The Practical Approach Series

SERIES EDITORS

D. RICKWOOD
*Department of Biology, University of Essex
Wivenhoe Park, Colchester, Essex CO4 3SQ, UK*

B. D. HAMES
*Department of Biochemistry and Molecular Biology,
University of Leeds, Leeds LS2 9JT, UK*

Affinity Chromatography
Anaerobic Microbiology
Animal Cell Culture (2nd edition)
Animal Virus Pathogenesis
Antibodies I and II
Biochemical Toxicology
Biological Membranes
Biomechanics—Materials
Biomechanics—Structures and Systems
Biosensors
Carbohydrate Analysis
Cell Growth and Division
Cellular Calcium
Cellular Neurobiology
Centrifugation (2nd edition)
Clinical Immunology
Computers in Microbiology
Crystallization of Nucleic Acids and Proteins
Cytokines
The Cytoskeleton
Diagnostic Molecular Pathology I and II
Directed Mutagenesis

DNA Cloning I, II, and III
Drosophila
Electron Microscopy in Biology
Electron Microscopy in Molecular Biology
Enzyme Assays
Essential Molecular Biology I and II
Fermentation
Flow Cytometry
Gel Electrophoresis of Nucleic Acids (2nd edition)
Gel Electrophoresis of Proteins (2nd edition)
Genome Analysis
HPLC of Macromolecules
HPLC of Small Molecules
Human Cytogenetics I and II (2nd edition)
Human Genetic Diseases
Immobilised Cells and Enzymes
In Situ Hybridization
Iodinated Density Gradient Media

Molecular Plant Pathology

Volume II

A Practical Approach

Edited by

SARAH JANE GURR

and

MICHAEL J. McPHERSON

Centre for Plant Biochemistry and Biotechnology
Department of Biochemistry and Molecular Biology
University of Leeds

and

DIANNA J. BOWLES

Centre for Plant Biochemistry and Biotechnology
Department of Biochemistry and Molecular Biology
and Department of Pure and Applied Biology
University of Leeds

OXFORD UNIVERSITY PRESS
Oxford New York Tokyo

Oxford University Press, Walton Street, Oxford OX2 6DP
Oxford   New York   Toronto
Delhi   Bombay   Calcutta   Madras   Karachi
Kuala Lumpur   Singapore   Hong Kong   Tokyo
Nairobi   Dar es Salaam   Cape Town
Melbourne   Auckland   Madrid
and associated companies in
Berlin   Ibadan

Oxford is a trade mark of Oxford University Press

A Practical Approach ⑥ is a registered trade mark
of the Chancellor, Masters, and Scholars of the University of Oxford
trading as Oxford University Press

Published in the United States
by Oxford University Press Inc., New York

A catalogue record for this book is available from the British Library

Library of Congress Cataloging in Publication Data
(Revised for vol. 2)
Molecular plant pathology : a practical approach.
(The Practical approach series)
Includes bibliographical references and indexes.
1. Plant diseases—Molecular aspects—Handbooks,
manuals, etc.   2. Phytopathogenic microorganisms—
Molecular aspects—Handbooks, manuals, etc.
3. Plant-pathogen relationships—Molecular aspects—
Handbooks, manuals, etc.   4. Molecular microbiology—
Technique—Handbooks, manuals, etc.   5. Plant molecular
biology—Technique—Handbooks, manuals, etc.
I. Gurr, Sarah Jane.   II. McPherson, M. J.   III. Bowles,
Dianna J.
SB732.65.M65   1992   581.2   91–24886
ISBN 0–19–963103–4 (v. 1 : hbk.)
ISBN 0–19–963102–6 (v. 1 : pbk.)
ISBN 0–19–963352–5 (v. 2 : hbk.)
ISBN 0–19–963351–7 (v. 2 : pbk.)

Typeset by
Footnote Graphics, Warminster, Wilts
Printed in Great Britain by
Information Press Ltd, Eynsham, Oxford

Preface

Molecular plant pathology is one of the fastest moving and most exciting fields in biology and has directly benefited from advances in modern molecular techniques. These have been applied to both pathogen and plant, allowing us to develop some understanding of the organisms themselves and of the complex interactions leading to compatibility or incompatibility.

The intention has been to provide a comprehensive handbook describing both the latest molecular techniques and more 'classical' approaches. Coverage is also given to areas of research that we believe will become increasingly important in molecular plant pathology: the nature of the signalling molecules involved in the recognition between plant and pathogen, and the nature of the signal transduction pathways in the plant that lead from those initial recognition events to defence-related changes in gene expression and strategies for the isolation of disease-resistance genes.

The contents encompass the expertise of many research workers with a broad range of hands-on experience in plant pathology. Some contributors specialize in specific classes of pathogens, whilst others focus on the diverse responses of the plant to pathogen invasion.

Due to the sheer volume of information required to provide this handbook, it has proved necessary to produce two *Practical Approach* volumes. The contents follow a sequence incorporating techniques for specific pathogens, followed by plant responses at the levels of genes, proteins, and defence-related compounds, the elicitor molecules and signalling pathways, and finally the disease resistance genes. The contents of the books are organized into six sections the first three of which contain a number of chapters describing complementary methods. A broad range of protocols are provided from pathogen isolation and culture, through physiology and biochemistry, to molecular biology and techniques for the localization of genes and their products *in situ* within cells of the infected plant.

Within Volume I, Section 1 provides an introduction to four classes of pathogen, viruses, bacteria, fungi, and nematodes. The importance of the first three to plant pathologists is obvious. However, it is now increasingly recognized that nematodes are crucially important pathogens of a wide range of crop species, yet, at the molecular level, we are only just beginning to understand their interaction with their host plants. We hope the inclusion of nematodes in this handbook will lead to the wider recognition of these parasites in molecular plant pathology.

Whilst detailed methods for various molecular biology approaches are provided in Section 2, the reader is also referred elsewhere for further discussion of techniques in plant molecular biology (1), polymerase chain reaction applications (2, 3), and a range of other general techniques in gel electrophoresis (4, 5) and nucleic acid methodology (6, 7). Similarly, protocols for carbohydrate analyses are complementary to those found in (8). A chapter on the use of baculoviruses as expression vectors has been included to highlight their enormous potential for the study of gene expression in host–pathogen interactions.

Within Volume II, Section 1 provides detailed protocols for the analysis of gene products induced during defence responses. Sections 2 and 3 concentrate, respectively, on the preparation of elicitors and on the analysis of signal transduction pathways. Descriptions of methods for the determination of calcium levels, phosphorylation, and membrane inositide turnover should provide both a readily accessible set of techniques for use by plant pathologists and also be of use to other plant biologists who may not have extensive biochemical experience.

Volume II ends with Section 4, which provides a review of strategies for the isolation of disease-resistance genes; a challenge that represents the dominant research objective of many molecular plant pathologists. Whilst no resistance gene has yet been isolated, techniques such as those described in Chapter 13 should hold the key to success in this search; success that will represent a landmark in plant science.

We hope these volumes will be of use to the community of plant pathologists world-wide, and that they will provide a ready source of important methods and approaches that can be applied to the study of plant and pathogen, whatever the specialized interest of the individual investigator.

Leeds S. J. G.
August 1991 M. J. M.
  D. J. B.

References

1. Shaw, C. S. (ed.) (1989). *Plant molecular biology: a practical approach*. IRL, Oxford.
2. McPherson, M., Quirke, P., and Taylor, G. R. (ed.) (1991). *PCR: a practical approach*. IRL, Oxford.
3. Erlich, M. A. (ed.) (1989). *PCR technology, principles and applications for DNA amplification*. Stockton Press, New York.
4. Hames, B. D. and Rickwood, D. (ed.) (1990). *Gel electrophoresis of proteins: a practical approach* (2nd edn). IRL, Oxford.
5. Hames, B. D. and Rickwood, D. (ed.) (1990). *Gel electrophoresis of nucleic acids: a practical approach* (2nd edn). IRL, Oxford.
6. Sambrook, J., Fritsch, E. F., and Maniatis, T. (ed.) (1989). *Molecular cloning: a laboratory manual* (2nd edn). Cold Spring Harbor Press, New York.
7. Davis, L. G., Dibner, M. D., and Batten, J. F. (ed.) (1986). *Basic methods in molecular biology*. Elsevier, New York.
8. Chaplin, M. F. and Kennedy, J. F. (ed.) (1987). *Carbohydrate analysis: a practical approach*. IRL, Oxford.

Contents

6. Isoflavonoid phytoalexins and their biosynthetic enzymes   45

Robert Edwards and Helmut Kessmann

7. Analysis of terpenoid phytoalexins and their biosynthetic enzymes   63

David R. Threlfall and Ian M. Whitehead

SECTION 2.   ELICITORS AND OLIGOSACCHARINS

8. Preparation and characterization of oligosaccharide elicitors of phytoalexin accumulation

Michael G. Hahn, Alan Darvill, Peter Albersheim,
Carl Bergmann, Jong-Joo Cheong, Alan Koller, Veng-Meng Lò

11. Analysis of components of the plant
phosphoinositide system   195

Bjørn K. Drøbak and Keith Roberts

12. Metabolic changes following infection of leaves by fungi and viruses 223

Richard C. Leegood and Julie D. Scholes

SECTION 4.   DISEASE RESISTANCE GENES

13. Strategies for cloning plant disease resistance genes 233

Richard W. Michelmore, Richard V. Kesseli, David M. Francis, Marc G. Fortin, Ilan Paran, and Chang-Hsien Yang

Appendix

Index

Contents of Volume I

Contributors

PETER ALBERSHEIM
Complex Carbohydrate Research Center, University of Georgia, PO Box 5677, Athens, GA 30613, USA.

CARL BERGMANN
Complex Carbohydrate Research Center, University of Georgia, PO Box 5677, Athens, GA 30613, USA.

THOMAS BOLLER
Plant Physiology Department, University of Basel, CH-4056 Basel, Switzerland.

DOUGLAS S. BUSH
Department of Botany, 2017 Life Sciences Building, University of California, Berkeley CA 94720, USA.

JONG-JOO CHEONG
Complex Carbohydrate Research Center, University of Georgia, PO Box 5677, Athens, GA 30613, USA.

ALAN DARVILL
University of Georgia, Complex Carbohydrate Research Center, 200 Riverbend Road, Athens, GA 30602, USA.

ALAN P. DAWSON
School of Biological Sciences, University of East Anglia, Norwich NR4 7TJ, UK.

BJØRN K. DRØBAK
Department of Cell Biology, John Innes Institute, Colney Lane, Norwich NR4 7UH, UK.

ROBERT EDWARDS
Department of Biological Sciences, University of Durham, Durham DH1 3LE, UK.

IAN B. FERGUSON
Mount Albert Research Centre, Division of Scientific and Industrial Research, Auckland, NZ.

MARC G. FORTIN
Department of Plant Science, McGill University, St-Anne de Bellevue, Quebec H9X 1C0, Canada.

DAVID M. FRANCIS
Department of Vegetable Crops, University of California, Davis, CA 95616, USA.

JOHN FRIEND
Department of Applied Biology, University of Hull, Hull HU6 7RX, UK.

STEPHEN C. FRY
Centre for Plant Science, King's Buildings, University of Edinburgh, Mayfield Road, Edinburgh EH9 3JH, UK.

MICHAEL G. HAHN
Complex Carbohydrate Research Center, University of Georgia, PO Box 5677, Athens, GA 30613, USA.

KIM E. HAMMOND-KOSACK
Sainsbury Laboratory, John Innes Centre, Colney Lane, Norwich, NR4 7UH, UK.

RUSSELL L. JONES
Department of Botany, 2017 Life Sciences Building, University of California, Berkeley, CA 94720, USA.

HEINRICH KAUSS
Universität Kaiserslautern, Fachbereich Biologie, D-1075 Kaiserslautern, Germany.

RICHARD V. KESSELI
Department of Biology, University of Massachusetts, Boston, MA 02125-3393, USA.

HELMUT KESSMANN
Agricultural Division, Ciba-Geigy Ltd, Basle, CH4002, Switzerland.

ALAN KOLLER
Complex Carbohydrate Research Center, University of Georgia, PO Box 5677, Athens, GA 30613, USA.

RICHARD C. LEEGOOD
Robert Hill Institute and Department of Animal and Plant Sciences, University of Sheffield, Sheffield S10 2TN, UK.

VENG-MENG LÒ
Complex Carbohydrate Research Center, University of Georgia, PO Box 5677, Athens, GA 30613, USA.

RICHARD W. MICHELMORE
Department of Vegetable Crops, University of California, Davis, CA 95616, USA.

Contributors

ILAN PARAN
Department of Field and Vegetable Crops, Hebrew University of Jerusalem, Rehovot, Israel, 76100.

KEITH ROBERTS
Department of Cell Biology, John Innes Institute, Colney Lane, Norwich NR4 7UH, UK.

JULIE D. SCHOLES
Robert Hill Institute and Department of Animal and Plant Sciences, University of Sheffield, Sheffield S10 2TN, UK.

DAVID R. THRELFALL
Department of Applied Biology, University of Hull, Hull HU6 7RX, UK.

KATHRYN A. VANDENBOSCH
Department of Biology, Texas A and M University, College Station, TX 77843, USA.

IAN M. WHITEHEAD
Biotechnology Department, Fermenich SA, La Junction, 1 Rue de Jeunes, Geneva, Swizterland.

CHANG-HSIEN YANG
Department of Vegetable Crops, University of California, Davis, CA 95616, USA.

Abbreviations

ARS	autonomously replicating sequence
BMV	brome mosaic virus
BSA	bovine serum albumin
βME	2-mercaptoethanol
CAT	chloramphenicol acetyl transferase
CCMV	cowpea chlorotic mottle virus
cM	centiMorgan
CsCl	caesium chloride
cDNA	complementary DNA
CTAB	cetyl trimethyl ammonium bromide
DEPC	diethylpyrocarbonate
DIG	digoxigenin
DMSO	dimethylsulphoxide
dNTP	deoxyribonucleotide triphosphate
ds DNA	double-stranded DNA
ds RNA	double-stranded RNA
DTT	dithiothreitol
ECL	enhanced chemiluminescence
EDTA	ethylene diamine tetraacetic acid
EGTA	ethyleneglycol aminoethyl ether tetraacetic acid
ELISA	enzyme-linked immunonsorbent assay
EMS	ethyl methane sulphonate
EtBr	ethidium bromide
EtOH	ethanol
FCS	fetal calf serum
GLC	gas–liquid chromatography
GLC–ms	gas–liquid chromatography–mass spectrometry
GlcNAc	N-acetylglucosamine
GRA	gel retardation assay
GuHCl	guanidium hydrochloride
GUS	β-glucuronidase
Hepes	hydroxyethylpiperazine ethanol sulphonic acid
HeBS	Hepes buffered saline
HPLC	high performance liquid chromatography
HPRNI	human placental ribonuclease inhibitor
IEF	iso-electric focusing
IPCR	inverse polymerase chain reaction

IPTG	isopropyl-β-D-thiogalactoside
IR	infra-red
LMP	low melting point
MeOH	methanol
MES	N-morpholine ethanol sulphonic acid
MLOs	mycoplasma-like organisms
MLV	murine leukaemia virus
mRNA	messenger RNA
NIL	near-isogenic line
NaAc	sodium acetate
NMR	nuclear magnetic resonance
NPV	nuclear polyhedrosis virus
n.t.	nucleotide
ORF	open reading frame
PAGE	polyacrylamide gel electrophoresis
PBS	phosphate buffered saline
PCN	potato cyst nematode
PCR	polymerase chain reaction
PEG	polyethylene glycol
PFGE	pulsed-field gel electrophoresis
pfu	plaque forming unit
PMSF	phenylmethylsulphonyl fluoride
pV	pathovar
PVP	polyvinylpyrrolidone
RAPD	random amplified polymorphic DNA
RFLP	restriction fragment length polymorphism
RKN	root-knot nematode
rRNA	ribosomal RNA
RNase A	ribonuclease A
r.p.m.	revolutions per minute
RT	reverse transcriptase
SCAR	sequence characterized amplified region
SCN	soybean cyst nematode
SDS	sodium dodecyl sulphate
SDW	sterile distilled water
ssDNA	single-stranded DNA
SS phenol	salt-saturated phenol
STS	sequence tagged sites
TAE	Tris–acetate–EDTA
Taq	*Thermus aquaticus* DNA polymerase
TBE	Tris–borate–EDTA
TCA	trichloroacetic acid
TE	Tris–EDTA

TLC	thin layer chromatography
TMV	tobacco mosaic virus
tRNA	transfer RNA
UPGMA	unweighted pair group-method with averaging
UV	ultraviolet
WDV	wheat dwarf virus

1

Callose and callose synthase

HEINRICH KAUSS

1. Introduction: callose formation represents a rapid defence reaction

Over the past century numerous cytochemical studies have shown that after various types of stress the polysaccharide callose is deposited, at selected sites, on to the plant cell wall. Callose mainly consists of linear β-1, 3-glucanase (1–4), and one of its functions is to close the connections that unite proto-plasts of contiguous cells (plasmodesmata, sieve pores). This process occurs within minutes of mechanical injury. Following fungal invasion there is a rapid formation (within hours) of callose-rich papillae basipetal to the sites of penetration. Such a deposition of papillae contributes to the resistance re-sponse of the host and can begin about one hour before fungal penetration pegs become visible microscopically. Callose formation also appears to tighten those cell wall parts neighbouring virus-induced lesions, or it can line the walls of cells that have undergone the hypersensitive response.

2. Microscopic observation and quantification of callose

2.1 Fluorescence microscopy

Deposition of callose is normally very localized and can only be evaluated by careful microscopical observation. Various staining and fluorescence methods (5) can be used to more or less selectively contrast callose, but the most reliable procedure makes use of commercial aniline blue which contains, as an impurity, the fluorochrome, 'Sirofluor'.

Pure Sirofluor (Biosupplies) is costly, but has the advantage that it can be used in water (25 μg/ml) at neutral pH, and is reported to give less back-ground fluorescence (6). For most practical work, however, commercial water-soluble aniline blue is adequate, although material bought from differ-ent companies may vary in fluorescence efficiency. We use aniline blue WS obtained from Merck or Serva, while other groups use dye from Fisher. For microscopy, aniline blue is solubilized at 0.1% (w/v) in 1 M glycine/NaOH buffer, pH 9.5. The solution can be kept for months at 4°C although its blue colour fades rapidly.

The nature of the plant material dictates whether the aniline blue solution is added directly to the sample on a microscope slide or added directly to the tissue in a test tube. A suspension of cells, protoplasts, or small pieces of leaves may be examined *in vivo* without pre-treatment, thus allowing better simultaneous observation of cellular details by phase-contrast microscopy.

Leaves infected by powdery mildew fungus can be used to gain experience with the technique. Leaves (for example from barley, wheat, red clover, plantain, *Phlox*) are carefully wiped with a cloth or a cotton pad to remove the mycelium from the surface and 1 mm strips are then cut and wetted with the stain. Ring-like callose spots around the penetration hyphae termed 'collars' (ineffective papillae) on epidermal cells stain immediately. With increasing time after the addition of the aniline blue, anticlinal walls of infected cells may become partially callose-positive. Dead epidermal cells, which have undergone the hypersensitive response, may be impregnated with callose all around. In red clover, in both infected and uninfected leaves, small dots of callose appear in pairs on either side of the walls, suggesting that the spots correspond to closed plasmadesmata. Beware that residual fungal hyphae are also fluorescent, and that certain cells (for example leaf hairs) in healthy tissues may contain β-1,3-glucans endogenously or are mechanically induced to form callose by handling of the cut tissue. Certain cells (for instance ribs in the epidermis of grasses) are autofluorescent or fluoresce under aniline blue, but with a different colour from callose.

Alternatively, the tissue can be rendered chlorophyll or autofluorescence-free by boiling in ethanol, or by prolonged incubation in ethanol at room temperature. Such procedures may elucidate cell wall details better, while observation of cellular details within the tissue, or counting of, for example papillae, requires the use of fixed (0.1% (v/v) glutaraldehyde) and embedded sections (5, 6).

Observation is preferentially performed with an epifluorescence micro-scope. A yellow-white fluorescence of callose is observed using the Zeiss filter set 18 (exc. 390–420 nm, colour splitter 425 nm, secondary filter 450 nm). With living green cells, the Zeiss filter set 02 can be used (exc. 365 nm, colour splitter 395 nm, secondary filter 420 nm), which suppresses the red chloro-phyll fluorescence and provides a more blue-white callose fluorescence. Microscopy provides rapid information on the localization and on the time-course of callose deposition. Quantitative estimations, however, should be made very cautiously since fluorescence efficiency is highly dependent on the degree of polymerization and packing density of the β-1,3-glucan and can also be quenched by other materials such as polyphenols in papillae.

Removal of callose by incubating the ethanol-extracted plant tissue with β-1,3-glucanase (e.g. laminarinase from Calbiochem or Sigma, 5 mg/ml in 0.2 M acetate/NaOH buffer, pH 5.5, for 2–24 h) can aid in callose identification. However, it should be noted that commercial enzyme preparations usually contain other hydrolases, and precise identification of callose requires further purification and characterization of the enzyme preparation used.

2.2. Induction and fluorometric determination of callose in suspension cells

For experimental work on the regulation of callose deposition in suspensions of cells and protoplasts a quantitative protocol based on a cytochemical assay has been reported (7), and is described in *Protocol 1*. A critical analysis of this method is given in reference 4.

Protocol 1. Callose induction and extraction

A. *From suspension-cultured cells, for example soybean (7),* Catharanthus roseus *(8)*

 1. Suspend 300 mg, fresh weight of cultured cells in 5 ml nutrient solution or 10 mM TES/NaOH (pH 7.0) or 10 mM MES/BIS–Tris (pH 5.8), each containing 2% (w/v) sucrose and 4% (v/v) growth medium.

 2. Supplement suspensions with appropriate callose elicitor (1–3), e.g. digitonin (5–40 μM), amphotericin B (2–50 μM), or chitosan (0.2–1 mg/g cell fresh weight); all are commercially available.[a]

 3. Shake or roll cells for 1–4 h. Allow to settle and remove most of supernatant. Add ethanol to 80% (v/v) to stop reaction and to extract autofluorescent material.

 4. Collect cells after 1 h, by suction on to a paper filter or nylon net, and wash with water.

 5. Transfer 300 mg (fresh weight) portions to 3 ml 1 M NaOH and homogenize in a ground-glass or Teflon Potter homogenizer, or disrupt by sonication.

 6. Heat broken cells to 80°C, 15 min, to solubilize callose.

 7. Cool and centrifuge (1000 g, 10 min).[b]

 8. Assay supernatant for callose according to *Protocol 2*.

 9. Wet the ethanol-extracted cells on the filter support with 1–2 ml 2 M H_2SO_4, partially hydrolyse for 15 min at room temperature.

10. Remove the acid by washing several times with distilled water.[c]

B. *From protoplasts of cell suspensions (8) or tobacco leaves*

 1. Induce callose by treating 4×10^5 protoplasts/2 ml with 1–10 μM digitonin or 1–10 μg chitosan for 1–4 h. Use suspension solution containing glucose as an osmoticum (8) and 2 mM $CaCl_2$.

3

Protocol 1. *Continued*

2. Mix protoplast suspension with 3–4 volumes pure ethanol, centrifuge (1000 g, 10 min).

3. Wash pellet in 80% (v/v) warm ethanol to completely decolorize material.

4. Solubilize pellet in 0.2–0.5 ml 1 M NaOH[c] at 80°C for 15 min.

Callose derived from certain cell suspensions is insoluble, or poorly soluble, in 1 M NaOH. If this is the case follow steps 9 and 10 from A above.

[a] Digitonin is soluble at 250 μM in water (at 60°C). Amphotericin B is soluble in methanol or dimethylsulphoxide (DMSO), but the final concentration of solvents should be below 0.5% (v/v). Chitosan is solubilized in 20 mM acetic acid (2 mg/ml), freeze-dried, and immediately resolubilized in the original volume of water. Chitosan precipitates above pH 6.0.

[b] Microscopic examination of the water-washed (neutral) pellet should verify that the cells have disintegrated and callose has been extracted (see Section 2.1).

[c] Cells treated in this way are sticky. A glass-fibre filter should be used (e.g. Whatman GF/A, 25 mm diameter), and homogenized and extracted together with the cells. DMSO at 121°C is used to extract and quantify alkali-insoluble callose (4). In this case, use 50 μl of DMSO-solubilized callose plus 350 μl water for *Protocol 2*, omitting step 3. DMSO quenches fluorescence and should, therefore, also be used for the standard.

Protocol 2. Quantitative fluorometric callose determination

1. Mix 200 μl of alkali-extracted callose from *Protocol 1* with 400 μl of 0.1% (w/v) aniline blue in water. A violet-red colour is produced.

2. Set up a blank of 1 M NaOH and standards of laminarin or pachyman and treat as in step 1.

3. Add 210 μl 1 M HCl. The colour changes to deep blue as the solution becomes mildly acidic.

4. Add 590 μl 1 M glycine/NaOH, pH 9.5, mix the tubes vigorously and heat for 20 min at 50°C. Incubate for 30 min at room temperature to reduce the colour.

5. Monitor fluorescence in an instrument with excitation at 360–410 nm and emission above 460 nm, or set spectrofluorometer excitation at 393 nm and emission at 480 nm (check optima for each instrument).

6. Construct a calibration curve with pachyman (Calbiochem) in 1 M NaOH (linear range 0–2 μg), or laminarin from *Laminaria digitata* (Sigma, soluble in hot water, linear range 0–200 μg).[a]

[a] The difference between pachyman and laminarin, with their respective degrees of polymerization of 700 and 20, demonstrates that the assay gives only relative callose amounts.

---

3. Callose synthase (β-1,3-glucan synthase, GSII)

In many host–pathogen interactions callose appears to be locally deposited directly from the plasma membrane on to the adjacent cell wall. It has long been speculated that the plasma membrane-localized β-1,3-glucan synthase (called GSII by plant cytochemists) may play a role in this deposition of callose. GSII is abundant in plant homogenates but is latent in undisturbed living cells. *In vitro* the β-1,3-glucan synthase is strictly Ca^{2+} dependent (1–3); in the presence of 0.2 mM spermine or 4 mM Mg^{2+}, half-maximal activity is attained at about 0.8 μM Ca^{2+}. It is considered, therefore, that the enzyme remains latent in undisturbed cells as long as the resting level of free Ca^{2+} in the cytoplasm is about 0.1 μM (1). Localized deposition of callose may start when Ca^{2+}-uptake results from a perturbation of the plasma membrane. Indeed, in suspension cells callose synthesis is induced by agents interfering with plasma membrane integrity (see Section 2.2). It has, in addition to Ca^{2+}, other, as yet unknown, signals which must play a role in callose induction *in vivo* (2, 3). Nevertheless, an experiment demonstrating the activation of the β-1,3-glucan synthase by Ca^{2+} *in vitro* is described in *Protocol 4* (see references 10 and 11 for results). It is worth considering, however, that enzyme activity measured in this way is the same both in microsomes from control cells and from cells which were experimentally induced to form callose (7). This indicates that the effectiveness of callose in rapid defence reactions is not only related to allosteric regulation of the enzyme but additionally to the sensitivity of the host cells and to the timing of the host–pathogen interaction (2).

3.1 Microsomes as an enzyme source

To assay β-1,3-glucan synthase, crude homogenates from tissues or suspensions of cells may be used, but are difficult to handle. Microsomes prepared according to *Protocol 3* represent a more convenient enzyme source.

Protocol 3. Preparation of microsomes

1. Wash suspension culture cells by suction on a filter with cold 50 mM TES/NaOH, pH 7.0.

2. Resuspend 2 g washed cells in 4 ml 50 mM TES/NaOH, 0.25 M sucrose, 1 mM DTT, pH 7.0.

3. Cool the cells on ice and homogenize in a Potter homogenizer, by sonication (0.5 min), or grind in a precooled mortar and pestle.

4. Dilute the homogenate with 6 ml of buffer from step 2 and centrifuge at 0°C, 500 g, 5 min.

5. Discard the pellet and centrifuge supernatant at 5000 g, 5 min to remove mitochondria and nuclei.

Protocol 3. *Continued*

6. Pellet microsomes from the supernatant from step 5 at 50 000 g for 10–20 min.

7. Dry the walls of the centrifuge tube with absorbant paper and resuspend microsomal pellet in 1 ml 50 mM TES/NaOH, 1 mM DTT, pH 7.0.[a]

8. Wash the microsomes in 6 ml 50 mM TES/NaOH, 1 mM DTT, pH 7.0. Reprecipitate at 50 000 g for 10–20 min and resuspend in 1 ml of the same buffer.

[a] Clumps of microsomes can be dispersed by careful use of a Teflon Potter homogenizer or rounded glass rod.

3.2 Enzyme assay

β-1,3-glucan synthase is activated synergistically by Ca^{2+} and polyamines or Mg^{2+} according to *Protocol 4*. Assay of enzyme activity can then be performed with radiolabelled UDP–glucose as substrate (*Protocol 5*) or a fluorometric assay (*Protocol 6*). Alternatively, the enzyme can be assayed (11) at a $[Ca^{2+}]$ $<10^{-3}$ μM with activation by poly-L-ornithine (30–70 kDa, 2.5–25 μg per assay). Activation of the enzyme can also be effected by limited proteolysis and assayed in solution A (see *Protocol 4*) and in the **obligatory** presence of 4 mM Mg^{2+} (10).

Protocol 4. Synergistic activation of β-1,3-glucan synthase

1. Mix Buffer A[a] (50 mM TES/NaOH, 4 mM EGTA, pH 7.0) and Buffer B (solution A containing 4 mM $CaCl_2 \cdot 4H_2O$) as indicated below to give, approximately, the following free Ca^{2+} concentrations

Buffer (ml)		Ca^{2+} (μM)
A	B	
5	0	$<10^{-3}$
2.5	2.5	~0.4
0.5	4.5	2–6
0	5	60–90

2. Add glycerol gravimetrically to 16% (w/v) to 5 ml aliquots of each of the four solutions (step 1).

3. Add cellobiose to 20 mM and digitonin[b] to 0.04% (w/v), then adjust pH to 7.0.

4. Prepare 10 mM $MgCl_2$ and 0.5 mM spermine-HCl as activators.

5. Set up duplicate assay reactions for each $[Ca^{2+}]$; mix 50 μl Ca^{2+} solution, 40 μl of water, spermine, or $MgCl_2$, and 10 μl of microsomes from *Protocol 3*.

Set up two time zero controls and heat inactivate for 5 min in a 95°C water-bath before adding substrate (*Protocols 5* and *6*).

[a] Buffers containing EGTA and Ca^{2+} are used to define low Ca^{2+} concentrations (10).

[b] Digitonin permeabilizes plasma membrane vesicles.

Protocol 5. Enzyme assay with radiolabelled substrate UDP–glucose

1. Prepare a 20 mM UDP–glucose solution containing 250 kBq/ml (~50 000 c.p.m./5 µl) UDP–[^{14}C]glucose or UDP–[^{3}H])glucose[a].

2. Add 5 µl of substrate to assay mixture from *Protocol 4* and incubate at 25°C for 2–20 min[b].

3. Inactivate enzyme by placing tubes in a 95°C water-bath for 5 min.

4. Spot the total assay mixture on to a 2×3 cm Whatman 3MM strip.

5. Dry chromatography strips under a stream of moderately warm air.

6. To remove unreacted substrate swirl the strips for ~1 h in a solution containing 150 ml ethanol and 350 ml 0.5 M aqueous ammonium acetate, either using a slow shaker or occasionally agitating by hand.

7. Dry the strips and measure radioactivity by liquid scintillation counting.

[a] Lower substrate concentrations (5–30 µM) may be used to demonstrate the allosteric properties of the β-1,3-glucan synthase and to show the synergistic action of spermine and Ca^{2+}.

[b] The reaction is linear with time as long as no more than one third of the substrate is polymerized.

Protocol 6. Fluorometric enzyme assay

1. Prepare assay mixtures as in *Procotols 4* and *5* in round-bottomed 2 ml centrifuge tubes containing 5 µl 20 mM UDP–glucose, without radio-tracer, as substrate.

2. Incubate at 25°C for 0, 5, 10, or 20 min.

3. Add 1.5 ml ethanol (65°C) and hold tube for 5 min at 65°C.

4. Centrifuge tubes for 10 min at 1000 g in a swing-out rotor arm to collect pellet (scarcely visible).

5. Drain ethanol by inverting tubes, and solubilize pellet in 200 µl 1 M NaOH. Place at 80°C for 15 min.

6. Use the solution from step 5 in the fluorometric assay described in *Protocol 2*. If fluorescence is greater than the highest value on the calibration curve, solubilize the pellet (from step 5) in 500 µl 1 M NaOH and use 200 µl for fluorometry.

Acknowledgements

Financial support from the Deutsche Forschungsgemeinschaft and the Fonds der Chemischen Industrie are gratefully acknowledged.

References

1. Kauss, H. (1987). *Annu. Rev. Plant Physiol.*, **38**, 47.
2. Kauss, H. (1990). In *The plant plasma membrane—structure, function and molecular biology* (ed. Ch. Larsson and I. M. Møller). Springer-Verlag, Heidelberg, pp. 320–50.
3. Kauss, H., Waldmann, T., and Quader, H. (1989). In *Signal perception and transduction in higher plants.* (ed. R. Ranjeva and A. Boudet). Springer-Verlag, Heidelberg. NATO ASI Series, Vol. H47, pp. 117–31.
4. Kauss, H. (1989). In *Modern methods of plant analysis. New Series* (ed. H. F. Linskens and J. F. Jackson), Vol. 10, p. 127. Springer-Verlag, Heidelberg.
5. Hächler, H. and Hohl, H. (1982). *Botanica Helvetica*, **92**, 23.
6. Stone, B. A., Evans, N. A., Bonig, I., and Clarke, A. E. (1984). *Protoplasma*, **122**, 191.
7. Köhle, H., Jeblick, W., Poten, F., Blaschek, W., and Kauss, H. (1985). *Plant Physiol.*, **77**, 544.
8. Kauss, H., Jeblick, W., and Domard, A. (1989). *Planta*, **178**, 385.
9. Conrath, U., Domard, A., and Kauss, H. (1989). *Plant Cell Reports*, **8**, 152.
10. Kauss, H., Köhle, H., and Jeblick, W. (1983). *FEBS Lett.*, **158**, 84.
11. Kauss, H. and Jeblick, W. (1985). *FEBS Lett.*, **185**, 226.

Lignin and associated phenolic acids in cell walls

JOHN FRIEND

1. Introduction

Many types of phenolic compounds are found in cell wall preparations or as insoluble residues remaining after the extraction of metabolites soluble in organic solvents. This chapter describes procedures for the isolation and analysis of lignin and other phenolic compounds from plant material.

2. Isolation of soluble and insoluble phenolic compounds

Since most analytical methods for phenolic compounds are specific for phenolic hydroxy-groups, it is essential that all soluble phenolic compounds are extracted from the tissue during the preparation of the insoluble residue or cell wall fraction.

Many simple phenolic compounds may not be completely extracted by non-aqueous organic solvents. Indeed, the removal of water by acetone or methanol may cause hydrogen-bonding of phenolic compounds, such as chlorogenic acid, to the insoluble residue leading to spurious results.

Moreover, it is possible for phenolic compounds to be oxidized by endo-genous phenolases, during the extraction procedure. The oxidation products of compounds, such as chlorogenic acid, can then chemically react with free-NH_2 or -SH groups on amino acid side chains to yield complex oxidation products bound to protein. This may be cell wall protein itself, or protein which is hydrogen bonded to the cell wall during the extraction procedure.

The procedure, originally devised by Laird, Mbadiwe, and Synge (1) for tobacco leaf, as modified by Ampomah and Friend (2), seems to overcome most of these difficulties (see *Protocol 1*). It involves extraction of the soluble phenolic compounds with methanol–chloroform–water (MCW), extraction of the protein with phenol–acetic acid–water (PAW) and leaves an insoluble residue which is assumed to contain lignin and cell wall polysaccharides. Ampomah and Friend (2) have shown that, with potato tuber discs, both the protein extract (PAW extract) and the insoluble residue (PAW residue) contain phenolic compounds which are different in the two fractions. The phenolic material in the PAW extract did not yield any free phenolic acids after alkaline hydrolysis (*Protocol 2*) but contained quinic acid (see *Protocol 3*); it was, therefore, assumed to be oxidized polymerized chlorogenic acid. Both *p*-coumaric and ferulic acids were found in the alkaline hydrolysate of the PAW residues. It was assumed that they had been esterified to either (or both) cell wall carbohydrates and lignin.

Protocol 1. Extraction of soluble and insoluble phenolic compounds from potato tuber discs

A. *Extraction of soluble phenols in methanol–chloroform–water (MCW)*

1. Chop 8–10 g tuber discs into small pieces and drop into a stainless steel beaker in liquid nitrogen.
2. Triturate, for example in an 'Ultra-turrax' top drive homogenizer, at maximum speed for 2 min.
3. Slurry the powder in 24–30 ml methanol–chloroform to give a methanol: chloroform:water ratio of approx. 2:1:0.8 (v/v/v), taking into account the water content of potato tubers.
4. Bubble nitrogen through slurry for approx. 20 min until the temperature approaches room temperature.
5. Centrifuge at 2000 g at 1°C for 15 min.
6. Decant supernatant by filtering to remove any floating material and homogenize residue with 25 ml methanol–chloroform–water (2:1:0.8). Flush with nitrogen and centrifuge as in step 5.
7. Repeat steps 1–6 with two further batches of tissue.
8. Use residue for extraction of bound phenolics, as described in part B.
9. Use MCW extracts to estimate free phenolic compounds (*Protocols 2, 3,* and *4*).

B. *Extraction of bound phenolic compounds in phenol–acetic acid–water*

1. Slurry residue from part A, step 8 in 20 ml phenol–acetic acid–water (PAW) ratio of 1:1:1 (w/v/v) while flushing with nitrogen.
2. Centrifuge at 2000 g at 1°C for 30 min.
3. Decant supernatant and slurry residue in 20 ml PAW, flush with nitrogen and place at 1°C overnight to allow protein to swell and dissolve.
4. Centrifuge at 2000 g at 1°C for 30 min.
5. Decant supernatant and extract residue 2× with 20 ml PAW and centrifuge each time.

10

6. Wash residue with ethanol to remove residual PAW and air-dry. Retain PAW residue (*Protocol 2* B, step 1).

7. Combine the 4 PAW extracts from part B, steps 2, 4, and 5, and dialyse overnight against running tap water to remove phenol.

8. Centrifuge contents of dialysis tubing at 2000 g at 1 °C for 15 min.

9. Wash residue with ethanol. Air-dry and weigh residue.

Protocol 2. Alkaline hydrolysis of bound phenolic compounds in (A) PAW extracts and (B) PAW residue

A. *PAW extracts*

1. Weigh 20 mg aliquots of dry precipitate (from *Protocol 1* B, step 9) into dry Thunberg tubes.

2. Add 1 ml 2 M NaOH to the tube and 2 ml 0.5 M H_2SO_4 to the stopper-receptacle of the tube.

3. Flush tube with nitrogen and incubate at 70 °C for 16 h to effect alkaline hydrolysis.

4. Following hydrolysis (precipitate completely degraded) cool tubes and mix tube and stopper-receptacle contents, to neutralize. Retain neutralized extract for assay (*Protocol 3*).

B. *PAW residue*

1. Alkaline hydrolyse residues as in A (steps 1–4).[a]

2. Following neutralization filter contents of each tube through a No. 1 sintered glass funnel and wash tube with 2 ml distilled water.

3. Adjust pH of each filtrate to pH 7.5. Make up to 10 ml. Retain neutralized extract for assay.

[a] Residue remaining after PAW extraction contains brown material. The remaining phenolics are removed by alkaline hydrolysis.

Protocol 3. Thiobarbituric acid assay for quinic acid moiety of oxidized, polymerized chlorogenic acid and related components

1. Add 250 µl of sample from *Protocol 1* A, step 9, *Protocol 2* A, step 4, or *Protocol 2* B, step 3, add 250 µl 25 mM HIO_4 in 63 mM H_2SO_4.

2. Shake vigorously and allow to stand for 20 min at room temperature.

Protocol 3. *Continued*

3. Remove excess periodate by addition of 500 µl of a 2% (w/v) solution of sodium arsenite in 0.5 M HCl.

4. Leave for 2 min, add 2 ml of a 0.3% (v/v) solution of thiobarbituric acid.

5. Heat in a boiling water-bath for 10 min.

6. Cool and transfer pink-coloured product to 3 ml of a 1:1 (v/v) mixture of isoamyl alcohol and 12% HCl.

7. Measure absorbance, A_{549}, of organic layer against a water blank, treated as above.

8. The quinic acid concentration is determined from a calibration curve prepared with 0.01–0.08 µmoles/sample quinic acid.

2.1 Ionization difference spectra

Ionization difference spectra have been used as an indication of lignin formation (4, 5). Such spectra can be obtained using hydrolysed extracts from PAW and the PAW residues.

Protocol 4. Ionization difference spectra

1. Add 4 ml 0.1 M NaOH (pH 12) to 1 ml of extract from *Protocol 2* A or B.

2. Add 4 ml 0.1 M phosphate buffer (pH 7) to a separate 1 ml of extract.

3. Record the absorbance of the extract (in 1) at pH 12 against that at pH 7 (in 2) between 220 and 360 nm.

4. Express results as A_{350} per 100 mg starting material (precipitation from PAW extract or PAW residue).

2.2 Demonstration and estimation of hydroxy cinnamic acids in hydrolysates of PAW residues

Hydroxy cinnamic acid (*p*-coumaric and ferulic acids) are found in the hydrolysate of the PAW residue and not in the PAW extract from potato tubers, these can be analysed according to *Protocol 5*.

Commercial phenolic acids (caffeic acid, ferulic acid, sinapic acid and *p*-coumaric acid) are in their *trans* forms and to obtain the *cis* isomers they have to be exposed to UV radiation. Dissolve 5 mg each of the phenolic acids in 1 ml of methanol in glass tubes and irradiate the solutions with UV using a mercury-vapour-discharge lamp, having 95% of the radiation at 360 nm. A 3 hour exposure to UV should yield a mixture of *cis* and unchanged *trans* isomers.

Protocol 5. TLC and HPLC of hydroxy cinnamic acids in hydrolysates of PAW residues

A. *TLC*

1. Run TLC on either microcrystalline cellulose (MC) or silica gel (SG) plates.

2. Mix toluene–acetic acid–water in the ratio 6:7:3 (v/v/v) acid for MC plates. Mix toluene–acetic acid (9:1) (v/v) for SG plates.

3. Detect with:

   (a) UV fluorescence (blue colour),

   (b) Spray with diazotized *p*-nitroaniline (0.5 ml 5% (w/v) NaNO$_2$, 5 ml 0.5% (w/v) *p*-nitroaniline in 2 M HCl, 15 ml 20% (w/v) sodium acetate). Mix in given order from stock solutions kept at 4°C. Use immediately. Ferulic acid gives pink colour; *p*-coumaric acid, an orange colour.

   Approximate R_f values

	MC	SG
ferulic acid	0.77	0.41
p-coumaric acid	0.53	0.23

B. *HPLC*

1. Use a 25 cm × 4.6 mm column packed with Hypersil 5 ODS.

2. Use the HPLC column in conjunction with a gradient elution apparatus with a UV detector set at 280 nm. (The method is quantitative if used in conjunction with an integrator.)

3. Prepare mobile phases:

   (a) 5% acetic acid in 10% methanol (HPLC grade)

   (b) 5% acetic acid in 100% methanol (HPLC grade)

4. Start gradient at an initial concentration of 15% 3(b) in 3(a) and complete it at 100% 3(b) in 3(a).

5. Inject 2 μl of extracts and standards and chromatograph at 2000 p.s.i. at a flow rate of 1 ml/min.

6. Calibrate the column frequently using 2 mg of recrystallized samples of standard phenolic acids in ethyl acetate.

7. Estimate phenolic acid amounts from calibration curves. Peak areas of each phenolic acid are linear in the range 0–2 μg.

8. Identify phenolic acids in the prepared extracts by comparing retention times with authentic standards or by 'spiking' with individual standards.[a]

---

[a] Most of the phenolic acids observed will probably be in the *trans* form but some minute quantities may also be in the *cis* form. The *cis* isomers can be identified by spiking the samples with authentic phenolic acids which have been exposed to UV radiation.

Lignin can also be determined as lignin thioglycollic acid. This method is described by Hammerschmidt (7). It is specific for lignin; however, the method described above does allow the elucidation of several other phenolic fractions.

Recently, an excellent method has been described for the quantitative assay of induced lignification in wounded wheat leaves (8). Lignin in leaves was stained with a diazonium salt; the intensity of the staining was determined with a scanning densitometer.

References

1. Laird, W. M., Mbadiwe, E. I., and Synge, R. L. M. (1976). *J. Sci. Food Agric.*, **27,** 127.
2. Ampomah, Y. A. and Friend, J. (1988). *Phytochemistry*, **27,** 2533.
3. Levy, C. C. and Zucker, M. (1960). *J. Biol. Chem.*, **235,** 2418.
4. Friend, J., Reynolds, S. B., and Aveyard, M. A. (1973). *Physiol. Plant Pathol.*, **3,** 495.
5. Ride, J. P. (1978). *Ann. Appl. Biol.*, **89,** 302.
6. Hartley, R. D. and Buchan, H. (1979). *J. Chromatog.*, **180,** 139.
7. Hammerschmidt, R. (1984). *Physiol. Plant Pathol.*, **24,** 33.
8. Barber, M. S. and Ride, J. P. (1988). *Physiol. Mol. Plant Pathol.*, **32,** 185.

3

Preparation and analysis of intercellular fluid

KIM E. HAMMOND-KOSACK

1. Introduction

The apoplast represents a key domain within the plant. It is a continuum with both the external environment and the plant cell, and, therefore, its contents can potentially be modified by both internal and external stimuli. The molecular events occurring peripheral to the plasma membrane are of immense importance to studies of plant–pathogen interactions, cell-to-cell interactions and to cell wall metabolism. This section describes a rapid, simple, and effective method to selectively analyse the apoplastic domain. The technique involves the recovery of intercellular fluids (IF). Such fluids, which are retrievable from all plant tissues, contains both free and ionically-bound polysaccharides, glycoconjugates, proteins, and enzymes from the cell wall, the surface of the plasma membrane and the external aqueous layers.

2. Methods

The preparation of IF from leaf, cotyledon, petiole, and stem tissue is basically a three-step process:

- the initial removal of air from the apoplastic space by the application of a vacuum

- the subsequent flooding of this region with fluid, achieved by a gradual raising of air pressure

- the recovery of the infiltrated fluid from the tissue by low speed centrifugation (1–5)

Protocols 1 and *2* describe specific methods appropriate for the preparation of IF from these different plant tissues. Infiltration is best achieved by trapping the tissue below the infiltration liquid with a weighted plastic mesh. When air is readmitted into the chamber, the tissue will appear uniformly watersoaked. If watersoaking is patchy, the initial evacuolation step should be repeated.

Following flooding the tissues are easily damaged and all further manipulations should be undertaken with care.

The preparation of IF from roots involves a different approach because the mechanical strength of this tissue is too low to permit collection of IF by low speed centrifugation. Instead, the technique relies on arranging the roots into a column and then flowing water or buffer gently through the column to elute the apoplastic components. *Protocol 3* describes a general method based on that of Sijmons (5) for the isolation of an extracellular peroxidase isoenzyme. Specific details of the original protocol are provided in the footnotes.

Protocol 1. Isolation of IF from leaf tissue (modified from references 1, 2)

1. Cut leaves with an area greater than 20 cm^2 into 2 cm wide strips. Smaller leaves are left intact. Rinse the leaves for >15 min in water to remove any debris from the surfaces or cytoplasmic contaminants from the cut edges.

2. Vacuum infiltrate for 10–15 min at room temperature[a] with water or buffer.

3. Blot material dry and roll into the barrel of a 20 ml plastic syringe. Place syringe, hub down, in an appropriate size centrifuge tube and centrifuge at 800 g for 10 min. The IF collects at the bottom of the tube.

4. Analyse IF immediately or store at −20°C.

[a] Vacuum infiltration can also be undertaken at a lower temperature to preserve enzyme activity and reduce protease digestion.

Protocol 2. Isolation of IF from hypocotyl, petiole, and stem tissue (modified from references 3, 4)

1. Cut into appropriate length sections (2.5 cm), and then wash in water as in *Protocol 1* (1).

2. Tightly pack sections vertically into the barrel of a 20 ml plastic syringe, in the bottom of which is placed a porous nylon disc to prevent lateral shearing forces. Vacuum infiltrate with water for 10 min at room temperature (see footnote *a* to *Protocol 1*).

3. Remove surface water by momentarily applying suction to the top of each syringe. Place the syringe containing the sections inside a centrifuge tube and centrifuge at 1000 g for 10 min. Collect IF and store as in *Protocol 1* step 4.

Protocol 3. Isolation of IF from root tissue (modified from reference 5)

1. Carefully harvest 3–4 cm lengths of unbranched roots (i.e. the tip region) or main roots with the laterals trimmed away. (Use scalpel and forceps and place isolated tissue on ice until all roots are collected.)

2. Lay roots parallel to form a bundle with a diameter of ~1.5 cm and a length of 8 cm. (This is achieved by rolling the roots in a sheet of parafilm.) Cut the root bundle to a length of 5 cm with a sharp razor blade, and carefully fit into a disposable PD-10 column (Sephadex). Dead volume is minimized by the addition of pre-swollen Sephadex G-25 until a thin layer has settled on top of the roots. Gently remove trapped air bubbles by placing the entire column for 1–2 sec under a water suction-generated vacuum. Allow the G-25 material to resettle then place a nylon filter on top of the column.

3. Elute the column with either water or a buffer system appropriate for the re-isolation of the components of interest.[a] Adjust to a flow rate of 1 ml eluant/min and collect 10 ml fractions. (To determine the void volume required to wash out the contents of damaged cells, aliquots taken from the eluted fractions can be assayed directly for the presence of cytoplasmic marker enzymes.) Dialyse fractions of interest against buffer—where appropriate—lyophilize, and then solubilize in water or buffer, store at −20°C until required for analysis.

[a] Sijmons (5) initially washed the column of root with 50 ml 10 mM potassium acetate pH 5.0 before applying a linear gradient (100 ml, 0–0.5 M NaCl in potassium phosphate buffer, pH 5.0) when isolating a cell wall bound peroxidase isoenzyme specifically associated with tissue suberization.

3. Detection of cytoplasmic contaminants

Visual examination of the IF collected should reveal a yellow to purplish/russet coloured liquid. A greenish coloration indicates cell rupture due to an excessive centrifugation force or pump pressure. The presence of cytoplasmic contamination can be identified by undertaking assays for either malate dehydrogenase or glucose-6-P-dehydrogenase (6).

4. Efficiency of extraction/choice of infiltration fluid

No qualitative changes in the protein profiles or polysaccharide composition of extracts have been observed when water or specific buffers are employed (1, 7). Water infiltration appears suitable for several plant species. However,

Table 1. Composition of infiltration fluids used to recover IF from various plant species

Infiltration fluid	Plant species	Reference
1. Water	*Solanum tuberosum*	2, 8, 21
	Lycopersicon esculentum	28
	Lactuca sativa	
	Hordeum vulgaris	
2. 25 mM Tris–HCl, pH 7.8 0.5 M sucrose, 10 mM $CaCl_2$, 0.5 M phenylmethyl fluoride, and 5 mM β-mercaptoethanol.	*Chenopodium amanticolor C. quinoa, L. esculentum Nicotiana clevelandii, N. glutinosa, N. sylvestris, N. tabacum, Phaseolus vulgaris, S. tuberosum, Vigna unguiculata*	7, 9, 10
3. 50 mM Tris–$CaCl_2$, pH 7.5	*N. tabacum* (basic PR proteins)	9
4. 32 mM/84 mM phosphate/ citrate buffer pH 2.8 and 5 mM β-mercaptoethanol	*L. esculentum*	11
5. 100 mM sodium phosphate buffer, pH 6.0.	*Cucumis sativus*	12

the overall yields of certain proteins are lowest when water is used as the extraction fluid, in other species, as, for example, in the extraction of the acidic PR proteins in tobacco (7). The composition of several infiltration fluids applied to various plant species is given in *Table 1*. Buffers with high ionic strength or containing >300 mM sucrose are unsatisfactory as large amounts of intracellular proteins are present in IF extracts. Successive extractions of the tissue result in the recovery of an additional 1–10% of the initial yield.

5. Observations and data analyses

The greatest yields of IF fluids are derived from leaves, the least from roots. From each organ type, the largest volumes are obtained after tissue expansion is complete. Typically, the yields are in the range 50–75 µl per g root, 125–150 µl per g hypocotyl, petiole, and stem, and 250–1000 µl per g leaf. The IF contains both water soluble polysaccharides and proteins. Total polysaccharide levels reported are in the range 300 to 500 µg/ml; total protein varies from 50 µg/ml to 1000 µg/ml (1, 3, 5, 7, 8).

Standard gas chromatography analysis of the carbohydrate component (13) can be undertaken if IFs are first passed through Whatman GF/A glass fibre paper, and the extracts then boiled for 5 min, vacuum-dried overnight at 47°C, and finally precipitated overnight with 80% ethanol.

Total protein determinations (14) and enzyme activity analyses can be undertaken directly on IF. Standard isoelectric focusing (IEF) and one-dimensional electrophoretic analyses typically require the sample to be lyophilized and resolubilized in the appropriate buffer or dye reagent. It is recommended to initially TCA precipitate the proteins prior to two-dimensional IEF/SDS–PAGE analysis (15). Such analyses are normally undertaken with a constant unit volume to allow analysis of *both* qualitative and quantitative changes in the composition of the apoplast.

6. Comments

The interpretation of results must be guarded, particularly where cellular integrity is questionable, as, for example, following the preparation of IF from tissue infected with a pathogen or responding hypersensitively. Direct immunocytochemical localization is usually undertaken to positively identify the extracellular locality of specific proteins (see Chapter 2 and ref. 16 this chapter).

Changes in the fluid composition potentially reflect both modifications to the secretory pathway and changes in extractability as caused by alterations in membrane and cell wall structure. The proportion of a particular protein in an extracellular location, is estimated by extraction of the tissue for total soluble protein after the removal of IF. The amount of specific protein remaining is compared with that recovered from the intact plant organ material. This approach would normally require a specific antibody or an appropriate enzyme assay to be available.

Purification of the more abundant products from IF is a relatively simple task utilizing standard column and electrophoretic fractionation techniques. Many secreted plant proteins have cyclocized *N*-termini and to gather amino acid sequence data the generation of peptide fragments and analysis of the resultant cleavage products is necessary. Peptide fragments are created either following digestion with different proteases (17) or by chemical cleavage according to established procedures (18).

7. Applications

7.1 Recognition events between host and pathogen

In studies on primary recognition events involving extracellular pathogens the analysis of IF is a particularly attractive tool. It has been most productively applied to a study of the interaction between *Cladosporium fulvum* and the tomato plant. DeWit's group (2) have identified the presence of several race-specific elicitors of necrosis and chlorosis by obtaining IF from compatible interactions between *Cladosporium fulvum* and tomatoes and then injecting this IF into the healthy leaves of cultivars containing different Cf resistance genes. Only in combinations which are genetically incompatible are necrotic or chlorotic symptoms induced 1–5 days later. Recently a product that appears to correspond to the fungal avirulence gene (*av. 9*) has been purified from IF (19), and a protein that may have a function in basic compatibility has been characterized (20).

The retrieval of IF could prove useful, in this context, in studies involving plant–bacterial associations where again pathogen colonization is restricted to the apoplast. In a study involving a fungus that penetrates cells during colonization this type of approach has proven unsuccessful (21). It is presumed that the key recognition events in this interaction occur in the vicinity of the haustorium, inside host cells, and are inaccessible to study in this manner.

7.2 Defence response proteins and enzymes

Most pathogens (excluding viruses) invade the apoplast or penetrate the cell wall at some stage during host colonization. In many plant species the products of several genes induced in the defence response are secreted to the apoplast. Defence gene products already identified in IF include chitinase and β-1,3-glucanases (8, 16, 22, 23, 24), peroxidases (24), and hydroxyproline-rich glycoproteins (25), as well as numerous additional 'pathogenesis-related proteins' of unknown function (26). These same gene products are also frequently identified in plant cell suspension cultures (27) as these culture media are often considered, experimentally, as an extension to the plant cells apoplastic domain, albeit a rather modified one.

7.3 Cell wall metabolism

Various workers attempting to ascribe specific roles to different extracellular acidic peroxidases during the oxidative polymerization of lignin precursors have analysed IF (24). Yields of these enzymes are found to be low when whole tissues are extracted. This is presumed to be due to the peroxidase isoenzymes non-specifically binding to other cell components. Other cell wall modifying enzymes, their specific substrates, and the general polymerization status of the cell wall could all readily be explored by an IF approach.

References

1. Hogue, R. and Asselin, A. (1987). *Can. J. Bot.*, **65**, 476.
2. DeWit, P. J. G. M. and Spikeman, G. (1982). *Physiol. Plant Pathol.*, **21**, 1.
3. Terry, M. E. and Bonner, B. A. (1980). *Plant Physiol.*, **66**, 321.
4. Cosgrove, D. J. and Cleland, R. E. (1983). *Plant Physiol.*, **72**, 326.
5. Sijmons, P. C. (1986). In *Molecular and physiological aspects of plant peroxidases*

(ed. H. Greppin, C. Penel, and Th. Gaspar), p. 221. University of Geneva, Switzerland.

6. Bergmeyer, J. and Grassl, M. (ed.) (1983). *Methods of enzymatic analysis Vol. III, Enzyme 1: oxidoreductases, transferases.* Verlag Chemie, Weinheim.

7. Parent, J. G. and Asselin, A. (1984). *Can. J. Bot.*, **62**, 564.

8. Kombrink, E., Schroder, M., and Hahlbrock, K. (1988). *Proc. Natl. Acad. Sci. (USA)*, **85**, 782.

9. Parent, J. G., Hogue, R., and Asselin, A. (1988). *Can. J. Bot.*, **66**, 199.

10. Hooft van Huijsduijnen, R. A. M., Alblas, S. W., deRijk, R. H., and Bol, J. F. (1986). *J. Gen. Virol.*, **67**, 2135.

11. Christ, U. and Mosinger, E. (1989). *Physiol. Mol. Plant Pathol.*, **35**, 53.

12. Smith, J. A. and Hammerschmidt, R. (1988). *Physiol. Mol. Plant Pathol.*, **33**, 255.

13. Albersheim, P., Nevins, D. J., English, P. D., and Karr, A. (1967). *Carbohydr. Res.*, **5**, 340.

14. Bradford, M. (1976). *Anal. Biochem.*, **72**, 248.

15. Hames, B. D. (1983). In *Gel electrophoresis of proteins: a practical approach* (ed. B. D. Hames and D. Rickwood), p. 1. IRL Press, Oxford.

16. Mäuch, F. and Staeheim, L. A. (1989). *Plant Cell.*, **1**, 447.

17. Cleveland, D. W. (1983). *Meth. Enzymol.*, **96**, 222.

18. Aitken, A., Geisow, M. J., Findlay, J. B. C., Holmes, C., and Yarwood, A. (1989). In *Protein Sequencing: a practical approach* (ed. J. B. C. Findlay and M. J. Geisow), p. 72. IRL Press, Oxford.

19. Schottens-Toma, I. M. J. and DeWit, P. J. G. M. (1988). *Physiol. Mol. Plant Pathol.*, **33**, 59.

20. Joosten, M. A. J. and DeWit, P. J. G. M. (1988). *Physiol. Mol. Plant Pathol.*, **33**, 241.

21. Crucefix, D. N., Mansfield, J. W., and Wade, M. (1984). *Physiol. Plant Pathol.*, **24**, 93–106.

22. Kauffman, S., Legrand, M., Geoffroy, P., and Fritig, B. (1987). *EMBO J.*, **6**, 3209.

23. Legrand, M., Kauffman, S., Geoffroy, P., and Fritig, B. (1987). *Proc. Natl. Acad. Sci. (USA)*, **84**, 6750.

24. Greppin, H., Penel, C., and Gaspar, Th. (1986). *Molecular and physiological aspects of plant peroxidases.* University of Geneva, Switzerland.

25. Cassab, G. I. and Varner, J. E. (1988). *Annu. Rev. Plant Physiol. Plant Mol. Biol.*, **39**, 321.

26. Van Loon, L. C. (1985). *Plant Mol. Biol.*, **4**, 111.

27. Ohashi, Y. and Matsuoka, M. (1987). *Plant Cell. Physiol.*, **28**, 573.

28. Holden, D. W. and Rohringer, R. (1985). *Plant Physiol.*, **79**, 820.

4

Biochemical analysis of chitinases and β-1,3-glucanases

THOMAS BOLLER

1. Introduction

Plant hydrolytic enzymes play an important role in host–pathogen interactions (1). Chitinase and β-1,3-glucanase are of particular interest for three reasons:

- The activities of both enzymes increase markedly in many plant tissues in response to pathogen invasion, to treatment with fungal cell wall components (so-called elicitors), or to the presence of the plant stress hormone, ethylene indicating a possible role in the plant stress response to a given pathogen (1, 2).

- Both enzymes strongly inhibit the growth of many fungi in culture, indicating that they may have a direct antibiotic function in plant–fungus interactions (3–5).

- Several of the so-called pathogenesis-related proteins (PR), i.e. proteins accumulating in hypersensitively-reacting tissue, have been identified as chitinases and β-1,3-glucanases (2, 6, 7). This chapter describes rapid, sensitive, and well-proven assay techniques for the two hydrolytic enzyme activities in extracts from plant tissues.

2. Analysis of chitinase

Chitinase activity can be measured spectrophotometrically, viscosimetrically, or radiochemically. The radiochemical assay is rapid and comparatively simple and is described in detail in Section 2.1 and *Protocol 1*. In Section 2.2, other chitinase assay methods are briefly presented and compared with the radiochemical assay.

2.1 Radiochemical method

The radiochemical assay of chitinase described here is based on the liberation of soluble radioactive fragments from insoluble radioactive chitin, first described by Molano *et al.* (8). Radioactive chitin is prepared by reacetylation of deacetylated chitin (chitosan) with [³H]acetic anhydride (8). According to the original method (8), the resulting 'regenerated' chitin is cut into small pieces in a mixer and used directly as a substrate. However, many investigators have found it difficult to obtain a suspension of regenerated chitin sufficiently fine to be pipetted. In *Protocol 1*, and as described in Section 2.1.1, the regenerated chitin is dissolved in HCl and re-precipitated in methanol, according to the method of Berger and Reynolds (9), to yield a finely dispersed, easily pipettable colloidal suspension of radioactive chitin.

2.1.1 Preparation of the substrate, colloidal radioactive chitin

The substrate is prepared by acetylating commercially available chitosan with [³H]acetic anhydride according to a modification of the procedure of Molano *et al.* (8) as described in *Protocol 1* (steps 1–6). The initial addition of a small amount of highly radioactive acetic anhydride yields about 15% reacetylation of the chitosan (*Protocol 1*, step 3), and is followed by the addition of a saturating amount of unlabelled acetic anhydride (*Protocol 1*, step 4). This modification results in an approximately five-fold increase in isotope incorporation compared with the original method (~30% instead of 5–6% of the ³H is present in the chitin). The theoretically possible maximal yield of 50%, which corresponds to one acetyl group incorporated per molecule of acetic anhydride, is not reached with the commercially available [³H]*N*-acetic anhydride.

Commercially available chitosan preparations, even batches from the same supplier, differ in their solubility after partial reacetylation (probably due to differences in the completeness of deacetylation). It is essential that the partially reacetylated chitosan remains soluble after step 3 of *Protocol 1* and so it is a good idea to do a test run of the entire procedure with non-radioactive chitin. (The resulting non-radioactive regenerated chitin can be dissolved in HCl (*Protocol 1*, step 7), and mixed in different proportions with the similarly dissolved radioactive preparation to yield colloidal chitin of differing specific activity.)

The regenerated radioactive chitin, finely dispersed in a mixer, is washed, dissolved in HCl and re-precipitated in methanol (9) to yield a colloidal suspension of [³H]chitin, as described in *Protocol 1*, steps 7–10. The resulting radioactive substrate, supplemented with azide, to prevent bacterial growth, can be stored for several months in a refrigerator. A portion of the preparation is dried and weighed (before the addition of azide) to determine the specific activity of the [³H]chitin. In a successful preparation, 1 mg of dry chitin, containing approximately 5 μmol GlcNac units, should contain approximately 19 kBq (0.5 μCi) of radioactivity, corresponding to a specific activity of approximately 3.8 MBq (0.12 mCi) per mmol GlcNac unit. The preparation initially contains approximately 0.04% of the radioactivity in TCA-soluble form, while after 3 months storage, the TCA-soluble part of the radioactivity may increase to ~0.1%. The TCA-soluble activity, which increases the substrate blank value in the enzyme assay, can be reduced by repeated centrifugation and resuspension if necessary. It is advisable to divide the preparation into several aliquots in fresh, sterile containers as the most frequent reason for spoilage of the substrate is contamination by minute amounts of chitinase.

Protocol 1. Preparation of radioactive, colloidal chitin

1. Soak 2.5 g chitosan in 50 ml 10% acetic acid in a 500 ml glass beaker. Stir for about 2 h to obtain a gluey mass.

2. Add 225 ml methanol, continue to stir for about 1 h until the chitosan is completely dissolved, yielding a viscous solution. Filter the solution through a 50 μm mesh nylon net.

3. Mix 0.2 ml acetic anhydride containing 5 mCi [^{3}H]acetic anhydride with 10 ml of a 9:2 (v:v) mixture of methanol and 10% acetic acid. Add immediately, with vigorous stirring, to the chitosan solution. Stir for about 1 min, then incubate at room temperature for approximately 2 h.

4. Add 3.5 ml acetic anhydride, stirring vigorously, and continue stirring vigorously until the liquid gels (approx. 2 min). Leave the regenerated chitin standing for 1 h.

5. Cut the gel block of radioactive regenerated chitin into pieces with a spatula. Add 100 ml water and homogenize in a mixer at full speed for 2 min.

6. Pour the regenerated chitin over a 50 μm mesh nylon net. Wash with 2 × 500 ml water and with 2 × 500 ml acetone. (Dispose of them highly radioactive washing fluid properly.)

7. Resuspend the regenerated chitin in 250 ml ice-cold concentrated HCl. Stir for approx. 10 min until the particles of chitin are completely dissolved.

8. Add the solution to 2 litres of vigorously stirred 50% ethanol. Continue stirring vigorously until a homogeneous milky suspension has formed.

9. Collect the colloidal chitin by centrifugation (e.g. Sorvall SS34 rotor, 6000 r.p.m., 5 min). Wash 3 times by resuspending in water and centrifuging as above. Adjust the pH of the suspension to 5.5 with 3 M sodium acetate, wash 3 times with water again.

10. Adjust the final suspension to 247.5 ml, add 2.5 ml 2% sodium azide, and homogenize the suspension in a mixer at full speed for 30 sec. Keep in a refrigerator; *do not freeze*.

2.1.2 Radiochemical assay of chitinase

The radiochemical assay of chitinase (8, 10) is described in *Protocol 2*. The amount of [^{3}H]chitin used per assay corresponds to approx. 0.3–0.6 μCi.

Each new preparation of radioactive chitin and each new chitinase preparation, requires the establishment of a calibration curve relating the amount of enzyme to the amount of radioactive product released (8, 10, 11). One nanokatal (nkat) of enzyme activity is defined as the amount of enzyme that catalyses the release of soluble chito-oligosaccharides corresponding to 1 nmol GlcNac in 1 sec at infinite dilution. The initial slope of the calibration curve is used to calculate the units. This calibration curve is typically non-linear for plant chitinases and reaches a plateau at about 15–25% of the radioactive chitin hydrolysed.

Protocol 2. Chitinase assay

1. Prepare 250 μl reaction mixtures in 1.5 ml microcentrifuge tubes in an ice-bath by adding the following:
   - 50 μl of 100 mM sodium acetate buffer, pH 5.0
   - 5–100 μl of crude enzyme extract, using e.g. 0.1 M sodium citrate, pH 5.0, as extraction buffer. (Or 5–100 μl of enzyme extract passed over a Sephadex G-25 column, using e.g. 10 mM sodium acetate, pH 5.0, as column buffer.)
   - extraction buffer (or column buffer) to bring the total volume to 150 μl
   - 100 μl of colloidal [^{3}H]chitin

2. Close the tubes and mix contents, then incubate the tubes in a shaking water-bath at 37°C for 30 min.
3. Put the tubes back in an ice bath, add 250 μl 10% TCA to each, and mix.
4. Centrifuge in a swing-out rotor at 3000 g for 10 min.
5. Carefully open the tubes, remove 250 μl of the supernatant, taking care not to disturb the sediment, mix with 4 ml scintillation liquid in a scintillation vial and measure radioactivity by liquid scintillation counting.
6. Determine chitinase activity with the aid of the calibration curve, taking the appropriate controls into account.

2.2 Other methods

Non-radioactive colloidal chitin can be used as a substrate when a sufficiently specific and sensitive assay is used to determine the hydrolysis products (11). Colorimetric methods based on the production of reducing sugars, although frequently used in microbiological work, are unsuitable for crude plant enzyme preparations which usually contain large amounts of reducing sugars. A more specific assay is based on the determination of GlcNac with p-dimethylaminobenzaldehyde according to Reissig et al. (12). Since the assay is specific for the GlcNac monomer, it is essential to enzymatically hydrolyse the chito-oligosaccharides formed in the enzyme reaction (13) before the assay.

Glycol chitin is a water-soluble chitin derivative, and as a highly viscous solution it can be used as a substrate in a sensitive viscosimetric enzyme assay (14). However, viscosimetric assays are time-consuming and unsuitable for routine work.

3. Analysis of β-1,3-glucanase

β-1,3-Glucanase can also be measured colorimetrically, viscosimetrically, or radiochemically. A colorimetric assay is described in detail in Section 3.1, while other methods are briefly presented and compared to the colorimetric assay in Section 3.2.

3.1 Colorimetric method

β-1,3-Glucanase is most commonly measured by colorimetric determination of the reducing groups formed in the enzyme reaction (15). Since crude plant extracts contain high levels of reducing sugars, they must be dialysed or passed over a molecular sieve column before the assay. The colorimetric method described here is based on the reduction of the neocuproin complex (16).

3.1.1 Preparation of the substrate, reduced laminarin

Commercially available laminarin is the predominantly β-1,3-linked storage glucan of the seaweed *Laminaria digitata*, and it is a useful substrate because of its solubility at ambient temperature. A disadvantage is the relatively high substrate blank of laminarin, due to its relatively short chain lengths. To eliminate this problem, the terminal groups of the laminarin chains are reduced with $NaBH_4$ (17), as described in *Protocol 3*.

Protocol 3. Prepration of reduced laminarin

1. Dissolve 3 g laminarin in 150 ml water. Stir at 80 °C for 20 min.
2. Add 1 g of solid sodium borohydride, continue stirring at 80 °C for 60 min.
3. Cool to 40 °C and adjust to pH 5.5 with glacial acetic acid.
4. Add a mixed-bed ion exchange resin (Amberlite MB3) and stir for 30 min to remove salts.
5. Filter the suspension. Wash the ion exchange beads by resuspending in 100 ml water. Combine the filtrates and adjust to 300 ml to obtain a 1% solution of reduced laminarin. Freeze aliquots at −20 °C.
6. Redissolve by heating to 60–80 °C.

3.1.2 Colorimetric assay for β-1,3-glucanase

The colorimetric assay of β-1,3-glucanase is described in *Protocol 4*. For each series of measurements with new, unknown enzyme extracts, include the following controls with the enzyme assay mixtures (E) in *Protocol 4*.

- substrate blanks (SB), containing column or dialysis buffer instead of enzyme extract
- reagent blanks (RB), containing column buffer instead of enzyme extract and 50 μl water instead of the laminarin solution
- reagent blanks with internal standard (RI), containing column buffer instead of enzyme, and 50 μl 3 mM glucose instead of laminarin
- enzyme blanks (EB), containing 50 μl water instead of laminarin
- enzyme blanks with internal standard (EI), containing 50 μl 3 mM glucose instead of laminarin

The amount of reducing sugar formed in the assay is calculated as follows:

$$\left[\frac{A_{450}(E) - A_{450}(EB)}{A_{450}(EI) - A_{450}(EB)} - \frac{A_{450}(SB) - A_{450}(RB)}{A_{450}(RI) - A_{450}(RB)}\right]150\,\text{nmol.}$$

Frequently, the presence of enzyme does not affect the reducing sugar assay, i.e.

$$A_{450}(EI) - A_{450}(EB) = A_{450}(RI) - A_{450}(RB).$$

In this case, the internal standards in the enzyme solutions can be omitted.

For each new enzyme preparation, establish a calibration curve with an enzyme dilution series in order to relate the amount of enzyme to product formation. These calibration curves are often, but not always, linear in the useful range of the assay (10–150 nmol glucose equivalents liberated). Use the appropriate calibration curve or conversion factor for calculation of the enzyme activity. One nkatal is defined as the amount of enzyme that catalyses the release of 1 nmol reducing sugar per second at infinite enzyme dilution.

Protocol 4. β-1,3-Glucanase assay

1. Prepare 250 μl reaction mixtures in 10 ml reagent tubes in an ice bath as follows:
- 50 μl of 100 mM sodium acetate buffer, pH 5.5
- 5–150 μl of an enzyme extract passed over a Sephadex G-25 column, using e.g. 10 mM sodium acetate, pH 5.0, as column buffer.
- column buffer to bring the total volume to 200 μl
- 50 μl of 1% reduced laminarin

Close the tubes and mix.

2. Incubate the tubes in a water-bath at 37°C for 30 min.

3. Return the tubes to the ice-bath and add 2 ml of the basic copper reagent (40 g Na_2CO_3, 16 g glycine, 0.45 g $CuSO_4 \cdot 5H_2O$ in 1000 ml water) and 2 ml of a fresh neocuproin solution (0.12 g neocuproin–Hcl in 100 ml water).

4. Incubate in a boiling water-bath for 12 min. Cool, add 3 ml water, mix and measure A_{450} in the spectrophotometer.

5. Determine β-1,3-glucanase activity with the aid of a calibration curve, taking the appropriate blanks and internal standards into account (see text).

3.2 Other methods

A radioactive assay, similar to the one described for chitinase (Section 2.1), can be performed when radioactive [³H]NaBH₄ is used to reduce the laminarin (18). The resulting [³H]laminarin is labelled exclusively in the terminal glucitol. *p*-Dioxane is used to precipitate the unreacted laminarin and to separate it from the soluble products. Laminarin is a relatively small polyglucan which differs little in solubility from the oligomers formed as products; maintenance of constant precipitation conditions is, therefore, critical.

Carboxymethylpachyman, a highly viscous soluble derivative of pachyman (an insoluble β-1,3-glucan, from the fruiting bodies of the basidiomycete fungus, *Poria cocos*), is a good substrate for a viscosimetric assay of β-1,3-glucan (19). This assay is somewhat time-consuming and tedious, but is highly sensitive to endoglucanases (as opposed to exoglucanases).

References

1. Boller, T. (1987). In *Plant–microbe interactions*, Vol. 2 (ed. T. Kosuge and E. W. Nester), p. 385. Macmillan, New York.
2. Boller, T. (1988). In *Oxford surveys of plant molecular and cell biology*, Vol. 5 (ed. B. J. Miflin), p. 145. Oxford University Press.
3. Schlumbaum, A., Mauch, F., Vögeli, U., and Boller, T. (1986). *Nature*, **324**, 365.
4. Mauch, F., Mauch-Mani, B., and Boller, T. (1989). *Plant Physiol.*, **88**, 936.
5. Ludwig, A. and Boller, T. (1990). *FEMS Microbiol. Lett.*, **69**, 61.
6. Legrand, M., Kauffman, S., Geoffroy, P., and Fritig, B. (1987). *Proc. Natl. Acad. Sci. (USA)*, **84**, 6750.
7. Kauffman, S., Legrand, M., Geoffroy, P., and Fritig, B. (1987). *EMBO J.*, **6**, 3209.
8. Molano, J., Duran, A., and Cabib, E. (1977). *Anal. Biochem.*, **83**, 648.
9. Berger, L. R. and Reynolds, D. M. (1958). *Biochim. Biophys. Acta*, **29**, 522.
10. Boller, T., Gehri, A., Mauch, F., and Vögeli, U. (1983). *Planta*, **157**, 22.
11. Boller, T. and Mauch, F. (1988). In *Methods of enzymology*, Vol. 161 (ed. W. A. Wood and S. T. Kellogg), p. 430. Academic Press, San Diego.

12. Reissig, J. L., Strominger, J. L., and Leloir, L. F. (1955). *J. Biol. Chem.*, **217**, 959.
13. Cabib, E. and Bowers, B. (1971). *J. Biol. Chem.*, **246**, 152.
14. Ohtakara, A. (1961). *Agr. Biol. Chem.*, **25**, 50.
15. Mauch, F., Hadwiger, L. H., and Boller, T. (1984). *Plant Physiol.*, **76**, 607.
16. Dygert, S., Li, L. H., Florida, D., and Thoma, T. A. (1965). *Anal. Biochem.*, **72**, 248.
17. Denault, L. J., Allen, W. G., Boyer, E. W., Collins, D., Kramme, D., and Spradlin, J. E. (1977). *Am. Soc. Brew. Chem. J.*, **36**, 18.
18. Keefe, D., Hinz, U., and Meins, F., Jr. (1990). *Planta,* **182,** 43.
19. Clark, A. E. and Stone, B. E. (1962). *Phytochemistry*, **1,** 175.

5

Localization of proteins and carbohydrates using immunogold labelling in light and electron microscopy

KATHRYN A. VANDENBOSCH

1. Introduction

Interest in inducible and tissue-specific genes has spawned a need for visualization of gene expression within complex, differentiated tissues. Three available techniques address this need by combining molecular probes with anatomical methods. These include *in situ* analysis of gene expression using reporter genes, *in situ* hybridization for identification of messenger RNA species, and immunogold labelling for localization of gene products. Together, the three approaches overlay molecular information on an anatomical outline, and are rapidly increasing the sophistication of our knowledge of tissue composition and organization on a cellular level.

The development of post-embedding labelling has enabled immunogold labelling to become routine practice for the demonstration of antigens in tissues (1). In this procedure, tissues are chemically fixed and embedded in plastic (with slight modifications from these protocols used for routine ultra-structural preservation). Tissue sections are then incubated sequentially in a primary antibody, specific for an antigen of interest, followed by a colloidal gold conjugate that binds to the primary antibody. The bound gold particles reveal the location of the antigen within the tissue. The combination of high specificity of immunological detection and good resolution by electron microscopy make immunogold labelling a powerful technique. This technique may also be used in combination with light microscopy where ultrastructural resolution is not required.

This chapter will detail immunogold labelling protocols which have been successfully applied to plant tissues; in particular emphasizing tissue preparation, choice of probes, and labelling methods.

2. Antibody production and screening

The success of the immunogold labelling technique rests largely upon the specificity of the antibody probes, and, therefore, careful preparation and screening of the antibodies are of critical importance. Both monoclonal and polyclonal antibodies have been used successfully in labelling protocols. However, both have their strengths and limitations, and these should be considered before preparing the probes for immunogold labelling purposes. Production of polyclonal and monoclonal antibodies are amply reviewed elsewhere (2–4), and need not be considered in detail here.

A high titre, monospecific polyclonal antiserum is the best choice for general purpose immunogold labelling, because it is a strong probe which reacts with a single macromolecule in the tissue to be studied. Polyclonal antisera to be used for labelling are usually raised in rabbits against purified proteins or glycoproteins. However, even when the immunogen is highly pure, a polyspecific antiserum may result. Major sources of spurious cross-reactivity of the antiserum can be due to components of the animal's diet, to infections, or other environmental factors. A sample of pre-immune serum, taken immediately before the first immunization, is an important control for both the initial screening of the antibody and for later gold-labelling controls. Non-immune serum, obtained from another animal, is not an adequate substitute. Cross-reactivity can also arise if the immunogen itself carries epitopes that are also carried on other macromolecules in the study tissue. Particularly problematic are glycoproteins with commonly occurring oligosaccharide side-chains. Western blots carried out before gold labelling will:

- detect non-specific binding
- help determine the appropriate dilution of the antibody for subsequent use

Monoclonal hybridoma technology overcomes many of the problems of non-specificity associated with polyclonal antisera, since monoclonal antibodies react only with a single epitope of a macromolecule. This attribute allows partially purified or unpurified material to be used to immunize either rats or mice; and antibodies of interest can then be selected during screening of the hybridoma cell lines. A major drawback to this technique is that production and screening of monoclonals is extremely labour-intensive. That a monoclonal antibody recognizes a single epitope is both problematic and advantageous. The epitope recognized by a given monoclonal may exist on a number of unrelated macromolecules, and monoclonal antibodies must be carefully screened by Western blotting. Furthermore, because monoclonals recognize a single epitope, they may fail to recognize an antigen vulnerable to the denaturation or conformational changes that may be caused by fixing and processing of tissues for immunolabelling. Monoclonal antibodies present an opportunity to examine tissue-specific isozymes or developmental modification of proteins, problems not easily approached with polyclonal antisera.

Although monospecific polyclonal antibodies actually amplify the detection of a component, since the antiserum recognizes multiple epitopes within the macromolecule.

3. Tissue preparation

The first post-embedding immunolabelling experiments used material fixed and embedded as for conventional electron microscopy. This approach yields specific and ample labelling of certain abundant proteins but the antigenicity of some other proteins may not be sufficiently well preserved for immuno-labelling. Subsequent modifications of the fixation and embedding protocols have improved the preservation of antigenicity, while maintaining high quality ultrastructural preservation of the tissue. *Protocol 1* describes a typical fixation and embedding schedule.

3.1 Fixation

Material prepared for conventional electron microscopy is first fixed in buffered glutaraldehyde, followed by osmium tetroxide post-fixation. Glutaraldehyde is a strong fixative, yielding excellent ultrastructural preservation. Many immunocytochemists prefer to either reduce the concentration of glutaral-dehyde, or to substitute with paraformaldehyde, a milder fixative. Sensitivity of the epitope to aldehyde fixation varies with the antigen. It is not always necessary to sacrifice the superior ultrastructural preservation afforded by glutaraldehyde in order to maintain immunoreactivity. Sensitivity of the anti-gen to aldehydes may be pre-screened using Western or dot blotting, or ELISA assays.

Post-fixation with osmium masks the antigenic site and hence greatly reduces immunolabelling. However, this masking effect can be overcome by treating the sections with a saturated solution of sodium metaperiodate, or other highly oxidative compounds (5). Treatment with periodate destroys certain carbohydrate epitopes, and, therefore, should not be used when detecting carbohydrates. Osmium post-fixation also renders tissues totally opaque, a condition not compatible with UV polymerization of embedding resin. For these reasons, and for simplicity's sake, most investigators prefer to omit osmium tetroxide from their fixation protocols.

3.2 Embedding

Epoxy resins, such as Epon/Araldite or Spurr's resin, which are highly favoured for conventional ultrastructural studies, result in sparse particle binding to the sections. More polar acrylic resins yield higher numbers of bound gold particles without sacrificing specificity or ultrastructural preserva-tion. The Lowicryl resins were specifically developed for immunolabelling purposes, although many plant scientists have found Lowicryl difficult to infiltrate into the tissues, and difficult to section. A popular substitute is LR White, which is easy to work with and yields a strong signal. Both LR White and Lowicryl have a low viscosity and may be infiltrated into tissues at low temperatures and polymerized with UV light instead of heat. The low temperature protocol is thought to maintain antigenicity better, although this advantage has not been documented in many cases. *Protocol 1* uses heat to polymerize LR White, but Bradley *et al.* (6) describe a low temperature protocol for the same resin.

Protocol 1. Fixation and embedding plant tissue for immunolabelling protocols

Fixation and embedding may be carried out at room temperature in a fume hood.

1. Prepare fixative in buffer at an appropriate osmotic strength for the tissue of interest.[a]

2. Dissect tissue pieces in fixative. Pieces should not exceed 1 mm in diameter. Fix tissue pieces for 1–2 h.

3. After fixation, dehydrate the tissue by passing through a graded ethanol series (10, 30, 50, 70, 90, and 100% ethanol), allowing 10 min for each step. To completely dehydrate the specimen, make 3 changes in 100% ethanol.

4. Following dehydration, infiltrate LR White resin monomer into the tissue as follows;

   - 50% LR White in dry ethanol for 30 min to 1 h
   - 75% LR White for 30 min to 1 h
   - 100% LR White for 30 min to 1 h
   - 100% LR White overnight

5. The following morning, change to fresh resin. After 6 h, make a final change of resin and place the tissue pieces into embedding moulds, such as BEEM or gelatin capsules. Oxygen inhibits polymerization of LR White, and so the moulds should be closed to air.

6. Place the moulds into an oven at 55°C and allow polymerization to proceed for 24 h.

[a] Fixative is 4% paraformaldehyde and 1% glutaraldehyde in 25–100 mM potassium phosphate buffer, pH 7.0. Paraformaldehyde should be freshly prepared from solid; glutaraldehyde should be EM grade.

3.3 Section preparation

Following embedding, thin sections (70–90 nm in thickness) are cut with a glass or diamond knife, and collected on to nickel or gold mesh grids.

Lowicryl and LR White resins are sensitive to damage by the electron beam, and so the sections should be supported on grids coated with a formvar or parlodion and carbon coating.

4. Immunolabelling for electron microscopy

4.1 Choice of gold probes

Early immunocytochemical protocols used ferritin as an electron dense marker for an antigen in tissue. Colloidal gold, introduced over a decade ago, has now supplanted ferritin. It offers several advantages, such as particle uniformity, the ability to make colloidal gold in a range of discreet particle sizes and multiple antigens can be labelled on a single section. Although it is not difficult to prepare gold conjugates (7) there are now so many commercial sources, that most investigators elect to purchase their probes. When selecting a source, the investigator should ascertain that the gold particles from a particular supplier are uniform in size and monodisperse (i.e. that there are few doublets or larger aggregates of particles).

Colloidal gold particles can be stabilized with a variety of different proteins that can be used as probes. Lectin–gold labelling and enzyme–gold labelling are now widespread techniques. For immunogold labelling, the investigator has the choice of colloidal gold conjugated to protein A, protein G, a secondary antibody, or streptavidin as a probe for detecting a primary antibody bound to tissue sections. Protein A–gold is widely used, especially when used with a polyclonal antiserum raised in rabbits.

However, protein A is not a suitable probe for all assay systems since it has low affinity for some isotypes and for antibodies raised in some species, notably rats and mice. Protein G, a cell wall protein from group G streptococci, resembles protein A in its ability to bind to the Fc region of antibodies, but demonstrates high affinity for a broader range of antibodies of different isotypes and from various animal species. For this reason, protein G has been suggested as a possible substitute for protein A in assay systems when a monoclonal antibody is used for the primary antibody (8). Alternatively, a secondary antibody–gold conjugate could be used in lieu of protein A–gold. Using primary antibodies raised in different species (such as rabbits and mice) enables a simple double-labelling scheme, where specific secondary antibodies are employed in the gold conjugates. Streptavidin–gold, in concert with a biotinylated primary antibody, can also be used in single or double-labelling schemes.

4.2 Procedures for labelling

Although specimen preparation can be time-consuming and somewhat tedious, the immunolabelling protocol itself can be carried out in less than three hours. *Protocol 2* describes a typical labelling procedure.

Choosing an appropriate dilution of the primary antibody will help to maximize the signal-to-noise ratio. For a given antibody, the best dilution for immunogold labelling will be similar to that chosen for other immunoassays, such as ELISA assays or Western blotting. The best dilution can be easily chosen in a pilot experiment.

4.3 Controls

In order to assure specificity in labelling, well-chosen controls must be run in parallel with the immunolabelling experiment. As with Western blotting, substitution of pre-immune for immune serum is the best control. When using monoclonal antibodies for labelling, a superfluous antibody (one known to recognize an epitope that is absent from the experimental tissue) may be substituted. An additional, useful control is to pre-absorb the antibody with its antigen before using it for immunolabelling. This pre-treatment should effectively eliminate all non-specific labelling. Finally, when examining the distribution of a protein in transgenic tissue, comparable non-transformed tissue should also be immunolabelled to demonstrate the absence of a signal in the absence of the protein.

Protocol 2. Gold labelling of thin section for electron microscopy observations

All solutions should be filtered through a 0.2 μm membrane filter before use.

1. Place grids with attached thin sections into a drop of blocking buffer for 15 min.

   Blocking buffer:

   - 1% (w/v) bovine serum albumin
   - 0.02% sodium azide
   - 0.05% Tween 20 in Tris-buffered saline (TBS; 10 mM Tris–HCl, pH 7.4 plus 150 mM NaCl)

   This and subsequent steps should be carried out in a covered dish to minimize evaporation.

2. Remove grids from blocking buffer and place in individual droplets (about 20 μl) of primary antibody appropriately diluted in blocking buffer. Incubate in primary antibody 1 h to overnight.

3. Rinse unbound antibody from sections by passing the grids through a series of drops of TBS. Remove excess buffer from the grids with filter paper.

4. Place grids into individual droplets of secondary antibody–gold or protein A–gold, diluted in blocking buffer (Janssen Auroprobes for electron microscopy require a 1:10 to 1:40 dilution). Incubate for 1 h.

5. Rinse the grids in TBS, as above, followed by a brief wash in distilled water.

6. Post-stain the sections, for immediate observation, for 10 min in 2% aqueous uranyl acetate, followed by 1 min in Reynold's lead citrate (for less contrast in the specimen, omit staining in lead citrate). Alternatively, allow the grids to dry for later observation.

---

5. Immunogold labelling for light microscopy

Colloidal gold probes were first developed as markers for electron microscopy. In recent years, immunogold labelling has become increasingly popular for light microscopic detection of antigens, as an alternative to immunofluorescence or immunoperoxidase assays. Colloidal gold has been used to label microtubules in mitotic spindles of endosperm cells (9), where it gives an intense red signal when viewed with bright-field illumination. However, in most cases, the protein of interest is at too low a concentration to be detected with gold labelling without further amplification.

Routine application of gold labelling for light microscopic visualization became practical with the introduction of silver amplification techniques (10, 11). This approach, which uses the ability of gold particles to catalyse the deposition of metallic silver, was modified from techniques for the electron microscopic detection of heavy metals. In principle, gold-labelled specimens are incubated in a silver developer, containing a silver salt and a reducing agent. Metallic silver precipitates on to the gold particles, encasing the antigenic site in an opaque shell of silver visible under the light microscope.

Silver enhancement of gold labelling has several advantages over alternative methods; the signal is:

- more intense than that obtained with immunoperoxidase assays;
- more permanent than immunofluorescence.
- A fluorescence microscope is not required, and the problem of autofluorescence, common in many plant tissues, is avoided.
- Silver enhancement of immunogold labelling may be carried out on tissue fixed and embedded as for electron microscopy, affording correlative localization studies using light and electron microscopy (12).

Procedures for tissues preparation for immunolabelling and silver enhancement are outlined in *Protocol 3*. This protocol uses resin embedded samples (as for electron microscopy) and yields excellent antigen and cellular preservation. Silver enhancement of gold labelling can also be used on paraffin-embedded or cryosectioned tissues prepared for light microscopy (13). These methods may actually result in an amplification of the signal since the embedding medium is removed before the sections are incubated with the antibody.

The original method for silver enhancement (10, 11) can be successfully used without modification. However, several properties of the developer, which contains hydroquinone and silver lactate in a citrate buffer, make it difficult to employ:

- The developer has a very short life, so that after about 15 minutes, spontaneous silver precipitation leads to a high background.
- The developer is extremely sensitive to contaminating ions, such as chloride from the incubation buffer, and this again leads to a high background.
- The developer is light sensitive, so enhancement must be carried out under darkroom safelight conditions or in total darkness.

Recently, Janssen Life Sciences Products have developed an alternative developer for silver enhancement which alleviates the difficulties encountered with the conventional developer. It is stable for a longer period of time, is relatively insensitive to water quality, and may be used under room light conditions. These qualities make the use of the silver enhancement technique an even simpler matter.

Protocol 3. Gold labelling and silver enhancement of semi-thin sections

All procedures are carried out at room temperature.

1. Embed tissue in LR White acrylic resin, as described in *Protocol 1*.
2. Cut semi-thin sections (0.5–1.0 μm) of embedded tissue using a glass knife on an ultramicrotome. Remove sections to a drop of water on a glass slide, and allow to dry at room temperature, or with slight heating (no more than 37°C). (Sections will adhere best if slides have been coated previously with 1% gelatin.) Store sections on slides in the dark before proceeding with labelling.
3. Apply blocking buffer (see *Protocol 2*) to sections and incubate for 15 min (100 μl of buffer covers 5–10 sections). Perform this and subsequent steps in a humid environment.
4. Remove blocking buffer by tipping the slide. Dry the area around the sections with a tissue. Never let the sections dry out at any time during immunolabelling.
5. Cover sections with a 50–100 μl drop of primary antibody, appropriately diluted in blocking buffer. Incubate 1 h to overnight.
6. Rinse sections with a steady stream of TBS (see *Protocol 2*) from a wash bottle. Place slides in a Coplin jar filled with TBS and wash 3× for 10 min each with constant agitation. Dry the area around the sections.
7. Cover sections with secondary antibody–gold or protein A–gold, diluted in blocking buffer (Janssen Auroprobes for light microscopy require a

1:10 to 1:40 dilution). Colloidal gold particles used for silver enhancement purposes should be 5 nm in diameter. Incubate for 1 h.

8. Rinse and wash sections in TBS as described in step 6.

9. Wash slides in distilled water, 3× 5 min each.

10. Proceed to silver enhancement (see text), or air dry and store slides for later development.

---

6. Examples of results

Immunogold labelling has been used to detect a wide variety of molecular components in plant cells, including proteins, carbohydrate side-chains on glycoproteins, and cell wall polysaccharides. The examples of immunolabelling of legume root nodules which follow illustrate the application of the techniques.

Figures 1 and *2* illustrate the localization of leghaemoglobin (Lb) in soybean nodules. Although Lb is the most abundant nodule protein, its location within the cell was disputed until it was localized in the infected cell cytoplasm (14). Further work, by VandenBosch and Newcomb (15), detected the protein in uninfected cells at a lower concentration (*Figure 3*). In this study, the nodule was embedded in LR White as described above, and probed with polyclonal antibodies against Lb, followed by protein A–gold.

Figure 2 depicts the localization of Lb in the same specimen, using immunogold labelling and silver enhancement. Semi-thin (0.5–1.0 μm) sections were labelled with the same antibody preparation, followed by goat anti-rabbit gold, and silver enhancement according to the method of Danscher and Nörgaard (11). Following counterstaining, the preparation was photographed using bright-field optics.

Pectin moieties were identified in the cell walls of pea nodules (*Figure 3*), using a monoclonal antibody against polygalacturonic acid (16). The tissue was prepared by low-temperature embedding (6). Following incubation in the primary antibody, the sections were labelled with goat anti-rat gold.

7. Future developments

Although the sensitivity of immunogold labelling has been enhanced by the development of modified fixation protocols and new embedding resins, some cellular components remain difficult to label because they are not well preserved by these techniques. An especially intractable labelling problem has been f-actin, which is poorly preserved by aldehyde fixation. Recently, Lancelle and Hepler (17) succeeded in labelling actin microfilaments by avoiding chemical fixation altogether. Instead, the tissue was preserved by rapid freezing in a hyperbaric freezer or by plunge freezing in liquid propane, followed by freeze-substitution and embedding in LR White resin. Until

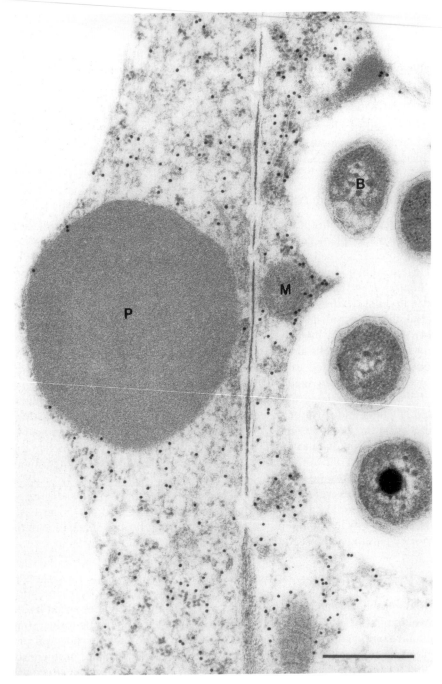

Figure 1. Immunogold localization of leghaemoglobin in two cells of a soybean nodule. See text for details of labelling protocol. Note the lack of labelling over the cell wall, organelles, and bacteria. ×49 000; bar, 0.5 μm; B, *Bradyrhizobium* bacteroid; M, mitochondrion; P, peroxisome.

40

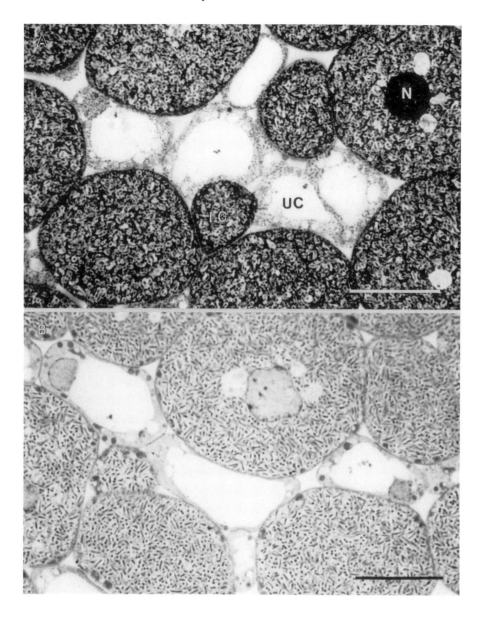

Figure 2. A. Immunohistochemical localization of leghaemoglobin, using silver enhancement of gold labelling. The metallic silver precipitate appears black in the micrograph, and demonstrates the presence of leghaemoglobin in the cytoplasm and nuclei (N) of infected cells (IC). The label is less dense in uninfected cells (UC), where it appears as a grainy precipitate. B. Silver enhancement of gold labelling, following incubation in pre-immune serum. No label is discernible. Sections were counterstained in azur II and methylene blue. ×1200; bar, 20 μm.

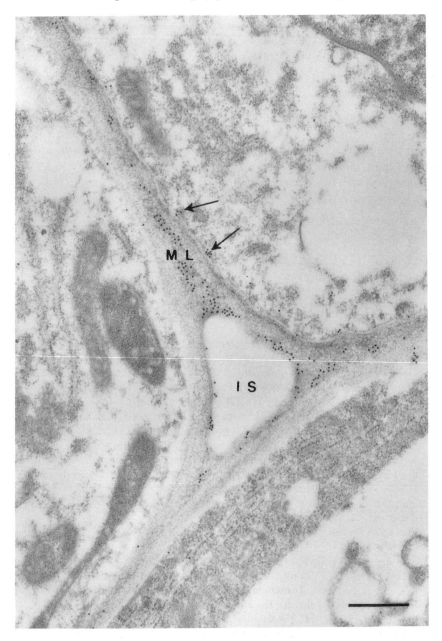

Figure 3. Immunogold localization of pectin in a pea nodule. Pectin is detectable in the middle lamella (ML) between adjacent cells, and lining an intercellular space (IS) at a three-way junction between cells. Some label is also present at the plasma membrane (arrows), probably indicating the presence of newly secreted pectin molecules. ×39 000; bar, 0.5 μm.

recently, rapid-freeze fixation was not applicable to most complex tissues because of cell disruption due to ice crystal damage, but the hyperbaric freezer has extended cryofixation techniques to complex tissues. The immunolabelling of actin suggests that rapid-freezing techniques will be efficacious not only for superior structural preservation, but also for retention of antigenicity.

An advantage of immunogold labelling, as stated above, is that several components can be co-localized in the same specimen. Use of multiple labelling techniques has, so far, been restricted to immunogold identification of two or more antigens. Recently, *in situ* hybridization methods have been developed for electron microscopy, using gold probes as labels (18). This advance opens the exciting possibility that transcripts, and the proteins they encode, could be detected within single cells.

References

1. Roth, J., Bendayan, M., and Orci, L. (1978). *J. Histochem. Cytochem.*, **26**, 1074–81.
2. De Mey, J. and Moeremans, M. (1986). In *Immunocytochemistry: modern methods and applications* (ed. J. M. Polak and S. Van Noorden), p. 3. John Wright and Sons, Bristol.
3. Ritter, M. A. (1986). In *Immunocytochemistry: modern methods and applications* (ed. J. M. Polak and S. Van Noorden), p. 13. John Wright and Sons, Bristol.
4. Galfre, G. and Butcher, G. W. (1986). In *Immunology in plant science* (ed. T. L. Wang), p. 1. Cambridge University Press, Cambridge.
5. Bendayan, M. and Zollinger, M. (1983). *J. Histochem. Cytochem.*, **31**, 101–9.
6. Bradley, D. J., Wood, E. A., Larkins, A. P., Galfre, G., Butcher, G. W., and Brewin, N. K. (1988). *Planta*, **173**, 149–60.
7. Slot, J. W. and Geuze, H. J. (1985). *Eur. J. Cell Biol.*, **38**, 87.
8. Bendayan, M. and Garzon, S. (1988). *J. Histochem. Cytochem.*, **36**, 597–607.
9. De Mey, J., Lambert, A., Bajer, A., Moermans, A., and De Brabander, M. (1982). *Proc. Natl. Acad. Sci. (USA)*, **79**, 1898–902.
10. Holgate, C. S., Jackson, P., Cowen, P. N., and Bird, C. C. (1983). *J. Histochem. Cytochem.*, **31**, 938.
11. Danscher, G. and Nörgaard, J. O. R. (1983). *J. Histochem. Cytochem.*, **31**, 1394–8.
12. VandenBosch, K. (1986). *J. Microscopy*, **143**, 187.
13. O'Brien, T. P. and McCully, M. E. (1981). *The study of plant structure: principles and selected methods*. Termarcarphi Pty Ltd., Melbourne, Australia.
14. Robertson, J. G., Wells, B., Bisseling, T., Farnden, K. J. F., and Johnson, A. W. B. (1984). *Nature*, **311**, 254–6.
15. VandenBosch, K. A. and Newcomb, E. H. (1988), *Planta*, **175**, 442–51.
16. VandenBosch, K. A., Bradley, D. J., Knox, J. P., Perotto, S., Butcher, G., and Brewin, N. J. (1989). *EMBO J.*, **8**, 335–42.
17. Lancelle, S. A. and Hepler, P. K. (1989). *Protoplasma*, **150**, 72.
18. Binder, M., Tourmente, S., Roth, J., Renaud, M., and Gehring, W. J. (1986). *J. Cell Biol.*, **102**, 1646–53.

6

Isoflavonoid phytoalexins and their biosynthetic enzymes

ROBERT EDWARDS and HELMUT KESSMANN

1. Introduction

Phytoalexins are low molecular weight antimicrobial compounds which accumulate in plant tissues after exposure to micro-organisms. In the case of isoflavonoid phytoalexins , the rates of synthesis and accumulation of these compounds in leguminous plants can be correlated with the resistance of the host to the pathogen (for recent reviews see references 1–3). The biosynthesis of isoflavonoid phytoalexins (for example medicarpin) proceeds via a multi-step pathway which begins with phenylalanine (*Figure 1*). The pathway may also be extended to produce more complex isoflavonoid phytoalexins, such as the prenylated pterocarpans phaseollin and the glyceollins. In this review we will use the biosynthetic route to medicarpin to illustrate the methods used to analyse isoflavonoids and to measure the activities of the enzymes involved in their synthesis.

2. Isolation of phytoalexins

2.1 Induction of phytoalexin biosynthesis

The exact methods of treatment for induction of phytoalexins will depend on the biological system used. Whole plant tissues may be elicited with suspensions of fungi (spores or mycelia) or bacteria either by surface application or by infiltration. Alternatively, abiotic elicitors such as heavy metal salts, may be used at a final concentration of 10^{-2}–10^{-3} M. However, biotic elicitors derived from micro-organisms or plants (see Chapter 8) are frequently the inducing agents of choice. When treating suspension-cultured plant cells the sterilized elicitor is added in aqueous solution with sterile water serving as a control treatment. In the case of crude elicitor preparations from fungal cell walls elicitation may be optimized in the range 10–100 mg glucose equivalents per litre of cell culture medium. Elicitor active fractions may be most conveniently obtained from yeast (4).

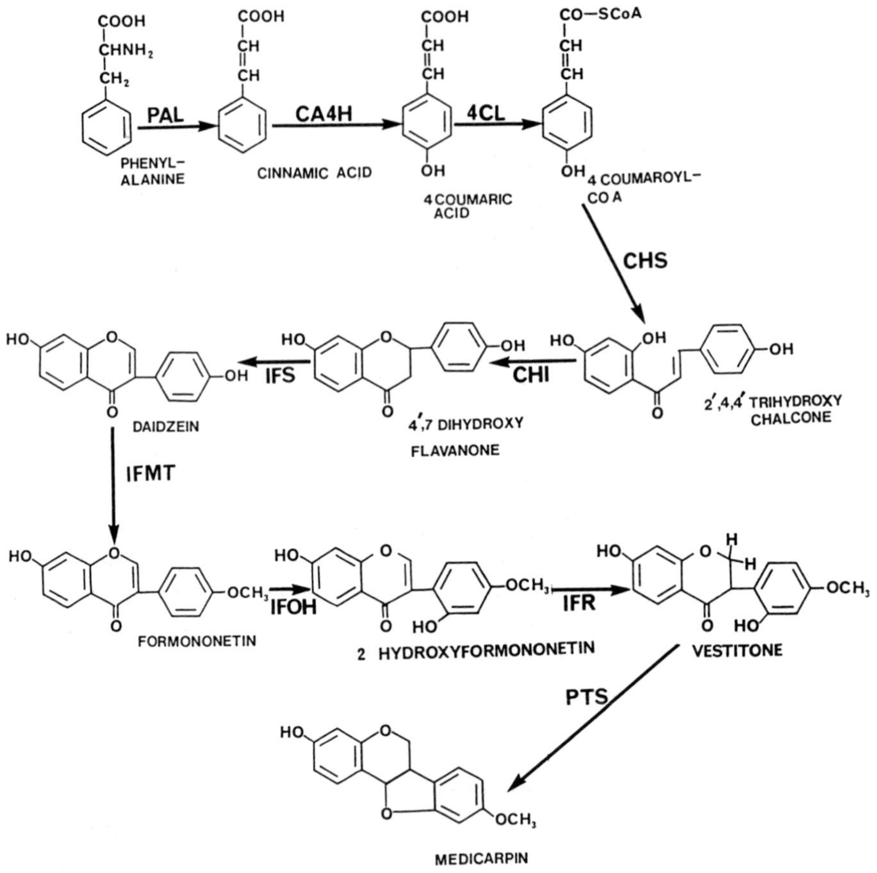

Figure 1. Biosynthesis of isoflavonoid phytoalexins in legumes. Enzyme abbreviations: PAL, L-phenylalanine ammonia lyase; CA4H, cinnamic acid 4-hydroxylase; 4CL, 4-coumarate CoA ligase; CHS, chalcone synthase; CHI, chalcone isomerase; IFS isoflavone synthase; IOMT, isoflavone *O*-methyl transferase; IFOH, isoflavone hydroxylase; IFR, isoflavone reductase; PTS, pterocarpan synthase.

2.2 Extraction of plant tissue

The lipophilic isoflavonoids and their related conjugates are readily extracted using organic solvents (acetone, methanol, or acetonitrile). The procedure given in *Protocol 1* is applicable to both cell cultures and whole plants (5).

Protocol 1. Extraction of plant tissue

A. *Cell cultures*

1. Separate the medium from the cells by vacuum filtration. Store cells and medium at −20°C until required.

2. Extract medium by partitioning twice with 2 volumes per volume diethyl ether or ethyl acetate (analytical grade solvents). Concentrate the organic phase under vacuum and redissolve in 1 ml methanol prior to analysis.

B. *Fresh or frozen plant tissue*

1. Extract in cold acetone (−20°C) to prevent hydrolysis of conjugates and peroxidative destruction of phenolics.

2. Homogenize 1 g tissue with 10 ml of cold acetone in a pestle and mortar containing 0.5 g acid-washed sand.

3. Vacuum filter the resulting liquid and re-extract the residue with 10 ml acetone followed by 10 ml acetone + methanol (1:1 v/v). Combine the filtrates and evaporate the sample under vacuum to 1–10 ml (depending on isoflavone content). Acetone should not be used to adjust the volume of the concentrate as it interferes with HPLC analysis.

2.3 HPLC analysis of isoflavonoids

The HPLC system described (5) may be used, with or without modification, to analyse plant extracts and to monitor the enzyme reactions described in subsequent sections. For isoflavone analysis a C-8 or C-18 reversed-phase column of dimensions 4 × 250 mm and particle size 5 μm is recommended. All solvents should be of HPLC grade and should be filtered and degassed prior to use. Solvents used are:

• Solvent A   1% aqueous phosphoric acid (85% w/v)
• Solvent B   acetonitrile

After equilibrating the system, the following elution programme is used at 0.8 ml/min: 80% solvent A, 20% solvent B to 40% solvent A, 60% solvent B in 45 min over a linear gradient. The eluant is monitored for UV absorbance depending on the absorption maximum of the products of interest (note: pterocarpans have lower molar extinction coefficients than do isoflavones). To distinguish between isoflavones which contain acidic functions from those which do not, the acid in solvent A may be omitted. Caution should be employed when identifying compounds in crude extracts solely on the basis of co-chromatography with standards (for sources of standards see *Table 2*). Confirmation of identity by further chromatography or by spectroscopy (6) is recommended.

2.4 TLC analysis of isoflavonoids

Cell extracts are prepared as detailed in Section 2.2 and applied as discreet bands to silica gel analytical TLC plates containing fluorescent indicator. Plates are developed in one of the solvents listed in *Table 1*. Unknowns are identified by co-chromatography with standards (best achieved using two-dimensional TLC) or by scraping the zones of interest and packing the silica gel into a Pasteur pipette plugged with glass wool. The isoflavone is eluted with spectroscopic grade methanol and analysed by scanning UV spectroscopy then compared with published spectra (6). Low concentrations of phenolic metabolites may be located on the TLC plate using one of the two following spray reagents.

2.4.1 Diazotized *p*-nitroaniline

Dissolve 12.5 mg *p*-nitroaniline in 2.5 ml 2 M HCl and then cool in an ice bath. Add 0.25 ml ice-cold 5% (w/v) $NaNO_2$ dropwise to the nitroaniline solution, followed by 7.5 ml of 20% (w/v) sodium acetate. Use immediately. An orange coloration denotes the presence of phenolics.

2.4.2 Fast blue salt

Dissolve 50 mg of Fast Blue B salt in 5 ml 0.2 M citrate–phosphate buffer (pH 4) and spray immediately. Plates are then heated to 110°C to develop coloration in the presence of *meta*-hydroxylated isoflavones.

Table 1. TLC solvents for isoflavone analysis

Solvent	v/v
1. Petroleum ether[a] + ethyl acetate + methanol	10:10:1 and 60:40:1
2. Toluene + ethyl acetate + methanol + petroleum ether[a]	6:4:1:3
3. Dichloromethane + methanol	15:1
4. Chloroform + methanol	50:1–10:1
5. Petroleum ether[a] + diethyl ether + acetic acid	25:75:1
6. Pentane + diethyl ether + acetic acid	75:35:3
7. Chloroform + acetone + 25% ammonia	90:10:5

[a] Petroleum ether is the 55°–65° B.pt fraction

3. Enzymes of isoflavonoid biosynthesis

In recent years most of the enzymes involved in isoflavonoid phytoalexin biosynthesis have been successfully assayed and a number purified to homogeneity. These advances have been achieved largely by improvements in the methods used to prepare enzyme extracts from plant tissues which contain large amounts of inhibitory phenolic material. In addition, the development of sensitive HPLC assays has enabled the researcher to monitor reactions for which there are no alternative procedures (see *Figure 2*). Before considering the individual assays in turn some important points are listed which should help ensure the successful determination of enzyme activity.

(a) Ensure that the enzyme of interest is stable following the freezing of plant material. For optimal storage freeze the tissue in liquid nitrogen and store at −70°C. Compare the enzyme activities before and after freezing to confirm stability.

(b) Phenolics may be removed from enzyme extracts by treating with insoluble polyvinylpyrrolidone (PVP) or with Dowex 1 ion-exchange resin. Dowex 1 should be washed extensively with distilled water and then equilibrated with the extraction buffer prior to use.

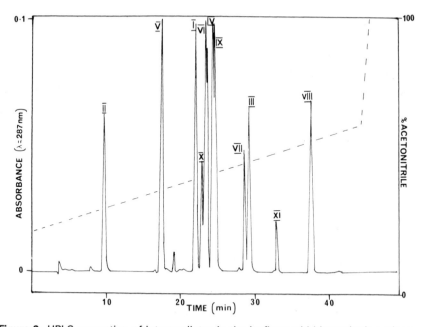

Figure 2. HPLC separation of intermediates in the isoflavonoid biosynthetic pathway. I, cinnamic acid; II, 4-coumaric acid; III, 2'4,4'-trihydroxychalcone; IV, naringenin; V, daidzein; VI, genstein; VII, formononentin; VIII, biochanin A; IX, 2'-hydroxyformononetin; X, vestitone; XI, medicarpin.

(c) Where the text refers to desalting by gel permeation chromatography two methods may be employed. For critical desalting pre-packed PD-10 columns (Pharmacia) are employed and used as recommended by the manufacturers. For smaller sample volumes spun-column elution is used. Briefly, a plastic syringe is plugged with glass wool and slurried Sephadex G-25 added such that the packed bed-volume is five to ten times the sample volume to be applied. The column is then pre-washed with one column volume of the final buffer and the excess solvent removed by centrifugation (70 g, 1 min). Finally the sample is applied and eluted by centrifugation (70 g, 1 min). Using spun-column desalting large numbers of samples may be routinely prepared for assay.

(d) When assaying these enzymes using the given procedures it should be borne in mind that the assay conditions may require adaptation. Thus in our experience of determining isoflavone *O*-methyl transferase activity the optimal pH for assay in alfalfa (pH 8.5), though only slightly different from that previously reported in chickpea (pH 8.8), resulted in significant differences in assay sensitivity. For this reason the assay conditions presented should be considered as guidelines only.

(e) Due to the low solubility in water of many of the phenolic substrates and products, binding of lipophilic compounds to the proteins present in the assay mixture may be encountered. For this reason the rate of product formation may only be directly proportional to protein concentration over a relatively narrow range. For crucial determinations this must be determined empirically for each assay. In any event meaningful results from controls containing no active enzyme can only be obtained with protein present.

3.1 Chemicals

3.1.1 Substrates

Table 2 describes the sources of all substrates and reference products used in this section. With the exception of 2'-hydroxyformononetin, vestitone, and medicarpin, the synthesis of the compounds which are not available commer-

Table 2. Sources of substrates

Substrate	Supplier/reference
Phenylalanine, 4-coumaric acid	Sigma Chemical Co.
Naringenin (4', 5, 7-trihydroxyflavone) daidzein, formononetin	Apin Chemicals, Ltd [a]
Cinnamic acid	Aldrich Chemical Co.
4-Coumaroyl CoA	Ref. 8 see *Protocol 2*
2',4,4'-Trihydroxychalcone	Ref. 8
2'-Hydroxyformononetin, vestitone, medicarpin	Refs. 9, 10

[a] Apin Chemicals Ltd.

cially is straightforward. For the synthesis of these isoflavones we recommend that the reader consults the references given or to consider isolating these compounds from a biosynthetic source. The purity of all substrates should be confirmed prior to use and, if necessary, improved to greater than 95% (w/w) by chromatography or recrystallization.

Because of the central role chalcone synthase (CHS) plays in the pathway and, therefore, the importance of its assay, the synthesis of the substrate 4-coumaryl CoA is described in *Protocol 2* (7).

Protocol 2. Synthesis of 4-coumaroyl CoA

1. Dissolve 2.64 g 4-coumaric acid in 80 ml ethyl acetate at 50°C. Filter the hot solution, cool to 30°C and add 1.73 g of *N*-hydroxysuccinimide with constant stirring. Slowly add 3.5 g of dicyclohexylcarbodiimide over 30 min and allow to stand for 48 h at room temperature.

2. Remove the insoluble dicyclohexylurea crystals by filtration, and partition the organic solvent 3 times against 1 M $NaHCO_3$ (1 volume per volume) and dry with Na_2SO_4 prior to evaporation under vacuum.

3. Purify the crude reaction product containing the succinimide ester by preparative TLC or flash column chromatography (silica gel 60) using chloroform + methanol (20:1 v/v) as developing solvent. Alternatively, recrystallization from chloroform + ethanol (8:2 v/v) gives similar results. The resulting yellow product has a melting point of 116–119°C.

4. Dissolve 210 mg $NaHCO_3$ in 2.5 ml distilled water and pass N_2 gas through the solution prior to adding 63 mg of coenzyme A (Li salt, Sigma). After 10 min, add 80 mg of the succinimide ester and evenly disperse the suspension with the dropwise addition of acetone. Allow to react for 2 h at room temperature under a N_2 atmosphere and analyse the reaction mixture by TLC in chloroform + methanol (20:1 v/v). An accumulation of UV absorbing material on the origin corresponds to the CoA ester.

5. Evaporate the acetone under a N_2 stream and dilute the aqueous phase. Partition against 1 vol. diethyl ether and then treat with 3 g of Dowex 50 W-X 8 after which time the solution should be colourless. After filtration the solution is acidified with formic acid and partitioned again with 1 vol. ethyl acetate.

6. Lyophilize the aqueous solution, and resuspend in 2 ml of 50 mM formic acid. Apply to a Sephadex G-10 column and elute with 50 mM formic acid. Test fractions for absorbance at 254 nm and 333 nm. After an initial minor peak of unreacted coenzyme A is observed the CoA ester elutes (max 254 and 333 nm) which may be quantified from its molar extinction coefficient at 254 nm log ε = 4.32.

7. Store the CoA ester at −70°C in water acidified to pH 3.5 with dilute HCl at a concentration of 0.55 mg/ml in 0.15 ml aliquots.

3.1.2 Preparation of radiolabelled substrates

[2-^{14}C]malonyl CoA (2.18 GBq/mmol) may be obtained from Amersham and adenosyl-L-methionine-S-[^{14}C-methyl] (17 GBq/mmol) from NEN. Following dilution to the required specific activity both of the labelled substrates should be stored at $-20\,°C$ in water which has been adjusted to pH 3.5 with dilute HCl. After thawing these substrates should be used immediately and should not be refrozen. For critical kinetic studies S-adenosyl methionine (SAM) should be treated with ion-exchange resin (11) to remove contaminating S-adenosyl homocysteine (SAH), a potent inhibitor of the transferase. Briefly, Dowex 1, which has been sequentially washed with 25:1 (v/w) 4 M NaCl, 10:1 (v/w) 0.01 M NaCl, and 40 (v/w) 0.5 M $NaHCO_3$, is packed into a small glass column and the diluted [^{14}C]SAM applied in water. Contaminating SAH binds to the column and the unretained SAM is acidified and stored. In practice, however, the use of stabilized salts of SAM, such as the *p*-toluene sulphonate form, is adequate for routine methyl transferase determination without further treatment.

3.1.3 Synthesis of [^{14}C]cinnamic acid

Radiolabelled cinnamic acid is no longer widely available although [3-^{14}C]cinnamic acid (1.76 GBq/mmol) can still be obtained from Cen Saclay. Alternatively [U-^{14}C]cinnamate can be synthesized from L-[U-^{14}C]phenyl-alanine (19 GBq/mmol, Amersham) by incubating the [^{14}C]phenylalanine with an active preparation of phenylalanine ammonia lyase (PAL). Prepara-tions of PAL are readily obtained by homogenizing elicited plant material in 0.1 M Tris–HCl (pH 8.5) containing 5 mM 1,10-phenanthroline, 1 mM phenylmethyl sulphonyl fluoride and 5% (w/v) PVP. Precipitate the cell free extract with $(NH_4)_2SO_4$ between 20–50% saturation and resuspend the pellet in 0.1 M Tris–HCl (pH 8.5) containing 2 mM 2-mercaptoethanol. Finally apply the extract to a Sephacryl S-300 column and elute the activity with the application buffer. Transfer the enzyme preparation (0.4 nkat in 4 mg pro-tein), in 3 ml, to a dialysis bag and incubate with [^{14}C]phenylalanine for 16 h at 37°C in 25 ml 50 mM Tris–HCl (pH 8.5) containing 0.1% (w/v) sodium azide. During the incubation the medium outside the dialysis bag is slowly circulated through a C18 SEP–PAK (Water Associates) to remove excess cinnamate from solution. Finally combine the dialysis bag contents and medium, and acidify the reaction mixture with HCl. Pump the acidified sample on to the SEP–PAK and wash the column with 0.01 M HCl. The bound radioactivity (largely cinnamic acid) is recovered by eluting with etha-nol. The [^{14}C]cinnamate may be further purified by TLC on silica gel in toluene + ethyl acetate + formic acid (25:75:1 (v/v/v)) to a final radio-chemical purity of 96.5% and a yield of 52%.

3.1.4 Synthesis of [^{14}C]naringenin

Radiolabelled naringenin (*Protocol 3*) can be obtained by a scaled-up pro-cedure of the chalcone synthase (CHS) assay outlined in Section 3.2.7. A reliable CHS preparation should be isolated as a $(NH_4)_2SO_4$ precipitate from an elicited plant source as described in Section 3.2.3.

Protocol 3. [^{14}C]naringenin synthesis

1. Desalt the protein precipitate in 50 mM potassium buffer (pH 8) containing 20 mM ascorbic acid, and adjust to a protein concentration of 1–2 mg/ml.

2. Incubate 100 μl of the enzyme extract with 58 μl of assay buffer, 30 μl 4-coumaroyl CoA, and 12 μl [3-^{14}C]malonyl CoA (2.18 GBq/mmol, 8.9 KBq total) at 35°C for 30 min (see Section 3.2.7 for assay details).

3. Partition the reaction contents twice against 1 ml ethyl acetate, and apply the concentrated organic phase to a silica gel TLC plate. After developing the plate in chloroform + ethanol (3:1 v/v), the pure [^{14}C]naringenin is eluted from the plate in methanol and stored as a methanolic solution (16–32 Bq/μl) at −20°C.

3.2 Assay of isoflavonoid biosynthetic enzymes

As discussed in Section 3 the specific conditions for assaying these enzymes may vary from those stated. Particular attention is drawn to the use of fresh tissue as opposed to frozen material. If in any doubt as to enzyme stability fresh tissue should be assayed, particularly when measuring the activity of membrane-bound enzymes, such as the P450's, and when assaying green tissue from whole plants.

3.2.1 Homogenization of samples

Frozen plant material may be homogenized by grinding with a pestle and mortar in liquid nitrogen. Thawed or fresh tissue is ground in a pestle and mortar in the presence of 0.4 (w/w) quartz sand or alternatively homogenized with an ice-cooled blade homogenizer, such as a Brinkman Polytron or Janke and Kunkel Ultraturrax set at half-maximal speed for 20 sec. For the individual enzyme assays the method of homogenization will only be specified if the results of using an alternative method are either detrimental or unknown. When using a blade homogenizer add PVP and Dowex 1 to the extract immediately after homogenization and not before.

3.2.2 Preparation of microsomes

Microsomes must be prepared to measure isoflavone synthase (IFS), isoflavone 2′-hydroxylase (IFOH), and to determine cinnamic acid 4-hydroxylase (CA4H) activity without using radiolabelled cinnamate. The procedure given in *Protocol 4* may be used to prepare microsomal fractions for all three assays.

Protocol 4. Preparation of plant microsomes

1. Homogenize 5 g tissue in 5 ml 0.1 M potassium phosphate buffer (pH 7.5) containing 0.4 M sucrose, 28 mM 2-mercaptoethanol, and 1.6 g Dowex 1 × 2 (200–400 mesh prewashed in phosphate buffer) in an ice-cold pestle and mortar with 5 g acid-washed sand for 5–10 min.

2. Decant the supernatant and rinse the pestle and mortar with a further 3 ml buffer. Centrifuge the combined extract at 8000 *g* for 15 min at 4°C.

3. Filter the supernatant through glass wool and rinse the filter with buffer such that the final filtrate volume is 10 ml. Centrifuge at 135 000 *g*, 80 min, 4°C.

4. Decant the cytosol and retain for enzyme assay if required (Section 3.2.3). Gently wash the pellet with 1 ml of the buffer and then blot-dry with filter paper.

5. Resuspend the microsomes in 0.5 ml 0.1 M potassium phosphate buffer (pH 7.5) containing 0.4 M sucrose and 3.5 mM 2-mercaptoethanol, either assay directly or store in 0.1 ml aliquots at −70°C (*note*: use fresh microsomes for assaying isoflavone synthase).

3.2.3 Assaying for multiple activities

Considerable time and effort may be saved if a single cell extract is assayed for multiple activities. Thus, when preparing microsomes it is perfectly feasible to measure soluble enzyme activities in the cytosolic fraction. If time does not allow the assay of the enzymes in fresh solution the preparations may be frozen or precipitated with $(NH_4)_2SO_4$ (80% saturation). Alternatively, plant tissue may be extracted with a general buffer, such as Tris–HCl, pH 7.5, and 14 mM 2-mercaptoethanol, and the cell-free supernatant divided into fractions which are independently precipitated with saturated solutions of $(NH_4)_2SO_4$ to 80% saturation. The precipitated pellets can then be stored at −70°C prior to desalting and analysis. The individual enzyme assays are now considered in order of reaction sequence (see *Figure 1*).

3.2.4 L-Phenylalanine ammonia lyase (PAL; EC 4.3.1.5)

The basis of this assay is the spectrophotometric measurement of formation of cinnamic acid (12; *Protocol 5*).

Protocol 5. Assay of phenylalanine ammonia lyase

1. Extract cells in 1–2 (v/w) 50 mM Tris–HCl (pH 8.5) containing 14 mM 2-mercaptoethanol and 5% (w/v) PVP and centrifuge at 10 000 *g*, 10 min, 4°C.

2. Incubate 0.1 ml samples of the supernatant at 40°C with 0.9 ml 12.1 mM L-phenylalanine in 50 mM Tris–HCl (pH 8.5) with a parallel incubation of sample with 0.9 ml 12.1 mM D-phenylalanine serving as control.

3. Monitor the formation of cinnamic acid by taking absorbance readings at 30 min intervals in quartz cuvettes at 290 nm up to 2 h.

4. Calculate PAL activity in μkats/kg protein as:

$$\frac{27780 \times (\Delta A_{290} \text{ L-Phe}/60 \text{ min} - \Delta A_{290} \text{ D-Phe}/60 \text{ min})}{\text{μg protein per incubation}}.$$

Cinnamic acid is a potent inhibitor of PAL activity and can give rise to non-linear kinetics. To remove endogenous cinnamic acid in the cell extract a 15 min treatment with Dowex 1 is recommended. To confirm the correlation between change in absorbance at 290 nm and cinnamic acid formation [U-^{14}C]L-phenylalanine may be added to the incubation to a final specific activity of 9.25 MBq/mmol. The formation of [^{14}C]cinnamic acid may then be followed by spotting 20 μl samples on to a TLC plate at the 30 min intervals used for the absorbance readings. The plates are then developed in toluene + ethylacetate + formic acid (50:50:5 v/v/v) and the [^{14}C]cinnamate quantified by scintillation counting.

3.2.5 Cinnamic acid 4-hydroxylase (CA4H; EC 1.14.13.11)

The basis of the assay is the chromatographic identification of 4-coumaric acid using radioisotopes (13) or HPLC. CA4H may be assayed in crude supernatants using radiolabelled cinnamic acid (*Protocol 6* A). Alternatively microsomal fractions may be prepared and the enzyme assayed after separating out the product by HPLC (*Protocol 6* B).

Protocol 6. Assay of cinnamic acid 4-hydroxylase

A. *Crude extracts*

1. Extract tissue in 1 (v/w) 50 mM Tricine–KOH (pH 7.5) containing 14 mM 2-mercaptoethanol and 5% (w/v) Dowex 1. Centrifuge the extract (10 000 g, 10 min, 4°C) and apply 0.3 ml of the supernatant to a G-25 spun column equilibrated with 50 mM Tricine–KOH buffer (pH 7.5) (*Note:* 2-mercaptoethanol is omitted as it has been reported that thiol reagents can interfere with the assay).

2. Prepare an NADPH generating system in 0.05 M potassium phosphate buffer as follows:

   • *Glucose-6-phosphate dehydrogenase.* Prepare a 67 nkat/ml solution in buffer containing 1% bovine serum albumin. Store in 0.2 ml aliquots at −70°C.

Protocol 6. *Continued*

- *NADP and glucose-6-phosphate.* Make up 15 mM stock solutions and store separately in 0.1 ml aliquots at −70°C.

3. Prepare an 11.1 mM stock solution of [^{14}C]cinnamic acid (3-^{14}C or U-^{14}C labelled) in ethanol to a specific activity of 45 MBq/mmol.

4. Pre-incubate 10 μl NADP, 10 μl glucose-6-phosphate, 5 μl of glucose-6-phosphate dehydrogenase, and 45 μl of enzyme at 30°C for 5 min prior to initiating the reaction with 5 μl [^{14}C]cinnamate. Incubate at 30°C for 30 min.

5. Stop the reaction with 75 μl of an ice-cold ethanolic solution of 4-coumaric acid (1 mg/ml). Apply 75 μl to a 3 cm wide band on a silica gel TLC plate and develop in toluene + ethyl acetate + formic acid (50:50:1 v/v/v). Resolve the [^{14}C]coumaric acid from cinnamic acid and quantify the zone by scraping off the silica gel into a scintillation vial containing 0.5 ml methanol. Assay the eluted radioactivity in a scintillation counter.

B. *HPLC*

1. Pre-incubate 50 μl of microsomal preparation (see *Protocol 4*) with 480 μl 0.1 M potassium phosphate buffer (pH 7) and 20 μl cinnamic acid (2 mM in acetonitrile) for 5 min at 30°C. Initiate the reaction by adding 50 μl of 20 mM NADPH and incubate for 30–60 min at 30°C.

2. Stop the reaction with 40 μl 6 M HCl and partition twice with 0.6 ml ethyl acetate (water-saturated), removing 0.45 ml of the organic phase after each extraction.

3. Evaporate the 0.9 ml of ethyl acetate under vacuum and redissolve the residue in 100 μl acetonitrile + water (1:1 v/v), analyse by HPLC at 290 nm following calibration with 4-coumaric acid (10–200 μM) (see Section 2.3 for HPLC system).

3.2.6 4-Coumarate: CoA ligase (4CL; EC 6.2.1.12)

The basis of this assay is the derivitization of activated product with hydroxylamine (14; *Protocol 7*).

Protocol 7. Assay of 4-coumarate: CoA ligase

1. Prepare the following reagents in 0.2 M potassium phosphate buffer (pH 7.5).

- Solution containing 50 mM MgSO$_4$ and 50 mM ATP, adjusted to pH 7.5 with KOH.[a]

- 10 mM CoA, adjusted to pH 7.5 with KOH.[b]

- 100 mM 4-coumaric acid in dimethylsulphoxide.[b]
- 1.33 M hydroxylamine–HCl.[b]

2. Prepare an enzyme precipitate as detailed in Section 3.2.3. Alternatively, extract tissue with 0.2 M KH_2PO_4 (pH 7.5) containing 14 mM 2-mercaptoethanol prior to precipitation with $(NH_4)_2SO_4$.

3. Desalt the protein pellet in 0.2 M KH_2PO_4 (pH 7.5) containing 12 mM dithiothreitol, then add 0.1 ml of the enzyme to 0.1 ml $MgSO_4$/ATP, 0.1 ml CoA solution, and 0.3 ml hydroxylamine.

4. Pre-incubate for 5 min at 30°C then initiate the reaction by adding 10 μl of 4-coumaric acid. Incubate reactions for 60 min at 30°C.

5. While incubating the samples, prepare an acidic $FeCl_3$ reagent from solutions of 5% $FeCl_3$ in 1 M HCl, 12% w/v trichloroacetic acid and 3 M HCl. Mix the three solutions at a ratio of 1:1:1 (v/v/v) and add 0.4 ml of the combined reagent to the samples to stop the reaction. After standing on ice, centrifuge the samples (10000 g, 10 min, 4°C) and measure the absorbance of the coumaroyl hydroxamate at 546 nm (log ε = 6.188).

[a] Store in aliquots at −70°C.
[b] Prepare fresh solutions.

3.2.7 Chalcone synthase (CHS; EC 2.3.1.74)

The basis of this assay is to follow the incorporation of radiolabelled malonyl CoA into a non-polar product (15; *Protocol 8*). The assay of the reductase involved in the formation of the deoxychalcone is described by Welle and Grisebach (16).

Protocol 8. Assay of chalcone synthase

1. Assay the enzyme in crude extracts after homogenization in 50 mM KH_2PO_4 (pH 8) containing 20 mM ascorbic acid. Alternatively, precipitate the enzyme with $(NH_4)_2SO_4$ as detailed in Section 3.2.3 and then desalt in 50 mM KH_2PO_4 (pH 8) with 20 mM ascorbate.

2. Prepare co-substrates and co-factors previously and store frozen at −70°C as follows:
   - 4-coumaroyl CoA (see *Protocol 2*)
   - dithiothreitol (DTT) 5.6 mM in 0.36 M KH_2PO_4 (pH 8)
   - [2-^{14}C]malonyl CoA as described in Section 3.1.2 (specific activity: 1.1 GBq/mmol).

3. Add 50 μl of enzyme to 10 μl of [^{14}C]malonyl CoA, 15 μl 4-coumaroyl CoA, and 25 μl of DTT. Incubate at 35°C for 30 min and stop the reaction by adding 20 μl of a 1.5 mg/ml solution of naringenin. After partitioning with

Protocol 8. *Continued*

0.2 ml of ethyl acetate (water saturated) centrifuge $10\,000\,g$, 3 min, 25°C or freeze at -20°C to clear the emulsion.

4. Assay the chalcone formed by either directly counting 75 μl of the ethyl acetate phase or, more satisfactorily, by separating naringenin from other products by TLC on silica gel plates in chloroform + ethanol (3:1 v/v). *Note*: When calculating product formation bear in mind that 3 moles malonyl CoA are required to synthesize 1 mole naringenin.

3.2.8 Chalcone isomerase (CHI; EC 5.5.1.6)

The basis of this assay is to monitor the decrease in absorbance following closure of chalcone ring (17; *Protocol 9*). **IMPORTANT NOTE**: The assay described utilizes a high concentration of KCN to inhibit endogenous peroxidase activity. When handling KCN solutions use all recommended precautions and work in a fumehood where possible. The preparation of a cyanide antidote kit and neutralizer is strongly recommended, and should be kept in an obvious place close at hand at all times. Another person should be present during the assay.

Protocol 9. Assay of chalcone isomerase

1. Extract tissue in 50 mM Tris–HCl (pH 8.5) containing 1.4 mM 2-mercaptoethanol and 5% (w/v) PVP. Centrifuge ($10\,000\,g$, 3 min, 4°C).

2. Add 20 μl of a 1 mg/ml solution of 2′,4,4′-trihydroxychalone or 2′,4,4′,6′-tetrahydroxychalcone to 2.5 ml of 60 mM KH_2PO_4 (pH 8) containing 50 mM KCN in a disposable plastic cuvette. Pre-incubate at 30°C for 5 min and then initiate the reaction by adding 20 μl of enzyme extract. For the assay with the tetrahydroxychalcone the addition of 5–10 mg/ml bovine serum albumin has been shown to improve the accuracy of the assay.

3. Measure decrease in absorbance at 400 nm on a chart recorder. Subtract the non-enzymic rate from the enzymic rate for the tetrahydroxychalcone. Calculate enzyme activity from the molar extinction coefficient (trihydroxychalcone log ε = 4.50; tetrahydroxychalcone log ε = 4.52 at 400 nm, pH 8).

3.2.9 Isoflavone synthase (IFS)

The basis of this assay is to monitor the quantification of radiolabelled product following chromatography (18; *Protocol 10*). Due to the low activity and instability of the enzyme assays are run at low temperatures.

Protocol 10. Assay of isoflavone synthase

1. Incubate 20 μl of the freshly prepared microsomal preparation (see *Protocol 4*) with 69 μl 0.1 M potassium phosphate buffer (pH 8.5) containing 0.4 M sucrose and 10 mM 2-mercaptoethanol, 5 μl 20 mM NADPH and 6 μl [^{14}C]naringenin (7.94 GBq/mmol, 185 Bq per assay) for 30–60 min at 15 °C.
2. Partition twice with 0.3 ml ethyl acetate (water saturated) and remove 0.25 ml of the organic phase at each separation. Evaporate the ethyl acetate to dryness and redissolve the residue in 30 μl methanol.
3. Add 10 μl each of 1 mg/ml methanolic solutions of naringenin and genistein to the sample and apply the entire extract to a cellulose TLC plate and develop in water + acetic acid (85:15 v/v).
4. Quantitate the radioactivity present in the bands corresponding to naringenin and genistein by extracting the cellulose in methanol and counting the eluant.

3.2.10 Isoflavone *O*-methyl transferase (IFMT; EC 4.3.1.5)

The basis of this assay is to monitor the incorporation of radioactivity from [^{14}C-methyl]SAM into a non-polar product (19; *Protocol 11*).

Protocol 11. Assay of isoflavone *O*-methyl transferase

1. Extract tissue in 0.2 M Tris–HCl (pH 7.5) containing 14 mM 2-mercaptoethanol, 5 mM EDTA, and 10 mM diethyldithiocarbamic acid (diethylammonium salt) and treat with 5% (v/w) Dowex 1. Due to the presence of compounds which can serve as competing methyl acceptors, it is recommended that extracts are either precipitated with $(NH_4)_2SO_4$, as detailed in Section 3.2.1, or desalted on spun columns prior to assay. During desalting the buffer conditions are changed to 0.2 M Tris–HCl (pH 8.5) containing 5 mM EDTA and 10 mM 2-mercaptoethanol.
2. Dissolve the isoflavone substrate in dimethylsulphoxide at a concentration of 10 mM. Prepare the [^{14}C]SAM to a concentration of 30 mM (8.33 MBq/mmol) as described in Section 3.1.2.
3. Incubate 50 μl of the enzyme with 5 μl of the isoflavone and 5 μl of [^{14}C]SAM for 30 min at 30 °C, then stop the reaction by adding 40 μl of methylated product (1 mg/ml) in ice-cold ethanolic solution.
4. Separate radiolabelled products from the substrate by TLC on silica plates in petroleum ether + ethyl acetate + methanol (60:40:1 v/v/v) prior to quantification by scintillation counting. Alternatively partition the incubation mixture against 0.2 ml of hexane + ethyl acetate (1:1 v/v) and count 0.1 ml of the organic phase.

3.2.11 Isoflavone 2' and 3' hydroxylase (IFOH)

The basis of this assay is the HPLC determination of product (20; *Protocol 12*). The positions of isoflavone hydroxylation are dependent upon the plant species under study. In plants such as alfalfa, isoflavone hydroxylation occurs in the 2' position (to give 2'-hydroxyformononetin), whereas in pea and chickpea hydroxylation in both the 2' and 3' positions (to give calycosin) may be assayed.

Protocol 12. Assay of isoflavone 2' and 3' hydroxylase

1. Incubate 100 µl microsomal preparation (see *Protocol 4*) with 330 µl of 0.3 M potassium phosphate buffer (pH 8) containing 0.4 M sucrose, 150 µl 20 mM NADPH and 20 µl 2 mM formononetin in methanol at 30°C for 30–60 min.

2. Partition twice with 0.6 ml ethyl acetate (water saturated) removing 0.45 ml of the organic phase at each extraction.

3. Take the ethyl acetate phase to dryness and redissolve in 100 µl methanol.

4. Analyse by HPLC (Section 2.3) at 248 nm following calibration with hydroxylated isoflavones (1–10 µM).

3.1.12 Isoflavone reductase (IFR)

The basis of this assay is the HPLC determination of product (21; *Protocol 13*).

Protocol 13. Assay of isoflavone reductase (IFR)

1. Homogenize cells with a pestle and mortar in 1 (v/w) 20 mM sodium phosphate buffer (pH 7.5) containing 1 mM DTT and 10% (w/v) PVP.

2. Centrifuge the cell extract (10 000 g, 10 min, 4°C) and desalt on a Sephadex column (Pharmacia PD-10) in the extraction buffer.

3. Prepare a 5 mM stock solution of the 2'-hydroxyisoflavone (for example 2'-hydroxyformononetin) in methanol and a 10 mM stock solution of NADPH (reduced form) in distilled water just prior to the assay.

4. Incubate 50 µl of the enzyme extract with 170 µl of 0.1 M Tris–HCl (pH 8.5), 25 µl of NADPH and 5 µl of 2'-hydroxyformononetin at 30°C for 15 min.

5. Stop the reaction by partitioning with 1 ml ethyl acetate and 0.75 ml of the organic phase taken to dryness prior to redissolving in 150 µl of methanol. Analyse the extract by HPLC (see Section 2.3) with the detector set at 278 nm following calibration with the product (for vestitone log $\varepsilon = 4.23$).

3.2.13 Ptercarpan synthase (PTS)

The basis of this assay is the HPLC determination of product (22; *Protocol 14*).

Protocol 14. Assay of ptercarpan synthase

1. Prepare an extract for PTS assay as described for IFR (*Protocol 13*).
2. Incubate 70 µl of the enzyme extract with 0.37 ml 0.2 M sodium phosphate buffer (pH 6), 50 µl of 10 mM NADPH, and 10 µl of 2.5 mM vestitone dissolved in methanol.
3. Analyse the product, after 15 min at 30°C, as for the IFR assay (*Protocol 13*) with the HPLC detector set at 287 nm (medicarpin, log ε = 3.9).

---

4. Concluding remarks

Using the methods described in this chapter it has been possible to monitor all of the biosynthetic steps involved in isoflavonoid phytoalexin biosynthesis in alfalfa. The stereochemistry of the resulting isoflavonoid phytoalexins may be determined by polarimetry as described by Kessmann *et al.* (23). The reader is encouraged to consult the original references for further details of the procedures if problems are encountered.

References

1. Dixon, R. A. (1986). *Biol. Rev.*, **61**, 239.
2. Barz, W., Daniel, S., Hinderer, W., Jacques, U., Kessmann, H., Koester, J., Otto, C., Tiemann, K. (1988). In *Applications of plant cell and tissue culture*, p. 178. (CIBA Foundation Symp. 137), Wiley, Chichester.
3. Hahlbrock, K. and Scheel, D. (1989). *Annu. Rev. Plant Physiol. Mol. Biol.*, **40**, 347.
4. Schumacher, H.-M., Grundlack, H., Fiedler, F., and Zenk, M. H. (1987). *Plant Cell Rep.*, **6**, 410.
5. Koester, J., Zuzok, A., and Barz, W. (1983). *J. Chromatog.*, **270**, 392.
6. Mabry, T. J., Markham, K. R., and Thomas, M. B. (1970). *The systematic identification of flavonoids.* Springer-Verlag, Heidelberg.
7. Stoeckigt, J. and Zenk, M. H. (1975). *Z. Naturforsch.*, **30c**, 352.
8. Moustafa, E. and Wong, E. (1967). *Phytochemistry*, **6**, 625.
9. Dewick, P. M. (1975). *J.C.S. Chem. Comm.*, 656.
10. Dewick, P. M. (1977). *Phytochemistry*, **16**, 93.
11. Shapiro, S. K. and Ehninger, D. J. (1966). *Anal. Biochem.*, **15**, 323.
12. Lamb, C. J., Merrit, T. K., and Butt, V. S. (1979). *Biochem. Biophys. Acta*, **582**, 196.
13. Bolwell, G. P., Robbins, M. P., and Dixon, R. A. (1985). *Eur. J. Biochem.*, **148**, 571.

14. Dixon, R. A. and Bendall, D. S. (1978). *Physiol. Plant Pathol.*, **13,** 295.
15. Lawton, M. A., Dixon, R. A., Hahlbrock, K., and Lamb, C. J. (1983). *Eur. J. Biochem.*, **129,** 593.
16. Welle, R. and Grisebach, H. (1988). *FEBS Lett.*, **236,** 221.
17. Dixon, R. A., Dey, P. M., and Whitehead, I. M. (1982). *Biochem. Biophys. Acta*, **715,** 25.
18. Kochs, G. and Grisebach, H. (1986). *Eur. J. Biochem.*, **155,** 311.
19. Dalkin, K., Edwards, R., Edington, B., and Dixon, R. A. (1990). *Plant Physiol.*, **92,** 440.
20. Hinderer, W., Flentje, U., and Barz, W. (1987). *FEBS Lett.*, **214,** 101.
21. Tiemann, K., Hinderer, W., and Barz, W. (1987). *FEBS Lett.*, **213,** 324.
22. Bless, W. and Barz, W. (1988). *FEBS Lett.*, **235,** 47.
23. Kessmann, H., Tiemann, K., Janssen, J. R., Reuscher, H., Bringmann, G., and Barz, W. (1988). In: *Plant cell biotechnology* (ed. M. S. S. Pais, F. Mavituma, and G. H. Novais), p. 211. Springer-Verlag, Heidelberg.

7

Analysis of terpenoid phytoalexins and their biosynthetic enzymes

DAVID R. THRELFALL and IAN M. WHITEHEAD

1. Introduction

Terpenoid phytoalexins and terpenoid stress metabolites (some of which have been ascribed phytoalexin status) are produced by members of nine plant families (1, 2). Their accumulation both *in planta* and in plant cell cultures can be induced by a variety of biotic and abiotic agents. The class of terpenoid phytoalexin produced is uniform within a plant family. The largest class comprises the sesquiterpenoid phytoalexins which, in common with the pentacyclic triterpenoid phytoalexins, are formed from (2*E*, 6*E*)-farnesyl pyrophosphate (FPP) (see *Figure 1*). The only other documented class of terpenoid phytoalexin is made up of the diterpenoid phytoalexins, which are formed from (all *E*)-geranylgeranyl pyrophosphate (GGPP) (see *Figure 1*). It should also be noted that the terpenoid precursor dimethylallyl pyrophosphate (DMAPP), provides the hemiterpenoid (dimethylallyl) residue(s) for many isoflavonoid phytoalexins produced by members of the Leguminosae (*see Figure 1*).

The phytopathology, elicitation, enzymology, and the regulation of accumulation of terpenoid phytoalexins have been studied most extensively in the cases of the acarbocyclic furanosesquiterpenoids produced by *Ipomoea batatas* L. (sweet potato, Convolvulaceae) (3), the carbocyclic sesquiterpenoids produced by members of the Solanaceae (see references 1, 4–10 and references therein), the mono and dimeric aromatic sesquiterpenoid aldehydes produced by *Gossypium* spp. (cotton, Malvaceae; 11, 12), the macrocyclic diterpenoid produced by *Ricinus communis* L. (castor bean, Euphorbiaceae; 13, 14) and the pentacyclic triterpenoids produced by *Tabernaemontana divaricata* R. Br. ex Ruem (Apocynaceae; 12, 15).

In Sections 2 and 3, we describe some of the general procedures used for the extraction and estimation of terpenoid phytoalexins, their characterization as phytoalexins, and the assay of some of the enzymes which are required for their biosynthesis.

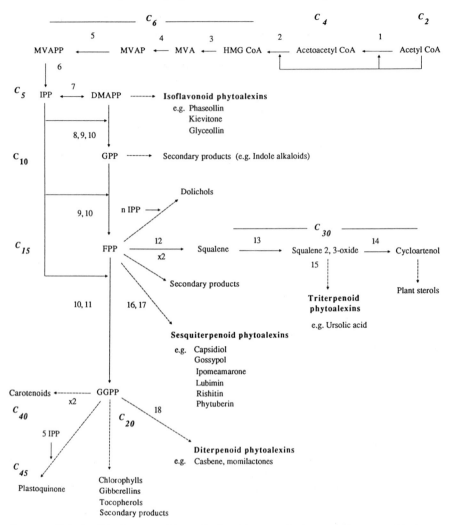

Figure 1. Relationship between biosynthetic pathways of phytosterol, terpenoid phyto-alexin, isoflavonoid phytoalexin, and other terpenoids.

1.1 Reagents, substrates, and markers

All reagents should be AnalaR or the best grade available. Organic solvents, after drying if necessary, should be glass redistilled (diethyl ether and di-isopropyl ether over reduced iron) before use, since even AnalaR grades contain appreciable amounts of non-volatile material which can interfere with the analysis of small amounts of phytoalexins. Glass or organic solvent re-sistant apparatus must be used for the extraction, purification, and storage of terpenoid phytoalexins. Contact of extracts and solutions of extracts with rubber, Parafilm, plastics, silicone seals, etc. must be avoided since these compounds are all excellent sources of lipid–soluble contaminants. We routinely store all samples at or below −20°C.

ADP, ATP, creatine kinase, creatine phosphate, dithiothreitol (DTT), FAD, glucose-6-phosphate, glucose-6-phosphate dehydrogenase, insoluble polyvinylpyrrolidone (PVP), NADP, NADPH, Triton X-100, and Tween 80 may be purchased from Sigma Chemical Co.

Table 1 lists the radiolabelled and unlabelled substrates needed for the enzymic assays (Section 3), either obtained from the commercial source shown or prepared by the method described in the reference quoted. The source of all other compounds are referred to in the text.

Allylic pyrophosphates are best stored in 0.1 M Tris–HCl buffer, pH 8.0, at

Table 1. Source of substrates needed for enzyme assays

Substrate (abbrev.)	Source or ref.
Unlabelled	
Dimethylallyl pyrophosphate (DMAPP)[b]	16, 17
(2*E*, 6*E*)- Farnesyl pyrophosphate (FPP)[b]	16, 17
(2*E*)-Geranyl pyrophosphate (GPP)*sb*	16, 17
(*S*)-3-Hydroxy-3-methylglutaryl coenzyme A (HMG-CoA)	Sigma
Isopenentyl pyrophosphate (IPP)[a,b]	16, 17
(*R,S*)-Mevalonic acid (MVA) lactone	Sigma
Squalene	Sigma
(*R,S*)-squalene 2,3-oxide	21
Radiolabelled	
[14C] Dehydroipomeamarone	23
[2-14C] Farnesyl pyrophosphate (FPP)	Amersham Int.
[1-3H2] Farnesyl pyrophosphate (FPP)[c]	10
[4,8,12-14C3] Farnesyl pyrophosphate (FPP)	18
[4,8,12,16-14C4] Geranylgeranylpyrophosphate (GGPP)[d]	19
(*S*)-[3-14C]-Hydroxy-3-methylglutaryl coenzyme A (HmG–CoA)	Amersham Int.
[14C]Ipomeamarone	24
[1-14C]Isopentyl pyrophosphate (IPP)	Amersham Int.
(*R,S*)-[2-14C] Mevalonic acid (MVA) lactone	Amersham Int.
(*R*)-[5-14C]MVA 5-phosphate (MVAP)	Amersham Int.
(*R*)-[1-14C]-MVA 5-pyrophosphate (MVAPP)	Amersham Int.
(*R*)-[5-14C]MVA 5-pyrophosphate (MVAPP)	Amersham Int.
[14C]Squalene	20
[4-3H]Squalene	21
(*R,S*)-[4-3H]-Squalene-2,3-oxide	21
(*R,S*)-[11,12-3H2]Squalene-2,3-oxide	22

[a] In the first phosphorylation step described in reference 16, the pH of the aqueous phase should be adjusted to pH 12 with LiOH and not NH4OH.
[b] The alcohols needed for this synthesis are available from Aldrich Chemical Co.
[c] Except that *E,E*-farnesol is now available from Aldrich Co.
[d] All *E*.

$-70\,^\circ$C. MVA lactone needs to be hydrolysed to the acid before use. This is done by carefully removing the solvent (benzene) that the labelled or unlabelled sample is delivered in, under a stream of nitrogen gas, followed by the addition of water containing an equimolar amount of $NaHCO_3$.

2. Extraction, detection, characterization, and estimation of terpenoid phytoalexins

The biological sample suspected of containing terpenoid phytoalexins is extracted with a suitable organic solvent (Section 2.1). The lipid extract obtained by this procedure is then examined for the presence of terpenoid phytoalexins and related stress metabolites by TLC, TLC-bioassay (in the case of a new biological system) and, usually, GLC. In many cases, the phytoalexin content of a lipid extract can be estimated directly without recourse to purification procedures, whereas in others a TLC purification step may be required to remove interfering lipids. The most common method of estimation is GLC although HPLC and colorimetric procedures are also used.

A satisfactory identification of the phytoalexins obtained from known systems can usually be achieved by TLC followed by GLC, HPLC, or UV analysis, especially if authentic samples are available for use as marker compounds. Unequivocal characterization of a terpenoid phytoalexin requires elemental analysis, and optical rotary dispersion (ORD) (in the case of optically active compounds), infra-red (IR), GLC mass spectrometry (GLC–MS), and most importantly high-field ^{1}H- and ^{13}C-nuclear magnetic resonance (NMR) measurements. The only terpenoid phytoalexins to give characteristic UV/visible spectra are gossypol, hemigossypol, and their derivatives (by virtue of the aromatic ring system) (25). The large scale extraction of the milligram amounts of material required for some of these analyses often requires column chromatography and/or preparative TLC of the lipid extract, the descriptions of which are beyond the scope of this chapter although examples can be found in references 26 and 27.

It should be noted that as a result of parasite-mediated reactions, modified phytoalexins may be isolated from some plant–pathogen interactions.

2.1 Extraction of terpenoid phytoalexins

It is not possible to recommend any particular procedure for the isolation of terpenoid phytoalexins and related stress metabolites from plant tissues, as these compounds vary so widely in their polarity and solubility, and in their accessibility *in vivo* to extraction solvents. The procedure outlined in *Protocol 1* A is suitable for the extraction of most sesquiterpenoid phytoalexins from non-photosynthetic tissues and can be modified to make it suitable for the extraction of gossypol and its derivatives, and triterpenoid phytoalexins (*Proto-*

col 1). The liquid–liquid partition step (*Protocol 1* A, step 4(b)) which forms part of this procedure is particularly important in the analysis of some tissues since it removes non-lipids and more polar lipids from the extract which simplifies subsequent analysis.

Protocol 1. Extraction of terpenoid phytoalexins from non-photosynthetic tissues

A. *Extraction of sesquiterpenoids from tuber tissue, tuber discs, callus material, and cells from suspension cultures* (5, 7, 30)

1. Harvest treated plant material. The analysis can often be simplified by taking only those areas of tissue which are known to contain phytoalexins, i.e. infected/elicitor treated tissue and immediately adjacent areas. Harvest cells from cell suspension cultures by vacuum assisted filtration of the cultures through a disc of Miracloth (Calbiochem Corp.) or nylon (pore size 1 µm) in a Buchner funnel and wash the retained cells with water (approx. 30–100 ml). Retain the initial filtrate plus washings for analysis (*Protocol 3*, Method 1).

2. Homogenize (mortar and pestle with acid-washed sand, microhomogenizer, or Ultraturrax) tissue with 5–10 ml methanol or chloroform ($CHCl_3$)–methanol (1:1, 2:1)/g fresh weight of tissue. Centrifuge homogenate at 1000 g for 5 min or vacuum filter through glass-fibre paper (Whatman GF/A). Re-extract pellet/residue (× 2) with 2.5–5 ml solvent/g fresh weight tissue.

3. Combine the supernatants/filtrates and follow steps 4(a) or 4(b).

4. (a) Take combined supernatants/filtrates to dryness in a rotary evaporator at 40°C with the addition of a few ml of ethanol to remove the last traces of water. Transfer the extract to a small vial with several 1 ml aliquots of a suitable solvent, for example acetone, then proceed to step 5.

   (b) Reduce combined supernatants/filtrates to ¼–⅛ volume in a rotary evaporator, transfer to a separatory funnel with washings (use 60% (v/v) methanol), dilute with an equal volume of water. Extract (2 or 3 × equal vol.) with either diethyl ether or ethyl acetate. Disperse emulsions by the addition of a few drops of ethanol. Combine solvent phases, and either take to dryness in a rotary evaporator at 40°C with the addition of ethanol to remove the last traces of water or dry over anhydrous Na_2SO_4 for several hours, filter through a sintered glass funnel and then take to dryness in a rotary evaporator at 40°C. Transfer the extract to a small vial with several 1 ml aliquots of a suitable solvent, for example acetone, then proceed to step 5.

5. Evaporate the sample to dryness under a stream of N_2 gas on a hot plate

Protocol 1. *Continued*

(40–50°C). Dissolve extracts in the smallest volume possible (but not less than 100 µl) of acetone and examine and/or purify by TLC (Section 2.2 and *Protocol 4*) prior to estimation by GLC (Section 2.3 and *Protocol 5*), HPLC (Section 2.4), or colorimetry (Section 2.5 and *Protocol 6*).

B. *Extraction of gossypol from cotton cotyledons and stele tissue* (25)

1. Homogenize tissue sample with 2 vol. (w/v) of 80% (v/v) aqueous methanol. Remove residue as described in part A, step 2 above. Re-extract residue twice.

2. Combine the supernatants/filtrates and dilute with water to 40% methanol.

3. Extract the diluted supernatants/filtrates with diethyl ether (3 × equal vol.). Combine diethyl ether phases, wash with water (3 × equal vol.) and dry over anhydrous Na_2SO_4 for several hours. Remove Na_2SO_4 by filtration through a sintered glass funnel and take diethyl ether phase to dryness in a rotary evaporator at 40°C. Transfer the extract to a small vial with several 1 ml aliquots of a suitable solvent. Evaporate sample to dryness under a stream of N_2 gas on a hot plate (40–50°C).

4. Dissolve extracts in the smallest volume possible (but not less than 100 µl) of 60% (v/v) aqueous methanol, examine by TLC (Section 2.2, *Protocol 4* and *Table 2*) and estimate by colorimetry (*Protocol 6* B).

C. *Extraction of pentacyclic triterpenoids from cultures of* T. divaricata (2)

1. Harvest cells, weigh, and extract 1 g by the procedures described above in part A, steps 1–3 (use methanol as the extractant).

2. Follow step A5 but add a few drops of H_3PO_4 to the diluted extract before extracting with diethyl ether. This acidification step ensures that the carboxylic acid substituent of the pentacyclic triterpenoids is fully associated which facilitates their extraction into the non-polar solvent.

3. Dissolve the lipid extract in a small, known vol. (approx. 1 ml) of $CHCl_3$–methanol (1:1) and examine by TLC (Section 2.2 and *Protocol 4*) and, after derivatization, GLC (Section 2.3 and *Protocol 5*).

Facilitated diffusion (*Protocol 2*) is the method of choice for photosynthetic tissues since it gives extracts which are low in leaf lipids and, more importantly, photosynthetic pigments which can interfere with the detection of phytoalexins on TLC plates. Method 2 (*Protocol 2*) has also been used to advantage with some non-photosynthetic tissues (28). However, it is not a very efficient method for the extraction of non-polar phytoalexins. Freeze-dried material can be extracted with CH_2Cl_2 (29).

Table 2. TLC of lipid extracts containing terpenoid phytoalexins: systems, chromogenic reagents, and example results

Adsorbents

Silica gel G

Silica gel GF_{254}

Rhodamine 6G-impregnated silica gel G

Solvents systems[a]

Furanosesquiterpenoids

| n-Hexane–ethyl acetate | (1:1, 9:1) |

Carbocyclic sesquiterpenoids

Ethyl acetate–cyclohexane	(1:1)
Ethyl acetate–propan-2-ol	(9:1)
n-Hexane-acetone	(3:1)
Chloroform–methanol	(19:1)
Carbon tetrachloride–acetone	(1:1, 1:2)
Diethyl ether	

Aromatic sesquiterpenoids

Benzene–methanol	(19:1)
Benzene–ethyl formate–formic acid	(75:24:1)
Toluene–ethyl formate–formic acid	(5:4:1)
Chloroform	
Ethyl acetate–hexane	(1:3)

'Hemigossypols' (R_f 0.3–0.5) are readily separated from 'gossypols' (R_f 0.70–0.75) by TLC on polyamide (Merck/Anachem) plates using:

| Chloroform–acetone–formic acid | (95:4:1) |

Diterpenoids

Momilactones

| Cyclohexane–dioxane | (7:3) |
| Chloroform–ethanol | (97:3) |

Casbene: this compound is present in insufficient amounts to detect by conventional TLC procedures. TLC of [^{14}C]casbene synthetase activity is given in *Protocol 17*.

Pentacyclic triterpenoids

Ethyl acetate–cyclohexane	(1:1)
Toluene–ethyl acetate	(17:3)
Chloroform–methanol	(9:1)
CH_2Cl_2–n-hexane–acetic acid	(45:30:9)
Do-isopropylether–acetone	(5:2)

Appearance of compounds under UV_{254} light: Red/pink bands (or spots) against a yellow background on rhodamine 6G-impregnated (or oversprayed) plates, except in the case of

69

Table 2. *Continued*

aromatic and other UV absorbing compounds which appear as dark bands on these and UV_{254} plates.

General terpenoid chromogenic spray reagents[b]

A. *Vanillin–H_2SO_4*
   1 g vanillin in 30 ml methanol containing 0.2 ml conc. H_2SO_4

B. *Iodine vapour followed by A*

C. *Carr–Price*
   $SbCl_2$ (22% w/w) in chloroform

D. *Anisaldehyde–H_2SO_4*
   0.5 ml anisaldehyde in 9 ml ethanol containing 0.5 ml conc. H_2SO_4 and 0.1 ml glacial acetic acid

After spraying with one of these reagents, heat the plate either in an oven to 100°C for 5 min or preferably with a hot hair-dryer as this allows colour development to be observed.

Example results	R_f	Colour with spray reagent[d]
Furanosesquiterpenoid		
Ipomeamarone	0.70[c]	Blue (A)
Ipomeamaronol	0.20[c]	Blue (A)
Carbocyclic sesquiterpenoid		
Capsidiol	0.12[e]	Turquoise (A)
Debneyol	0.30[e]	Mauve (A)
Lubimin	0.30[e]	Blue (A)
Phytuberin	0.70[e]	Heliotroph (A)
Rishitin	0.20[e]	Blue (A)
Solvavetivone	0.62[e]	Buff (A)
Aromatic sesquiterpenoid		
Hemigossypol	0.31[f]	Magenta[g]
6-Methoxy-hemigossypol	0.47[f]	Deep rose[g]
Diterpenoid		
Casbene	see *Protocol 17*	
Momilactone A	0.55[h]	Purple (A)
Momilactone B	0.47[h]	Light blue (A)
Pentacyclic titerpenoid		
Monohydroxy-ursolic acids	0.32[i]	Purple (D)
Ursolic acid	0.47[i]	Green (D)

[a] Solvents containing polar components (for example water-miscible) cannot be used with rhodamine 6G-impregnated plates (see *Protocol 4*, step 1).
[b] Aldehyde and phenolic reagents are used for the detection of gossypol and related compounds (25). Ehrlich's reagent is used routinely for the detection of furanosesquiterpenoids (23, 24).
[c] *n*-Hexane–ethylacetate (9:1).
[d] The spray reagent used is shown in parenthesis.
[e] Ethyl acetate–cyclohexane (1:1).
[f] Polyamine system described above.
[g] Phloroglucinol reagent (25).
[h] Chloroform–ethanol (97:3).
[i] Toluene–ethyl acetate (17:3).

Protocol 2. Extraction of terpenoid phytoalexins from leaf tissue
(facilitated diffusion)

A. *Extraction of sesquiterpenoid phytoalexins*

Method 1 (31)

1. Steep leaf tissue in benzene or toluene (10 vol. (w/v)) for 12 h.

2. Separate solvent extract from tissue by decantation or filtration and remove solvent in a rotary evaporator at 45°C. Transfer sample to a small vial with several 1 ml aliquots of a suitable solvent.

3. Treat extracts as described in *Protocol 1*A, step 5.

Method 2 (32)

1. Submerge leaf tissue in 15 vol. (w/v) of 40% (v/v) aqueous ethanol in a glass-stoppered conical flask fitted with a side arm. Place under gentle vacuum for 10 min to infiltrate solvent into leaf spaces. Shake flasks for 5 h. Remove solvent by decantation or filtration and repeat extraction.

2. Combine extracts and reduce to 0.1 vol. in a rotary evaporator at 40°C.

3. Transfer the reduced extract to a separatory funnel and partition against diethyl ether (3 × equal vol.).

4. Combine the organic phases and take to dryness in a rotary evaporator at 40°C as described in *Protocol 1*A, step 4.

5. Treat extract as described in *Protocol 1*A, step 5.

B. *Extraction of gossypol from cotton tissue culture cells* (33)

1. Stir filtered plant cells overnight with 5 vol. methanol saturated with $NaHSO_3$.

2. Remove methanol phase by filtration, and concentrate to 3–6 ml *in vacuo* or under a stream of N_2 gas.

3. Estimate total sesquiterpene aldehydes by colorimetry (*Protocol 6* B), or separate gossypol and related compounds by TLC (*Protocol 4* and *Table 2*) prior to colorimetry (*Protocol 6*).

Terpenoid phytoalexins are easily extracted from aqueous solutions (for example diffusates from fruit cavities, infected wells in tubers and cored stems, or the media of plant cell suspension cultures) with water immiscible organic solvents (*Protocol 3*). In the case of diffusates, the phytoalexins can be extracted by the use of reverse phase SEP-PAK cartridges (*Protocol 3*, Method 2).

Protocol 3. Extraction of terpenoid phytoalexins from culture media and aqueous diffusates

Method 1 (10)

1. Obtain culture filtrates from cell suspension cultures as described in *Protocol 1 A*, step 1. Diffusates from fruits are also filtered in this way to remove seeds and other plant material.

2. Transfer the filtrate to a separatory funnel and extract with diethyl ether (2 or 3 × equal vol.). Disperse emulsions by the addition of a few drops of ethanol. Combine solvent phases and either take to dryness in a rotary evaporator at 40°C with the addition of ethanol to remove the last traces of water or dry over anhydrous Na_2SO_4 for several hours, filter through a sintered glass funnel and then take to dryness in a rotary evaporator at 40°C.

3. Transfer the lipid extract to a small vial by rinsing the flask with several millilitres of diethyl ether. Evaporate the sample to dryness under a stream of N_2 gas on a hot plate (40–50°C).

4. Analyse extracts by TLC and quantitate phytoalexins by a suitable method (Sections 2.2–2.5).

Method 2

1. Pass diffusate through a SEP-PAK C_{18} cartridge (Water Associates Ltd) which has been pre-equilibrated with methanol according to the manufacturer's instructions.

2. Elute the phytoalexins with 1 ml methanol.

3. Evaporate methanol on a hot plate (not above 60°C) under a stream of N_2 gas.

4. Analyse extracts by TLC and quantitate phytoalexins by a suitable method (Sections 2.2–2.5).

2.2 TLC analysis of terpenoid phytoalexins

TLC is used both for screening plant lipid extracts for the presence of known or putative terpenoid phytoalexins (analytical TLC), and, when necessary, the preparation (prep. TLC) of phytoalexin-enriched fractions for analysis by GLC, HPLC, or colorimetry (*Protocol 4*). Preparative TLC (or alternatively column chromatography) is a particularly important step in the case of extracts containing large amounts of extraneous lipids which, if not removed, can adversely affect the resolution and useful life of GLC and HPLC columns.

In the screening procedure, lipid extracts from both control and infected/elicited tissues are subjected to TLC and the separated compounds located by spraying the plate with a chromogenic reagent. However, this procedure is not infallible since some terpenoid phytoalexins give weak or only transient colour reactions which can be easily overlooked, and in crude extracts even strong colour reactions may be masked by pigments and/or other terpenoid or non-terpenoid compounds which react with the reagents (*Table 2*). Compounds which occur uniquely in the extract from infected/elicited tissue can be assayed for antibiotic activity by the TLC bioassays described in Section 2.6.

Protocol 4. TLC of lipid extracts containing terpenoid phytoalexins: methods

A. *Thin layer plates*

1. Prepare silica gel G and silica gel GF$_{254}$-coated plates (20 cm × 20 cm × 0.25–0.5 mm) from the appropriate silica gel (TLC grade) according to the manufacturer's instructions (or purchase, Anachem/Merck).

2. Prepare rhodamine 6G-impregnated silica gel G plates by replacing 12.5% (v/v) of the water used to prepare the gel with a (filtered) 0.4% (w/v) ethanolic solution of rhodamine 6G (BDH). These plates are particularly useful for preparative TLC, since on examination of the developed plate under UV, even non-UV absorbing compounds can be located with a high degree of accuracy. **Warning**: These plates are not suitable for use in systems employing polar (i.e. water miscible) solvents, either when developing TLC plates or recovering samples from the gel, as the rhodamine 6G is eluted. In these cases, plates can be oversprayed with a 0.04% (w/v) acetone solution of rhodamine 6G after development.

3. Prepare AgNO$_3$-impregnated plates by dissolving 10% (w/w silica gel) AgNO$_3$ in the water used to prepare the gel, before mixing with the silica. These plates are usually only used for analytical work as their loading capacity is greatly reduced (see part B, preparative TLC). They *must* be stored and run in the dark, and, in any case, are only usable for a few days.

B. *Application of samples*

Analytical TLC

1. Dissolve the lipid extract in a small volume of a suitable solvent (for example acetone).

2. Apply an aliquot as a spot or narrow band (1–1.5 cm long), 1.5 cm from the bottom of a TLC plate. When studying a system for the first time apply a series of spots/bands of increasing loading to ensure that sufficient amounts of the phytoalexin(s) are present to give a colour reaction. This also facilitates the detection of minor components.

3. Spot marker compounds (if available) to the left and right of the samples on the plate.

Protocol 4. *Continued*

Preparative TLC

1. Dissolve the extract in a small volume of a suitable solvent (see above).

2. Apply the whole sample to a TLC plate(s) as a narrow band 2–20 cm in length by means of a 100 μl syringe. The loading capacity of a TLC plate depends on several factors, including the adsorbent used, the thickness of the adsorbent layer, the resolution of the compounds of interest, and the purity of the sample, but is generally in the 5–50 mg range. The loading of plates with larger amounts of material often requires several overlaying applications to be made to the plate.

3. Allow the solvent to evaporate before the next portion of the sample is applied to ensure that the sample is loaded as a tight band. This can be aided by the use of a (hot) hair-drier in some cases.

C. *Development of the TLC plates*

1. Develop the TLC plate in a chromatography paper-lined TLC tank which has been pre-equilibrated with the developing solvent (*Table 2*). In our experience, better resolution can be obtained by the following method.

   • Allow the developing solvent to migrate *just* to the top of the applied sample, remove the plate and dry off the solvent. This helps to concentrate the sample as a tight band.

   • Develop the plate as normal. The resolution of closely running compounds can often be improved by redeveloping the plate (2–3 times) with the same solvent.

D. *Detection and recovery of material from the gel*

Analytical TLC

1. Spray the plate with a suitable chromogenic reagent (*Table 2*)

Preparative TLC

1. Examine the developed preparative TLC plate under UV_{254} light (silica gel GF_{254} and rhodamine 6G-impregnated plates) and/or mask all but one edge of plate (for example with a glass plate) and spray the exposed area with an appropriate chromogenic reagent (*Table 2*).

2. Scrape the area(s) of gel containing the compound(s) of interest into a small, sintered glass funnel(s) and elute the compound(s) from the gel(s) with a suitable solvent (for example diethyl ether).

2.3 GLC analysis of terpenoid phytoalexins

GLC analysis (*Protocol 5*) is used both as part of the identification procedure and for the quantification of most sesquiterpenoid and pentacyclic triterpenoid phytoalexins. It may be performed on crude extracts provided that they do not contain excessive amounts of extraneous lipids. However, in the case of tissues which have been extracted with polar solvents (*Protocol 1*), it is often necessary to prepare a phytoalexin-enriched fraction of the extract by TLC (*Protocol 4* B, preparative TLC), prior to attempting GLC analysis. In most studies, separation of the phytoalexins is effected on packed columns and compounds are detected by means of a flame ionization detector (FID). Sesquiterpenoids are normally analysed without derivatization (34), although this technique has been employed. The only procedure described for the GLC analysis of pentacyclic triterpenoids involves both methylation and acetylation of the samples prior to analysis (15).

Protocol 5. FID–GLC analysis of sesquiterpenoid and pentacyclic triterpenoid photoalexins

A. *Sesquiterpenoid phytoalexins* (34)

Column: 3% OV 225 on Gas Chrom Q (80–100 mesh, Phase Sep) or Supelcoport (100–120 mesh, Supelco Corp.) (2 m × 2 m i.d. Pyrex glass column)

Carrier gas: argon or nitrogen, 45 ml/min

Temperatures: injection port 235°C, detector 300°C, column 180°C

Sample preparation and injection: dissolve the lipid extract or material from preparative TLC in a small volume (approx. 100 μl) of acetone (containing internal standard[a] if used). Inject 1–5 μl of sample on to column

Example RR_t *values*: methyl stearate (R_t 4 min), 1.0; rishitin, 1.5; debneyol, 1.7; capsidiol, 2.1; lubimin, 2.4.

B. *Triterpenoid phytoalexins* (15)

Column: 3% OV 17 on Gas Chrom Q (80–100 mesh, Phase Sep) (2 m × 2 mm i.d. Pyrex glass column)

Carrier gas: argon or nitrogen, 45 ml/min;

Temperatures: injection port 360°C; detector 360°C; column 280°C for 5 min, 280–320°C in 5 min, 320°C for 20 min.

Sample preparation:

1. Transfer an aliquot (300 μl) of the $CHCl_3$–methanol solution containing the lipid extract (*Table 2* C) or material recovered from preparative TLC to a small vial.

2. Place vial on a hot plate (not above 60°C) and remove solvent under a stream of N_2 gas.

3. Add 0.5 ml of a solution of diazomethane in diethyl ether and leave the mixture to stand for 30 min.

Protocol 5. *Continued*

4. Remove the diethyl ether under a stream of N_2 gas.

5. Add 100 µl acetic anhydride and 20 µl of pyridine.

6. Leave the mixture to stand on a hot plate for 2 h or overnight at room temperature.

7. Add 200 µl ethanol, transfer the vial to a hot plate and remove all of the solvent under a stream of N_2 gas.

8. Dissolve the sample in 100–200 µl acetone and inject 1–5 µl on to the column.

Example RR$_t$ *values*:   methyl acetyl–ursolic acid (R_t 12 min), 1.0; steryl acetates, 0.5–0.79; methyl acetyl–oleanolic acid) 0.94; methyl diacetyl–monohydroxyursolic acids, 1.27.

[a] Response factors depend on the GLC instrument and operating conditions, and can vary from day to day.

The identification of compounds is best achieved by co-chromatography with authentic marker compounds rather than by reliance on published data, although retention times relative to known compounds are useful guides. In some cases, the more useful Kováts indices are available. The amount of a phytoalexin present is calculated from its integrated peak area [measured manually by triangulation (height × width at half-peak) or electronically with a computing integrator] by using either the peak area given by an aliquot of a stock solution of the phytoalexin, or by using a response factor calculated for the machine in use between an easily obtainable internal standard (for example methyl stearate or methyl arachidate, Sigma) and the stock solution of the phytoalexin. This second method is useful with samples containing several phytoalexins and when the internal standard is added to samples at the beginning of the analysis to correct for extraction errors. Alternatively, the amount of a phytoalexin can be expressed relative to a known weight of the internal standard.

2.4 HPLC analysis of terpenoid phytoalexins

HPLC has not been exploited to any great extent as a method for the purification, separation, and estimation of terpenoid phytoalexins. Reverse-phase columns have been used as a step in the purification of momilactones [Partisil 10/$CHCl_3$–ethanol–hexane in the ratio 50:1:49 (v/v/v), detc. 225 nm] (35), for the separation of gossypol and related compounds (Ultrasphere ODS 5 µM/60–67% methanol, detc. 254 mm) prior to their estimation by fluorimetry (12), and for the estimation of sesquiterpenoid phytoalexins in crude extracts of infected potato tubers (µ Bondpak C_{18}/70% methanol, detc. 200 mm) (36).

2.5 Colorimetric analysis of terpenoid phytoalexins

Colorimetry is used routinely for the estimation of ipomeamarone and related compounds, and of gossypol and its derivatives (*Protocol 6*). It was used for the estimation of rishitin prior to the development of GLC procedures.

Protocol 6. Colorimetric estimation of terpenoid phytoalexins

A. *Ipomeamarone and related furanoterpenoids* (37)

1. Dissolve the crude lipid extract (*Protocol 1* A) or TLC purified material in 5 ml ethanol.

2. Take an aliquot of the ethanol solution and make up to 2 ml with ethanol. Add 1 ml of a 10% (w/v) ethanolic solution of 4-dimethylaminobenzaldehyde (Sigma), and then 2 ml 40% (v/v) aqueous H_2SO_4.

3. Incubate mixture at 30°C for *exactly* 15 min.

4. Read absorbance at 527 nm while mixture is still warm against an appropriate blank.

5. Express results as 'ipomeamarone equivalents' by reference to a standard curve (0–10 mg) prepared from pure ipomeamarone.

B. *Gossypol and related aldehydes* (25)

1. Dissolve the crude lipid extract (*Protocol 1* B) or TLC purified material in a suitable volume of ethanol.

2. Take 1 ml of this solution and add 0.5 ml of a 5% (w/v) ethanolic solution of phloroglucinol and 1 ml HCl.

3. Leave mixture to stand for 30 min, then add ethanol to 10 ml.

4. Read absorbance at 550 nm against an appropriate blank.

5. Express results as 'gossypol equivalents' by referring to a standard curve (0–5 mg) prepared from pure gossypol.

2.6 Detection of phytoalexins in plant extracts

A variety of bioassays have been employed to assess the antibiotic activity of compounds suspected of being phytoalexins. The details of many of these techniques and some of their associated problems have been reviewed elsewhere (38).

The two techniques we have chosen to include here are TLC bioassays for antifungal and antibacterial activity. They are convenient and sensitive methods which, in many cases, can be used with an authentic test organism from a given host–pathogen interaction. They can be used with crude or partially purified plant extracts to provide a quick answer to the question: Are any phytoalexin-like compounds present in the extract? In general, plant pigments and other endogenous substances present in the lipid extracts do not interfere with the assays, but this should be confirmed in each system by running a control extract obtained from uninfected/unelicited plant material.

2.6.1 TLC bioassay of antibacterial activity

The method described here (*Protocol 7*) uses *Pseudomonas syringae* pv. *phaseolicola* as the indicator organism. Glycerol is added to the overlay medium to act as a carbon source and electron donor. This facilitates the reduction of 2,3,5-triphenyltetrazolium chloride to pink-coloured formazans by bacterial dehydrogenases. Where bacterial growth is inhibited a colourless zone is apparent against a pink background. The test, which is a modified version of published protocols (39), gives clear, reproducible results within 24 h and is sensitive to levels of 0.05–5 µg for purified samples of photo-alexins.

Protocol 7. Identification, by TLC bioassay, of compounds possessing antibacterial activity (40)

1. Prepare the indicator strain (in this case *Pseudomonas syringae* pv. *phaseolicola*) by incubating cultures overnight in 25 ml nutrient broth (Oxoid) containing 0.25 ml glycerol in a 250 ml Erlenmeyer flask on an orbital shaker at 25–28°C and 150–200 r.p.m. King's B broth has also been used successfully to culture indicator strains. (The absolute age and concentration of the indicator bacteria inoculum and the composition of the culture medium used to grow the indicator cells appears not to be critical.)

2. Apply aliquots of crude lipid extracts from infected/elicited and control tissues to a silica gel G TLC plate (*Protocol 4* B: analytical TLC) along with marker phytoalexins (if available) and develop in a suitable solvent system (*Table 2*).

3. Dry off the developed plate with a (hot) hair-dryer and leave it in a fume cupboard for at least 2 h to ensure that all of the solvent has evaporated. Place the plate on a *level* surface.

4. Prepare the overlay medium. For a single (20 × 20 cm) TLC plate this is 50 ml nutrient broth (Oxoid) containing 0.5 ml glycerol and 0.3–0.75 g (0.6–1.5% w/v, see step 6 below) Bacto-Agar (Difco). Autoclave this media at 121°C for 15 min, cool in a water-bath to 45°C, and add 50 mg 2,3,5-triphenyltetrazolium chloride (Sigma). Mix thoroughly and maintain at 45°C.

5. Collect the bacterial cells by centrifuging the prepared culture at 3000 g for 10 min at room temperature. Resuspend the pellet in 2 ml buffer (1.4 mM KH_2PO_4, 2.5 mM Na_2HPO_4).

6. Mix this bacterial suspension with the overlay medium. Take up this mixture *immediately* in a pre-warmed 50 ml pipette and apply evenly over the TLC plate. This is achieved by passing the pipette over the plate in a zig-zag fashion, first from top-to-bottom and then from left-to-right, whilst allowing the mixture to run out at a constant rate. Overlaying the TLC plate in this manner ensures an even thickness of bacterial suspension and gives uniform background colour development. It is very important to minimize the time *P. syringae* pv. *phaseolicola* cells are kept at 45°C since prolonged heat shock can lead to reduced colour development. The most even colour development is achieved with relatively high concentrations of agar (1.0–1.5% w/v), but with some agars premature gelling in the pipette necessitates their use at less than 1%.

7. Wait until the medium has solidified, place the TLC plate in a box lined with moist tissue paper, and incubate overnight (or longer if necessary) at room temperature.

8. Examine the plate for the presence of pale spots (areas containing anti-bacterial compounds) against a deep pink–red background.

2.6.2 TLC bioassay of antifungal activity

The method described here (*Protocol 8*) uses *Helminthosporium carbonum* as the test organism. A concentrated spore suspension of this organism in a glucose–mineral salts medium is sprayed directly on to a pre-developed TLC plate. Where fungal growth is inhibited, a pale zone is apparent against a background of black fungal growth. Hyaline (colourless) fungi may be visualized by the methods described in reference 41.

Protocol 8. Identification, by TLC bioassay, of compounds possessing antifungal activity (42, 43)

1. Prepare the suspension medium:
- 0.7 g KH_2PO_4
- 0.3 g $Na_2HPO_4 \cdot 2H_2O$
- 0.4 g KNO_3
- 0.1 g $MgSO_4 \cdot 7H_2O$
- 0.1 g Nacl
- in 100 ml distilled water

Make up a 30% (w/v) solution of glucose. Sterilize both solutions by autoclaving at 121°C for 15 minutes.

2. Apply aliquots of crude lipid extracts from infected/elicited and control tissues to a silica gel G TLC plate (*Protocol 4* B, analytical TLC) along

Protocol 8. *Continued*

with marker phytoalexins (if available) and develop in a suitable solvent system (*Table 2*).

3. Dry off the developed plate with a (hot) hair-dryer and leave it in a fume cupboard for at least 2 h to ensure that all of the solvent has evaporated.

4. Mix 10 ml of the glucose solution with 60 ml of the suspension medium.

5. Add a small volume of this mixture to a sporulating culture of *Helmintho-sporium carbonum*. Dislodge spores from the fungal mat with a sterile spatula to obtain a concentrated spore suspension (approx. 2×10^6 spores/ml).

6. Spray this solution directly on to the prepared TLC plate. Be careful not to overwet the plate.

7. Incubate the plate under moist conditions (*Protocol 7*, step 7) in the dark at 25–30 °C for 2–5 days.

8. Fungitoxic compounds are indicated by pale zones of inhibition against a background of fungal growth.

---

3. Assay of multienzyme systems and individual enzymes of terpenoid phytoalexin biosynthesis

All terpenoids are formed from acetyl-CoA by the conventional acetyl-CoA → MVA → polyprenyl pyrophosphate pathway (see *Figure 1*). However, the pathways leading from the polyprenyl pyrophosphates to the various terpenoid phytoalexins show considerable diversity.

The rapid and efficient synthesis of terpenoid phytoalexins (in response to elicitation) results from: the stimulation of some of the enzymes of the general terpenoid pathway; the induction of the specific enzymes needed for the synthesis of terpenoid phytoalexins from polyprenyl pyrophosphates, and significantly, the inhibition of the enzymes on directly competing branch pathways of terpenoid biosynthesis (6, 10, 15). In view of its key role in the regulation of mammalian steroidogenesis, HMG-CoA reductase (the enzyme which catalyses the formation of the first committed intermediate on the terpenoid pathway) has been the subject of many studies in plants. As yet, however, no clear regulatory role has been established for this enzyme in the biosynthesis of terpenoid phytoalexins.

The study of the enzymology and regulation of terpenoid phytoalexin biosynthesis is complicated by the following:

- There are at least three compartments which are capable of the biosynthesis of terpenoids in the plant cell, i.e. cytosol plus microsomes, plastids, and mitochondria (*Table 3*).

- Within each of these compartments some of the enzymes are present in the soluble phase whilst the remainder are membrane-bound.
- In order to carry out terpenoid biosynthesis, the mitochondrial compartment (and possibly the plastid compartment as well) needs to import IPP from the cytosolic–microsomal compartment (*Table 3*).
- In some cell-free assays more than one enzyme in the same compartment or enzymes in different compartments, may be able to utilize a test substrate.

Nevertheless, despite these complications, many of the enzymes of interest can be assayed with confidence in fairly crude cell-free preparations.

In several of the systems that we and others have studied, crude cell-free extracts prepared from either control or elicitor-treated suspension cultures/plant tissues are capable, when incubated with the appropriate substrate, co-substrates and co-factors, of catalysing an impressive number of the enzymic reactions described below. For example, crude cell-free extracts from elicitor-treated cultures of *Tabernaemontana divaricata*, in the presence of ATP, NADPH, and molecular oxygen convert MVA into mono-, di-, and tri-hydroxylated pentacyclic triterpenoids, which represents about a dozen enzymic steps! (15). This has two important consequences for investigators working with such systems. Firstly, it is possible to estimate the overall biosynthetic capabilities of a crude cell-free system by incubating preparations with readily obtainable substrates such as [14C]MVA or [14C]IPP. Although this procedure does not provide any absolute kinetic data it can be very informative when used to compare preparations from control and infected/elicited tissue(s) and when both are performed in the presence and absence of the co-substrates/co-factors mentioned above. Secondly, the assay of an enzyme may be complicated by the fact that its product acts as substrate for the next enzyme in the pathway. This means that the activity of the enzyme being assayed can only be estimated by isolating the product and any metabolites formed from the product during the reaction. However, in some cases, the omission of a co-substrate/co-factor (which effectively blocks part of the pathway) or the physical separation of enzymes, i.e. by dividing the crude cell-free extract into soluble and microsomal fractions, can help to circumvent this problem.

The procedures outlined below are straightforward and should be applicable to a variety of systems which synthesize terpenoids. The rationale employed will enable investigators to estimate the biosynthetic capacity of whole cells, the biosynthetic capabilities of crude cell-free preparations, and the activity of some of the individual enzymes, in either crude or only partially purified cell-free extracts.

3.1 Substrates

A major problem for researchers in all fields of terpenoid biochemistry is that only a few of the specialized substrates required for specific enzyme assays are

Table 3. Biosynthetic capabilities of the three compartments synthesizing terpenoids in a plant cell

Compartment	Starting substrate	Prenyl-pyrophosphates produced	Constitutive biosynthetic products	Phytoalexins produced
1. Cytosol-microsomes	Acetyl-CoA (from mitochondria)	FPP[a]	Phytosterols	Sesquiterpenoid Pentacyclic triterpenoid
		Dehydrodolichyl PP	Dolichol	
2. Mitochondria	IPP (from cytosol)	FPP Polyprenyl PP	Cytochrome[a] Ubiquinone	Non-reported
3. Plastid(s)	CO_2, acetyl-CoA or IPP[c]	GGPP[b]	Carotenoids, gibberellins, phylloquinones, tocopherols, and chlorophylls	Diterpenoid
		Nonaprenyl PP	Plastoquinone-9	

[a] It was reported that Z, Z-FPP is needed for the synthesis of hemigossypol and related compounds (11). More recent studies, however, suggest that E, E–FPP is the true precursor (44).

[b] In healthy tissues, the synthesis of GGPP is catalysed by GGPP synthetase (DMAPP + 3 IPP → GGPP). In infected/elicited tissues synthesizing diterpenoid phytoalexins, GGPP is formed by the tandem operation of FPP synthetase (DMAPP + 2 IPP → FPP) and FPp transferase (FPP + IPP → GGPP) (Section 3.6.3).

[c] Although most authorities believe that a plastid can synthesize its complement of terpenoids starting from either CO_2 (chloroplasts) or acetyl-CoA (non-photosynthesizing plastids), some workers are of the opinion that it lacks the early part of the terpenoid pathway and must import IPP from the cytosol.

available from commercial sources. Most assays require a single radiolabelled substrate, but the prenyl transferase assays also require unlabelled substrates for example dimethylallyl pyrophosphate (DMAPP). Unlabelled compounds are also required for adjusting the specific activities of labelled compounds and for use as markers. Companies such as Amersham International plc do provide a service to custom synthesize radiolabelled chemicals, but this is relatively expensive if only small amounts of material are required. However, surplus amounts are sometimes available at reasonable prices. Researchers have thus had to resort to the syntheses of these compounds by either chemical or biochemical methods. It is, however, sometimes possible to obtain small amounts of some of these compounds from other researchers in this field. A list of commercial sources and/or references for the synthesis (either chemical, biochemical, or both) of all the compounds in both radio-labelled and unlabelled form required for the work described in this section is given in *Table 1*.

3.2 Estimation of the capacity of cells in culture to synthesize terpenoids

Feeding experiments with the terpenoid precursor (R)-$[2$-$^{14}C]MVA$ are a useful way of estimating the capacity of plant cells in culture to synthesize extraplastid terpenoids (the plastid envelope is impermeable to MVA). MVA is the substrate of choice as

- it enters the terpenoid pathway after HMG-CoA reductase (a putative regulatory enzyme)
- it is the first committed precursor of terpenoid compounds
- it is taken up rapidly by the plant cell

These features mean that it can be administered at levels which are saturating for phytosterol synthesis (45). Thus, a careful chemical and radiochemical analysis of the amounts of various terpenoids (intermediates and products) formed from exogenously supplied MVA in both elicited and control cell cultures can yield useful information on rate-limiting steps within a pathway. By using this technique, we were able to obtain evidence for the inhibition of squalene synthetase and squalene 2,3-oxide:cycloartenol cyclase in elicitor-treated cell cultures of potato (5) and *Tabernaemontana divaricata* (15), respectively. The details of a typical analysis are beyond the scope of this article but can be found in the aforementioned references.

3.3 Enzyme assays: general procedures

All of the assays described below are based on the use of radiochemicals, the sources of which are described in *Table 1*. Spectrophotometric assays are possible for some of the enzymes, but require substantial amounts of enzyme-enriched fractions and thus are not practicable for physiological studies involving multiple assays of relatively crude cell-free preparations obtained from small amounts of plant material.

The identification of products formed in cell-free incubations is usually achieved by comparing the radioactive products with authentic standards if these are available. The methods used include radio-TLC (both normal and AgNO₃-TLC, Section 2.2, *Protocol 4* and *Table 1*), radio-GLC (Section 2.3 and *Protocol 5*), and recrystallization (to a constant specific radioactivity) of both the product and/or derivatives of the product. An unknown product needs to be isolated in milligram amounts to enable identification by GLC-MS and high-field ^{1}H- and ^{13}C-NMR spectroscopy. Fuller details and examples of these techniques are given elsewhere (10, 15, 46).

The incubation conditions described for each assay contain an indication of protein concentration, substrate concentration, length of incubation, and appropriate controls which enable measurements of enzyme activity to be made whilst the reaction rate is linear. These conditions should be treated as a guide only and must be re-established in any investigation. Protein values quoted are based on the Lowry method. Enzyme activities are expressed as μmol substrate utilized (or product formed h^{-1} mg protein^{-1}. The equation used for the calculation of specific activities is:

$$Sp.\ act. = d.p.m. \times V_{Ass}/V_{Aliq} \times 1/SA \times 1/(2.2 \times 10^6) \times 1/prot. \times 1/t \times x,$$

where d.p.m. = d.p.m. in aliquot counted; V_{Ass} = volume (ml) in which sample for counting was taken up; V_{Aliq} = volume (ml) of aliquot of V_{Ass} taken for counting; SA = specific radioactivity ((>Ci/>mol) of radiosubstrate; $prot$ = mg protein in assay mixture; 2.2×10^6 = dpm in $1\,\mu$Ci; t = period (h) of incubation; and x = correction for loss of label (for example squalene synthetase assay, *Protocol 14*) and/or incorporation of more than one molecule of substrate into the product (when results are expressed as μmol product formed).

3.4 Preparation of cell-free extracts

We have used the methods described in *Protocol 9* to obtain crude cell-free preparations, a soluble fraction (high-speed supernatant), and a microsomal fraction from both potato tuber tissue and cell suspension cultures of a number of plants (7, 10, 15). Minor differences and modifications exist between the methods described here and those used by other research groups.

Protocol 9. Preparation of cell-free extracts (7, 10, 15, 23, 24)

Perform all extraction and centrifugation procedures at 0–4°C

A. *Preparation of crude cell-free extract from suspension culture cells and potato tuber discs*

1. Harvest the cells from a control or elicitor-treated culture by vacuum-filtration through a disc of Miracloth (Calbiochem. Corp) in a Buchner funnel. Wash the retained cells with approx. 300 ml distilled water. Alternatively, remove the top 0.5 mm from discs which have been aged for 24 h, and then incubated with a suitable elicitor or SDW for an appropriate period of time.

2. Transfer freshly harvested cells (1–10 g) or tissue to a pre-cooled (−4°C) mortar and pestle. Add 0.5 ml extraction buffer/g of cells, 0.2 g acid washed sand/g cells, 0.1 g insoluble polyvinylpyrrolidone (PVP)/g of cells and homogenize this mixture for 30 sec.[a]

Extraction buffer

0.1 M potassium phosphate (K–Pi), pH 7.5, containing:

- 0.5 M sucrose
- 2 or 10 mM Na_2EDTA
- 10 mM 2-mercaptoethanol/g cells

3. Squeeze the homogenate through one layer of Miracloth or four layers of cheesecloth into a pre-chilled centrifuge tube.

4. Centrifuge at 4500 g for 15 min at 0–4°C. The resulting supernatant is designated **crude cell-free preparation** and usually contains between 0.2–1 mg protein/ml.

5. *Optional step[b]*

Apply 2 ml of the crude cell-free preparation to a pre-packed disposable Sephadex G-25 column (PD-10 column, Pharmacia) which has been pre-equilibrated with running buffer (0.05 M K–Pi, pH 7.5, containing 0.5 M sucrose, 1 mM Na_2EDTA, and 1 mM 2-mercaptoethanol). Elute the column according to the manufacturer's instructions.

B. *Separation of crude-cell free preparation into soluble and microsomal fractions*

1. Centrifuge the crude cell-free preparation or Sephadex G-25 eluate at 16 000 g for 40 min to remove intact and broken organelles (mitochondria, plastids, etc.).

2. Centrifuge the 16 000 g supernatant at 105 000 g[c] for 60 min to precipitate the microsomes. The supernatant from this step represents the **high-speed supernatant**.

3. Resuspend the microsomal pellet obtained in step 2 in a suitable volume of the running buffer described in part A, step 5 and re-centrifuge under the same conditions as step 2. Resuspend the washed microsomes in a suitable volume of running buffer (approx. 1/10 the starting vol.) to

Protocol 9. *Continued*

obtain the **microsomal fraction**. Readjust the pH to 7.5 with 1 M Tris if necessary.

[a] This method can be used with other tissues (for example cotyledons and seedlings) provided that the volume of extraction buffer used is increased. If the homogenate is to be passed through Sephadex (step 5) prepare extraction buffer with 10 mM Na_2EDTA.

[b] This step is particularly useful for removing low molecular weight compounds (for example endogenous co-factors, phytoalexins) which may interfere with the assay of certain enzymes.

[c] Higher speeds or longer times may be required for extracts prepared with viscous buffers (i.e. those containing large amounts of sucrose, glycerol, or soluble PVP).

3.5 Estimation of the biosynthetic capabilities of a crude cell-free preparation

This technique (*Protocol 10*) measures the incorporation of [1-^{14}C]IPP into terpenoids in a crude cell-free preparation. Preparations (*Protocol 9*) from both control and infected/elicited plant tissue are incubated both with and without an NADPH-generating system and/or molecular oxygen. [2-^{14}C]MVA can often be used in place of [1-^{14}C]IPP providing ATP or an ATP-generating system is included in the incubation mixture.

Protocol 10. Estimation of the capabilities of a crude cell-free preparation to biosynthesize terpenoids from [2-^{14}C]MVA or [1-^{14}C]IPP

1. Mix 0.5 ml crude cell-free preparation[a] (0.1–0.3 mg protein) and 0.5 ml assay buffer [50 mM K–Pi-buffer, pH 7.5, containing 0.5 M sucrose, 1 mM Na_2EDTA, 1 mM 2-mercaptoethanol and 10 mM $MgCl_2$ (added last to avoid precipitation)] in two thick-walled test tubes (13 × 120 mm).

2. Add 25 μl NADPH-generating system[b] to one tube and an equal volume of assay buffer to the other. If (*R*)-[2-^{14}C]MVA is used as substrate, add 25 μl of an ATP-generating system[c] to both tubes at the same time as the NADPH-generating system.

3. Equilibrate tubes at 30 °C for 5 min.

4. Add 4 μl 4.5 mM (*R*)-[2-^{14}C]MVA (18 nmol, 55 mCi/mmol) or 4 μl 0.9 mM [1-^{14}C]IPP (3.6 nmol, 55 mCi/mmol) to both tubes to start the reactions.

5. Incubate the tubes at 30 °C for 45 min.

6. Add 5 ml $CHCl_3$–methanol (1:2) to each tube to stop the reactions. Allow to stand for 30 min, at room temperature.

7. Centrifuge tubes at 600 *g* for 10 min to precipitate denatured protein.

8. Transfer the supernatant of each tube to a clean thick-walled test tube and add 5 ml distilled water.

9. Repeat centrifugation (step 7). Remove the water/methanol phase (upper phase) by means of a Pasteur pipette attached to a vacuum line.

10. Overlay the CHCl₃ phase with 5 ml distilled water.

11. Repeat centrifugation (step 7) and removal of the upper aqueous phase as for step 9.

12. Transfer the CHCl₃ phase of each tube to a small screw-topped vial and carefully evaporate off the solvent on a hot plate (20–40°C) under a stream of N₂ gas. Traces of water can be removed by adding a drop of ethanol and repeating the evaporation step.

13. Dissolve the CHCl₃-soluble lipid extract in 100 μl acetone containing 100 μg each of unlabelled markers, i.e. farnesol, squalene, and terpenoid phytoalexin(s) where appropriate. Assay radioactive content of 10 μl of the extract by liquid scintillation counting to estimate aliquot needed for TLC. (Store the remainder at −20°C unless analysed immediately by TLC.)

14. Apply a suitable aliquot, for example 10–20 μl[d] to a rhodamine 6G-impregnated silica G plate (*Protocol 4* B: analytical TLC) and develop the plate with ethyl acetate–cyclohexane (1:1)[e] or other suitable solvent (*Table 2*).

15. Dry off the solvent thoroughly with a (hot) hair-dryer and visualize the marker(s) under UV₂₅₄ nm light. Scan each lane for areas of radioactivity with a radio-TLC scanner or prepare an X-ray radioautograph.

16. Scrape areas of gel corresponding to markers and/or containing radio-activity directly into a scintillation vial. Add 10 ml of a suitable scintillation cocktail[f] and assay radioactivity by liquid scintillation counting or recover individual bands after preparative TLC of the extract and characterize products (Sections 2 and 3.3).

---

[a] Sephadex-treated preparations often give more clearcut results (*Protocol 9*, step 5).

[b] Freshly prepared solution of assay buffer containing 129 mM glucose-6-phosphate (sodium salt), 10.4 mM NADP (sodium salt), and 1.5 units of glucose-6-phosphate dehydrogenase/ml.

[c] Freshly prepared solution of assay buffer containing 541 mM creatine phosphate (disodium salt), 108 mM ADP (sodium salt), and 2800 units of creatine phosphokinase/ml.

[d] Typically we apply 6000–10000 d.p.m. of ¹⁴C or 100000–200000 d.p.m. of ³H per lane.

[e] Most hydrocarbon products will migrate in or near to squalene in this solvent system. To resolve these products repeat step 14 but use petroleum ether (40–60°C) as the developing solvent (*Protocol 4*, step 8). Farnesol and other oxygenated products remain at the origin.

[f] Any lipid-compatible scintillation cocktail can be used for the assay of the radioactivity content of samples dissolved in organic solvents. However, cocktails containing a polar organic solvent (for example 20% ethanol) must be used for the assay of the radioactivity content of TLC gels and aqueous samples.

3.6 Estimation of the enzymes of the central terpenoid biosynthetic pathway

Although it is possible to assay all of the enzymes of the general terpenoid pathway, it is not normally attempted during physiological studies with relatively crude cell-free preparations. The enzymes that are assayed are those which have a (putative) regulatory role (i.e. are always or in some circumstances, rate-limiting), for example HMGR, and those enzymes that are specifically induced or suppressed in response to elicitation. These enzymes are identified by feeding experiments with MVA (Section 3.2) and by studies on the biosynthetic capabilities of crude cell-free preparations (Section 3.5 and *Protocol 10*).

3.6.1 Microsomal HMG-CoA reductase (HMGR; EC 1.1.1.34, step 3 in *Figure 1*)

This assay (*Protocol 11*) measures the NADPH-dependent formation of (R)-[^{14}C]MVA from (R, S)-[3-^{14}C]HMG-CoA. The product is lactonized by treatment with acid prior to purification by TLC, recovery, and liquid scintillation counting (LSC). In the absence of ATP the product will not be further metabolized.

Protocol 11. Assay of microsomal HMG-CoA reductase (4, 47)

1. Mix 0.1 ml microsomal fraction[a] (0.1–0.4 mg protein) with 0.6 ml assay buffer (50 mM K–Pi, pH 7.5, containing 4.2 mM Na_2EDTA, 8.3 mM dithiothreitol [DTT], and 1 mg bovine serum albumin [BSA]) and 25 μl NADPH-generating system (*Protocol 10*, footnote *b*) in a 1.5 ml microfuge tube.

2. Equilibrate tube at 30°C for 5 min.

3 Add 5 μl of 0.42 mM (S)-[3-^{14}C]HMG-CoA (2.1 nmol, 0.5 mC2/mmol)[b] to start the reaction.

4. Incubate at 30°C for 30 min.

5. Add 0.1 ml 3 M HCl to stop the reaction and 20 μl of 5% (R, S)MVA lactone in ethanol as carrier.

6. Stand vials at room temperature for at least 30 min to ensure lactonization of labelled mevalonate.

7. Centrifuge at 4000 g for 10 min to remove precipitated protein.

8. Apply 100 μl of the reaction mixture to a silica gel G TLC plate (*Protocol 4 B*, analytical TLC) and develop in toluene–acetone (1:1).

9. Expose the plate to iodine vapour. The MVA lactone will be seen as a brownish-yellow band at an R_f of between 0.5 to 0.7.

10. Mark the band and allow it to decolorize.

11. Scrape the gel containing the lactone directly into a scintillation vial. Add 10 ml scintillant (*Protocol 10*, footnote *f*) and assay radioactivity by LSC.

[a] Changes in microsomal HMGR activities can be assayed in crude extracts as its activity is very much higher than that of chloroplast HMGR. It is, however, preferable to use a microsomal fraction (*Protocol 9* B).

[b] Prepared by mixing the appropriate amounts of (*S*)-[3-^{14}C]HMG-CoA and (*R,S*)-HMG-CoA.

3.6.2 Pyrophosphomevalonate decarboxylase (EC 4.1.1.33, step 6 in *Figure 1*)

This assay (*Protocol 12*) involves trapping the $^{14}CO_2$ evolved from (*R, S*)-[1-^{14}C]mevalonic acid in aqueous KOH. The evolution of $^{14}CO_2$ is non-linear for the first 5 minutes and, therefore, each assay requires a control to correct for this (37).

Protocol 12. Assay of pyrosphophomevalonate decarboxylase (37)

1. Set up two Warburg manometer flasks as follows:

Centre well: 0.1 ml 2-phenylethylamine (Aldrich) and a fluted paper wick

Side-arm: 0.1 ml crude cell-free preparation

Main compartment: 0.9 ml assay buffer—50 mM K–Pi, pH 7.5, containing:

- 5 mM ATP
- 5 mM $MgCl_2$
- 5 mM KF
- 0.25 μCi (*R, S*)-[1-^{14}C]MVA (4.2 nmol, 60 mCi/mmol)

2. Seal each flask tightly with a Subaseal rubber cap and equilibrate at 30°C for 5 min.

3. Start the reactions by mixing the contents of the main compartment and the side-arm of each flask.

4. Incubate both flasks at 30°C with shaking.

5. After 5 min inject 0.3 ml 6 M HCl into the reaction mixture in the main compartment of the control.

6. After 20 min inject 0.3 ml 6 M HCl into the reaction mixture in the main compartment of the second flask.

7. Continue to incubate both flasks for a further 1 h to ensure complete trapping of liberated $^{14}CO_2$.

8. Transfer contents of the central well to scintillation vials. Add 10 ml scintillant (*Protocol 10*, footnote *f*) and assay radioactivity by LSC.

9. Calculate activity from the difference in radioactivity present in the two samples.

3.6.3 Cytosolic and stromal (2*E*, 6*E*)-farnesyl pyrophosphate (FPP) synthetases (geranyl transferase, EC 2.5.1.1, step 9 in *Figure 1*) and stromal farnesyl transferase (a GGPP synthetase, EC 2.5.1.1, step 11 in *Figure 1*)

FPP synthetase catalyses the sequential addition of two molecules of iso-pentenyl pyrophosphate (IPP) to one molecule of dimethylallyl pyrophos-phate (DMAPP) to form one molecule of FPP. The cytosolic synthetase provides the FPP used in the synthesis of phytosterols and the sesquiter-penoid and triterpenoid phytoalexins, whereas the stromal synthetase provides the FPP needed by the stromal farnesyl transferase. The stromal farnesyl transferase catalyses the addition of one molecule of IPP to one molecule of FPP to form one molecule of geranylgeranyl pyrophosphate (GGPP). This GGPP is used in the synthesis of diterpenoid phytoalexins. (The stromal GGPP required for the synthesis of constitutive plastid diter-penoids and related compounds, *Table 3*, is made by a GGPP synthetase which catalyses the sequential addition of three molecules of IPP to one molecule of DMAPP to form one molecule of GGPP, step 10 in *Figure 1*.)

The assay of cytosolic FPP synthetase should be performed on stromal-free high-speed supernatants, whilst the assay of stromal FPP synthetase and farnesyl transferase should be performed on osmotically-shocked plastids. Alternatively, partly purified enzymes from crude cell-free preparations can be used. The assay of crude cell-free preparations or high-speed supernatants can provide a preliminary indication of changes in the activity of the cytosolic FPP synthetase since in most tissues (particularly non-photosynthetic tissues, for example tubers, etiolated seedlings, tissue cultures, etc.) the activity of GGPP synthetase is very low compared to the activity of this enzyme. The assays using cell-free preparations or high-speed supernatants will also show an increase in the total FPP synthetase activity (i.e. stromal and cytosolic FPP synthetase activity) and stromal FPP transferase activities between control and elicited tissues synthesizing diterpenoid phytoalexins.

These prenyl transferase assays (*Protocol 13*) use [1-^{14}C]IPP with an appro-priate prenyl pyrophosphate acceptor. The product, FPP (farnesyl synthetase) or GGPP (farnesyl transferase) is not isolated directly from the reaction mixture. Instead, the activity of the enzyme is calculated from the radioactivity extracted into diethyl ether after acid-hydrolysis of the reaction mixture. The diethyl ether-extract contains two isomeric alcohols formed from the prenyl pyrophosphate by acid hydrolysis (for example FPP + H$^+$ → farnesol + nerolidiol [3:1]) and also any diethyl ether-soluble products formed from the prenyl pyrophosphate during the assay. The radioactive substrate (IPP) is not an allylic pyrophosphate and is, therefore, not hydrolysed by the acid treat-ment. The methods use a control (lacking the added prenyl pyrophosphate acceptor) to correct for background levels of phosphatases, IPP isomerase, and possible extraction of [^{14}C]IPP.

Protocol 13. Assay of FPP synthetase and FPP transferase (GGPP synthetase)[a]

A. *FPP synthetase*

1. Mix 0.1 ml crude cell-free preparation or high-speed supernatant (*Protocol 9*), with 0.4 ml assay buffer (40 mM Tris malate, pH 6.8, containing 4 mM $MgCl_2$) in each of two tubes.

2. Add 10 µl of 2.4 mM DMAPP or GPP to one tube and an equal volume of buffer solution to the other (control) tube.

3. Equilibrate the tubes at 30 °C for 5 min.

4. Add 2 µl 1.2 mM [1-^{14}C]IPP (2.4 nmol,[b] 55 mCi/mmol) to both tubes to start the reaction.

5. Incubate at 30 °C for 30 min.

6. Add 0.5 m ethanol to each tube to stop the reactions.

7. Add 0.5 ml 6 M HCl to each tube and incubate at 30 °C for 15 min to hydrolyse allylic pyrophosphates to their corresponding free alcohols.

8. Extract the free alcohols with diethyl ether (2 × equal vol.). Combine the extracts in a scintillation vial and evaporate the solvent under a stream of N_2 gas.

9. Add 10 ml scintillant[c] to each vial and assay radioactivity by LSC.

10. Calculate farnesyl synthetase activity from the difference between the extractable radioactivity in the DMAPP/GPP containing and control assays.

B. *FPP transferase (GGPP synthetase)*

1. As part A, step 1, except that the assay buffer used contains 50 mM Tris–HCl, pH 8.5, containing 2 mM $MgCl_2$ and 25% (v/v) glycerol.

2. As part A, step 2, above except use 10 µl 2 mM FPP solution.

3. Follow part A, steps 3–9 above.[b]

4. Calculate farnesyl transferase activity from the difference between the FPP containing and control assays.

[a] See reference 48 and *Protocol 8*.

[b] Considerably smaller amounts of [1-^{14}C]IPP can be used as the K_m for the reaction is about 0.1 µM.

[c] See *Protocol 10*, footnote *f*.

3.6.4 Squalene synthetase (SQS; EC 2.5.1.21, step 12 in *Figure 1*)

This NADPH-dependent microsomal enzyme catalyses the tail-to-tail condensation of two molecules of FPP to form one molecule of squalene. In response to elicitation, SQS activity is rapidly suppressed in Solanaceous plants synthesizing sesquiterpenoid phytoalexins (10, 49). Conversely, however, SQS activity is stimulated in cell suspension cultures of *Tabernaemontana divaricata* synthesizing pentacyclic triterpenoid phytoalexins (15).

[2-^{14}C]FPP, [4,8,12-^{14}C$_3$]FPP, [1-^{3}H$_2$]FPP, or any other labelled species of FPP can be used in the assay (*Protocol 13*). If [1-^{3}H$_2$]FPP is used as the substrate, a correction must be made in the calculation of enzyme activity for the loss of one ^{3}H atom, which occurs in the formation of 1 molecule of ^{3}H$_3$-labelled squalene. The method involves inhibiting the activity of squalene 2,3-epoxidase by performing the assay, *Protocol 14*, under anaerobic conditions. Labelled farnesol, which is produced by non-specific phosphatases in the crude cell-free extract, is separated from squalene by TLC.

Protocol 14. Assay of squalene synthetase[a]

1. Degas all buffers with either He or N$_2$ gas prior to use and perform steps 2–4 under a blanket of one of these gases.[b]

2. Mix 0.5 ml crude cell-free preparation (0.1–0.3 mg protein) or 0.1 ml microsomal fraction made up to 0.5 ml with running buffer (*Protocol 9* B) with 0.5 ml of the assay buffer described in *Protocol 10* and 25 μl NADPH-generating system (*Protocol 10*, footnote *b*).

3. Equilibrate the tube at 30°C for 5 min.

4. Add 10 μl of 2.5 mM [1-^{3}H$_2$]FPP (25 nmol, 40 mCi/mmol), or other labelled species of FPP: Section 3.6.4, to start the reaction.

5. Incubate at 30°C for the appropriate time (5–45 min).

6. Add 5 ml CHCl$_3$–methanol (1:2) to stop the reaction. Allow to stand at room temperature for 30 min.

7. Follow *Protocol 10*, steps 7–12.

8. Dissolve the CHCl$_3$-soluble lipid extract in 100 μl acetone containing 100 μg each of unlabelled farnesol and squalene as markers. Count 10 μl of the extract to estimate aliquot needed for TLC. Store the remainder at −20°C unless analysed immediately by TLC.

9. Apply a suitable aliquot, for example 10 μl to a rhodamine 6G-impregnated silica gel G plate[c] (*Protocol 4* B, analytical TLC) and develop the plate with petroleum ether (40–60°C, *Protocol 10*, footnote *e*) in a TLC tank covered with a thick cloth (to protect it from draughts which can cause the plate to develop unevenly with this volatile solvent).

10. After development, dry off the solvent thoroughly with a (hot) hair-dryer and visualize the marker bands of squalene (bright pink, R_f approx. 0.30) and farnesol (bright pink, R_f 0.0) under UV$_{254}$ nm light. Other radioactive products can be detected on the plate by the methods mentioned in

Protocol 10, step 15. Scrape the areas of the gel containing the squalene into scintillation vials. Add 10 ml scintillant (*Protocol 10*, footnote *f*) and assay radioactivity by LSC.

[a] See reference 10; micro-assay has been reported (49).

[b] This is most easily achieved by attaching a series of Pasteur pipettes to a gas manifold and positioning the tip of a pipette in the assay tube just above the surface of each reaction mixture.

[c] A method utilizing a silica gel column is given in *Protocol 15*, steps 7–9.

3.6.5 Squalene epoxidase (EC 1.14.99.7, step 13 in *Figure 1*)

This microsomal mixed-function oxygenase catalyses the NADPH- and O_2-dependent epoxidation of squalene to form (*S*)-squalene 2,3-oxide. Feeding experiments with MVA have suggested that this enzyme may play a regulatory role during the synthesis of pentacyclic triterpenoid phytoalexins in cell suspension cultures of *Tabernaemontana divaricata* (15). The assay (*Protocol 15*) employs a control (minus enzyme) to correct for the possible non-enzymic epoxidation of squalene.

Protocol 15. Assay of squalene epoxidase (50)

1. Transfer 10 μl of toluene containing 1.5 mg Tween 80 and 10 nmoles [^{14}C]squalene (1 mCi/mmol) and evaporate the organic solvent under a stream of N_2 gas.

2. Add 0.2 ml assay buffer (20 mM Tris–HCl, pH 7.4, containing 0.1 mM FAD which has been pre-equilibrated at 30°C) to each tube.

3. Mix the contents of each tube with a vortex mixer to emulsify the substrate and detergent.

4. Add 0.8 ml crude cell-free preparation or 0.1 ml microsomal fraction (*Protocol 9* B) made up to 0.8 ml with resuspension buffer to one tube and 0.8 ml of the appropriate buffer (for example resuspension buffer in the case of microsomes) to the other.

5. Incubate the tubes at 30°C for 5 min.

6. Add 25 μl of NADPH-generating system (*Protocol 10*, footnote *b*) to each tube to start the reaction and incubate both tubes at 30°C for 30 min.

7. Add 1 ml 60% (w/v) KOH in methanol to each tube to terminate the reaction and leave the tubes in the cold (4°C) overnight.

8. Extract the substrate and products with petroleum ether (40–60°C, 3 × equal vol.). Combine the petrol extracts, reduce vol. to approx. 2 ml under nitrogen and apply the sample to a short column (approx. 2–3 cm in a Pasteur pipette) of silica gel (column grade) topped with a 1 cm layer of anhydrous Na_2SO_4 made up in petrol.

Protocol 15. *Continued*

9. Elute the column with approx. 5 ml petrol (this fraction contains un-reacted [^{14}C] squalene). Then elute the column with approx. 5 ml diethyl ether and collect the eluate in a scintillation vial (this fraction contains squalene 2,3-oxide and any metabolites formed from this product during the reaction).

10. Evaporate the diethyl ether under a stream of N_2 gas. Add 10 ml scintillant (*Protocol 10*, footnote *f*) and assay radioactivity by LSC.

3.6.6 Squalene 2,3-epoxide:cycloartenol cyclase (EC 5.4.99.7, step 14 in *Figure 1*) and squalene 2,3-epoxide:α/β-amyrin cyclases (step 15 in *Figure 1*)

These microsomal enzymes catalyse the cyclization of (3 *S*)-squalene 2,3-oxide to form cycloartenol and amyrins. During the synthesis of pentacyclic triterpenoid phytoalexins in cell suspension cultures of *Tabernaemontana divaricata* the activity of squalene 2,3-epoxide:cycloartenol cyclase is suppressed whilst the activities of the squalene 2,3-epoxide:α/β-amyrin cyclases are induced. The assays (*Protocol 16*) measure the formation of amyrins and/or cycloartenol from [^{3}H]squalene 2,3-oxide.

Protocol 16. Assay of squalene 2,3-epoxide:cycloartenol and squalene 2,3-epoxide:α/β-amyrin cyclases (51)

1. Transfer 10 μl of toluene containing 1.5 mg Tween 80 and 40 nmoles (*R*, *S*)-[^{3}H]squalene 2,3-oxide (0.22 μCi; 56 mCi/mmol) to each of two test tubes and evaporate solvent under a stream of N_2 gas.

2. Add 0.8 ml assay buffer (*Protocol 10*) (which has been pre-equilibrated to 30°C) to each tube and mix the contents with a vortex mixer to emulsify the substrate and detergent.

3. Add 0.2 ml microsomal fraction (*Protocol 9* B) which has been pre-equilibrated to 30°C for 5 min, to start the reaction.

4. Incubate at 30°C for 1 h.

5. Add 5 ml CHCl$_3$–methanol (1:2) to each tube to stop reaction. Allow to stand for 30 min at room temperature.

6. Follow *Protocol 10*, steps 7–12.

7. Follow *Protocol 10*, steps 13 and 14, but use lanosterol (a 4,4'-dimethylsterol) as the only marker.

8. Apply samples to a silica gel G plate (*Protocol 4* B: analytical TLC) and develop the plate with toluene–ethyl acetate (17:3).

9. Follow *Protocol 10*, step 15.

10. Elute the band containing the lanosterol marker and other 4,4-dimethyl sterols (R_f 0.4). Transfer the eluted material to a small vial with a suitable solvent, for example diethyl ether and take to dryness under a stream of N_2 gas. Acetylate the 4,4-dimethyl sterols by adding 100 μl of acetic anhydride and 20 μl of pyridine to the vial. Mix and leave on a hot plate at 60°C for 2 h or overnight at room temperature.

11. Add 200 μl ethanol to the vial and evaporate the solvents under a stream of nitrogen on a hot plate (temp. not above 60°C).

12. Apply the acetylated materials to an $AgNO_3$-impregnated silica gel G plate (*Protocol 4*, steps 1 and 2 and part B, analytical TLC) and develop the plate with toluene–petrol (1:1).

13. Follow *Protocol 10*, step 15.

14. Scrape the areas of gel corresponding to the markers and/or containing radioactivity (α-amyrin acetate, R_f 0.37; β-amyrin acetate, 0.37,[a] cyclo-artenyl acetate, R_f 0.32; 24-methylenecycloartanyl acetate, R_f 0.26) directly into a scintillation vial. Add 10 ml of a suitable scintillation cocktail[b] and assay radioactivity by LSC.

15. The activity of squalene 2,3-epoxide:cycloartenol cyclase is calculated by adding the amounts of radioactivity in the 24-methylene cycloartanyl acetate and cycloartenyl acetate fractions. The activity of squalene 2,3-epoxide:α/β-amyrin cyclases is calculated from the radioactivity in the amyrin acetate fraction.

---

[a] The method to separate α-amyrin and β-amyrin can be found in reference 52.
[b] See *Protocol 10*, footnote *f*.

3.6.7 Enzymes of the central terpenoid pathway not described

The assays for the following enzymes are not described. The step number corresponds to the arrow number in *Figure 1*: acetoacetyl-CoA thiolase (EC 2.3.1.9, step 1); HMG-CoA synthetase (EC 4.1.3.5, step 2); MVA kinase (EC 2.7.2.36, step 4); 5-phosphomevalonate kinase (EC 2.7.4.2, step 5); isopentenyl pyrophosphate isomerase (EC 5.3.3.2, step 7), and GPP synthetase (EC 2.5.1.1, step 8).

3.7 Estimation of the activation of some of the enzymes specific to the biosynthesis of terpenoid phytoalexins

Only a few of these enzymes have been assayed on an individual basis. The best characterized is the GGPP-carbocyclase casbene synthetase which produces the macrocyclic diterpenoid phytoalexin casbene in a single step from GGPP (13).

We have developed assays for 5-*epi*-aristolochene synthetase (EASY), the FPP-carbocyclase involved in the biosynthesis of the eremophilene phytoalexins capsidol and debneyol (10), and for the FPP-carbocyclase involved in the biosynthesis of the vetispirane lubimin (reference 7 and D. R. Threlfall and I. M. Whitehead, unpublished work). Others have characterized a reductase (Section 3.7.2) and a hydroxylase (Section 3.7.3) which are involved in the biosynthesis of the furanosesquiterpenes (23, 24).

Stromal FPP synthetase and stromal FPP transferase which technically belong in this section have been discussed already (see Section 3.6.3).

3.7.1 FPP and GGPP carbocyclases

i. 5-epi-Aristolochene synthetase (EASY, step 16 in Figure 1)

This assay (*Protocol 17*) measures the formation of 5-*epi*-aristolochene from [1-^{3}H$_2$]FPP (although other radiolabelled species of FPP can be used, Section 3.6.4) by an identical procedure to that used for the assay of SQS (*Protocol 14*) except that an NADPH-generating system is not required. If the cell-free preparation has been gel-filtered (*Protocol 9 A*, step 5) to remove endogenous amounts of NADPH there is no need to perform this assay under an inert atmosphere.

Protocol 17. Assay of 5-*epi*-aristolochene synthetase (EASY)[a]

1. Mix 0.5 ml crude cell-free preparation or 105 000 g supernatant (prepared from elicitor-treated *N. tabacum* cultures) with 0.5 ml of the assay buffer described in *Protocol 10*.

2. Follow SQS assay (*Protocol 14*, steps 2–8).

3. Locate areas of radioactivity by scanning the plate with a radio-TLC scanner.

4. Transfer the area of the gel containing 5-*epi*-aristolochene (R_f 0.60) to a scintillation vial. Add 10 ml scintillant (*Protocol 10*, footnote *f*) and assay radioactivity by LSC.

   [a] See reference 10; a micro-assay has been reported (49).

---

ii. Vetispira-1(10), 11-diene synthetase (step 17 in Figure 1)

This enzyme is assayed by the method described for EASY (*Protocol 17*). The product, which is probably vetispira-1(10), 11-diene (D. R. Threlfall and I. M. Whitehead, unpublished work) has not yet been fully characterized.

iii. Casbene synthetase (step 18 in Figure 1)

This assay (*Protocol 18*) measures the formation of [^{14}C]casbene from [2-^{14}C]GGPP. The product is isolated on a silica gel F TLC plate, the top part of which has been dipped into an AgNO$_3$ solution. This focuses the product at the $AgNO_3$ interface. Other diterpenes which may be formed in the reaction, migrate into the $AgNO_3$ region.

Protocol 18. Assay of casbene synthetase (53)

1. Prepare a crude cell-free preparation or soluble fraction by the methods outlined in *Protocol 9* A, except that the K–Pi buffer is replaced by 100 mM Tris–HCl, pH 8.5.
2. Mix 0.5 ml crude cell-free preparation or soluble fraction with 0.5 ml assay buffer (100 mM Tris–HCl, pH 9.0, containing 10 mM $MgCl_2$) in a test tube.
3. Equilibrate the test tube at 30°C for 5 min.
4. Start the reaction by adding 10 μl of 0.33 mM [2-^{14}C]GGPP (3.6 nmol, 2 mCi/mmol).
5. Incubate tube at 30°C for up to 30 min.
6. Add 1 ml 25% (v/v) ethanol in petroleum ether (40–60°C) to stop the reaction.
7. Mix layers vigorously on a vortex mixer for 1 min.
8. Remove organic layer and repeat extraction (3 × 1 ml petroleum ether (40–60°C).
9. Combine organic extracts and reduce volume to approx. 0.1 ml under a stream of N_2 gas.
10. Apply extract on to a silica gel F (Merck/Anachem) TLC plate (*Protocol 4*B), the top 13.5 cm of which has been immersed in a solution of 4% (w/v) $AgNO_3$ in acetonitrile–ethanol (9:1) and dried.
11. Develop plate in benzene–petroleum ether (40–60°C) (3:7).
12. Follow *Protocol 10*, step 15.
13. Scrape the area of silica from 6 to 7 cm (containing the origin of the $AgNO_3$) into a scintillation vial. Add 10 ml scintillant (*Protocol 10*, footnote *f*) and assay radioactivity by LSC.

3.7.2 Dehydroipomeamarone reductase

This microsomal enzyme catalyses the NADPH-dependent reduction of dehydroipomeamarone to ipomeamarone. The assay (*Protocol 19*) measures the formation of [^{14}C]ipomeamarone from [^{14}C]dehydroipomeamarone. The product is isolated by TLC.

Protocol 19. Assay of dehydroipomeamarone reductase (23)

1. Prepare a microsomal fraction by the method described in *Protocol 9* B.
2. Mix 50 μl of the microsomal fraction with 0.4 ml assay buffer (20 mM K–Pi pH 7.5, containing 0.5 M sucrose and 1 mM DTT).

Protocol 19. *Continued*

3. Add 12 μl NADPH-generating system (*Protocol 10*, footnote *b*) and equilibrate tubes at 30°C for 5 min.

4. Add 10 μl of 3.8 mM [^{14}C]dehydroipomeamarone (38 nmol, 15.2 mCi/mmol, in acetone) to start the reaction.

5. Incubate tube at 30°C for 1 h.

6. Add 5 ml CHCl$_3$–methanol (1:2) to each tube to stop reaction. Allow to stand for 30 min at room temperature.

7. Follow *Protocol 10*, steps 7–12.

8. Dissolve the lipid extract in 100 μl acetone and apply a suitable aliquot (*Protocol 10*, footnote *d*) to a silica gel G plate (*Protocol 4* B: analytical TCL).

9. Develop the TLC plate to a height of 8 cm with *n*-hexane–ethyl acetate (1:1). Dry off the solvent and then develop the plate to a height of 16 cm with *n*-hexane–ethyl acetate (4:1).

10. Locate the [^{14}C]ipomeamarone with a radio-TLC plate scanner (R_f approx. 0.70). Scrape this area of gel into a scintillation vial, add 10 ml scintillant (*Protocol 10*, footnote *f*) and assay radioactivity by LSC.

3.7.3 Ipomeamarone 15-hydroxylase

This microsomal mixed-function oxygenase catalyses the NADPH- and O$_2$-dependent hydroxylation of ipomeamarone to ipomeamaronol. The assay (*Protocol 20*) measures the formation of [^{14}C]ipomeamaronol (15-hydroxyipomeamarone) from [^{14}C]ipomeamarone. The product is isolated by TLC.

Protocol 20. Assay of ipomeamarone 15-hydroxylase (24)

1. Prepare a microsomal fraction as described in *Protocol 9* B, except that:

   The homogenization buffer is replaced by 50 mM Tris–HCl buffer, pH 8.5, containing:

   - 0.5 M sucrose
   - 1 mM Na$_2$EDTA
   - 1% (w/v) sodium isoascorbate
   - polyclar AT (0.2 g/g fresh weight)
   - 0.5% (w/v) BSA

   The equilibration buffer for Sephadex G-25 column is replaced by 50 mM Tris–HCl buffer, pH 8.5, containing:

   - 0.5 M sucrose

- 1 mM Na$_2$EDTA
- 1 mM dithiothreitol (DTT)

The resuspension buffer replaced by 50 mM Tris–HCl, pH 8.5, containing:

- 0.5 M sucrose
- 0.3 mM DTT

2. Mix 1 ml microsomal fraction (1–4 mg protein) with 0.5 ml assay buffer (50 mM Tris–HCl, pH 8.0, containing 0.5 M sucrose).

3. Add 25 μl of an NADPH-generating system (*Protocol 10*, footnote *b*).

4. Equilibrate tube at 30°C for 5 min.

5. Add 15 μl of 6.2 mM [^{14}C]ipomeamarone (93 nmol, 15 mCi/mmol, in acetone) to start reaction.

6. Incubate at 30°C for 1 h.

7. Add 5 ml CHCl$_3$–methanol (1:2) to stop the reaction. Allow to stand for 30 min at room temperature.

8. Follow *Protocol 10*, steps 7–12.

9. Follow *Protocol 19*, steps 8 and 9.

10. Locate the [^{14}C]ipomeamaronol with a radio-TLC plate scanner (R_f approx. 0.20). Scrape this area of gel into a scintillation vial, add 10 ml scintillant (*Protocol 10*, footnote *f*) and assay radioactivity by LSC.

11. The enzyme activity can also be calculated from the amount of [^{14}C]ipomeamarone (R_f approx. 0.7) recovered.

References

1. Brooks, C. J. and Watson, D. G. (1985). *Nat. Prod. Rep.*, 427.
2. van der Heijden, R., Verheij, E. R., Schripsema, J., Baerheim Svendsen, A., Verpoorte, R., and Harkes, P. A. A. (1988). *Plant Cell Rep.*, **7**, 51.
3. Ôba, K., Reiko, Y., Fujita, M., and Uritani, I. (1982). In *Plant infection: the physiological and metabolic basis* (ed. Y. Asada, W. R. Bushnell, S. Ouchi, and C. P. Vance), p. 157. Springer-Verlag, Berlin.
4. Stermer, B. A. and Bostock, R. M. (1987). *Plant Physiol.*, **84**, 404.
5. Brindle, P. A., Kuhn, P. J., and Threlfall, D. R. (1988). *Phytochemistry*, **27**, 133.
6. Chappell, J. and Nable, R. (1987). *Plant Physiol.*, **85**, 469.
7. Coolbear, T. and Threlfall, D. R. (1985). *Phytochemistry*, **24**, 1963.
8. Preisig, C. L. and Kúc, J. A. (1987). In *Molecular determinants of plant diseases* (ed. S. Nishimura, C. P. Vance, and N. Doke), p. 203. Springer-Verlag, Berlin.
9. Ôba, K., Kondo, K., Doke, N., and Uritani, I. (1985). *Plant Cell Physiol.*, **26**, 873.
10. Threlfall, D. R. and Whitehead, I. M. (1988). *Phytochemistry*, **25**, 2567.

11. Heinstein, P., Widmaier, R., Wegner, P., and Howe, J. (1979). *Recent Adv. Phytochem.*, **12**, 313.
12. Pierce, M. and Essenberg, M. (1987). *Physiol. Mol. Plant Pathol.*, **31**, 273.
13. Moesta, P. and West, C. A. (1985). *Arch. Biochem. Biophys.*, **238**, 325.
14. Dudley, M. W., Dueber, M. T., and West, C. A. (1986). *Plant Physiol.*, **81**, 335.
15. van der Heijden, R., Threlfall, D. R., Verpoorte, R., and Whitehead, I. M. (1989). *Phytochemistry*, **28**, 2981.
16. Cornforth, R. H. and Pöpjak, G. Nadeau, R. G., and Hanzlik, R. P. (1969). *Meth. Enzymol.*, **15**, 359.
17. Davisson, V. J., Woodside, A. B., and Poulter, C. D. (1985). *Meth. Enzymol.*, **110A**, 130.
18. Green, T. R. and Baisted, D. J. (1972). *Biochem. J.*, **130**, 983.
19. Oster, M. O. and West, C. A. (1969). *Arch. Biochem. Biophys.*, **127**, 112.
20. Pópjak, G. (1969). *Meth. Enzymol.*, **15**, 442.
21. Nadeau, R. G. and Hanzlik, R. P. (1969). *Meth. Enzymol.*, **15**, 346.
22. Elder, J. W., Benveniste, P., and Fonteneau, P. (1977). *Phytochemistry*, **16**, 490.
23. Inoue, H., Ôba, K., Ando, M., and Uritani, I. (1984). *Physiol. Plant Pathol.*, **25**, 1.
24. Fujita, M., Ôba, K., and Uritani, I. (1982). *Plant Physiol.*, **70**, 573.
25. Bell, A. A. (1967). *Phytopathology*, **57**, 759.
26. Masamune, T., Murai, A., Takasugi, M., Matsunaga, A., Nobukatsu, K., Sato, N., and Tomiyama, K. (1977). *Bull. Chem. Soc. Jpn*, **50**, 1201.
27. Ward, E. W. B., Unwin, C. H., Rock, G. L., and Stoessl, A. (1976). *Can. J. Bot.*, **54**, 25.
28. Henfling, J. W. D. M. and Kúc, J. (1979). *Phytopathology*, **69**, 609.
29. Farmer, E. F. and Helgeson, J. P. (1987). *Plant Physiol.*, **85**, 733.
30. Lyon, G. D. (1972). *Physiol. Plant Pathol.*, **2**, 411.
31. Bailey, J. A., Vincent, G. G., and Burden, R. S. (1976). *Physiol. Plant Pathol.*, **8**, 35.
32. Hammerschmidt, R. and Kúc, J. (1979). *Phytochemistry*, **18**, 874.
33. Heinstein, P. (1985). *J. Nat. Prod.*, **48**, 907.
34. Heisler, E. G., Siciliano, J., and Bills, D. D. (1979). *J. Chromatog.*, **154**, 297.
35. Cartwright, D. W., Langcake, P., Pyrce, R. J., Leworthy, D. P., and Ride, J. P. (1981). *Phytochemistry*, **20**, 535.
36. Heisler, E. G., Siciliano, J., Kalan, E. B., and Osman, S. F. (1981). *J. Chromatog.*, **210**, 365.
37. Ôba, K., Tatematsu, H., Yamashita, K., and Uritani, I. (1976). *Plant Physiol.*, **58**, 57.
38. Smith, D. A. (1982). In *Phytoalexins* (ed. J. A. Bailey and J. W. Mansfield), p. 218. Blackie, London.
39. Lyon, F. M. and Wood, R. K. S. (1975). *Physiol. Plant Pathol.*, **6**, 117.
40. Slusarenko, A. J., Longland, A. C., and Whitehead, I. M. (1989). *Bot. Helv.* **99**, 203.
41. Woodward, S. and Pearce, R. B. (1985). *Plant Pathol.*, **34**, 477.
42. Ingham, J. L. (1976). *Phytopath. Z.*, **87**, 353.
43. Homan, A. L. and Fuchs, A. (1970). *J. Chromatog.*, **51**, 327.
44. Essenberg, M., Stoessl, A., and Stothers, J. B. (1985). *J. Chem. Soc., Chem Commun.*, 556.
45. Goad, L. J. (1983). *Biochem. Soc. Trans.*, **11**, 548.

46. Sato, K., Ishiguri, Y., Doke, N., Tomiyama, K., Yagihashi, F., Murai, A., Katsui, N., and Masamune, T. (1978). *Phytochemistry*, **17**, 1901.
47. Suzuki, H. and Uritani, I. (1976). *Plant and Cell Physiol.*, **17**, 691.
48. Dudley, M. W., Dueber, M. T., and West, C. A. (1986). *Plant Physiol.*, **81**, 335.
49. Vögeli, U. and Chappell, J. (1988). *Plant Physiol.*, **88**, 1291.
50. Ono, T. and Imai, Y. (1985). *Meth. Enzymol.*, **110**, 375.
51. Duriatti, A., Bouvier-Nave, P., Benveniste, P., Schuber, F., Delprino, L., Balliano, G., and Cattell, L. (1985). *Biochem. Pharmacol.*, **34**, 2765.
52. Elder, J. W., Benveniste, P., and Fonteneau, P. (1977). *Phytochemistry*, **16**, 490.
53. Dueber, M. T., Adolf, W., and West, C. A. (1978). *Plant Physiol.*, **62**, 598.

8

Preparation and characterization of oligosaccharide elicitors of phytoalexin accumulation

MICHAEL G. HAHN, ALAN DARVILL, PETER ALBERSHEIM, CARL BERGMANN, JONG-JOO CHEONG, ALAN KOLLER, and VENG-MENG LÒ

1. Introduction

Plants utilize a variety of defence responses when confronted with invasive micro-organisms (1, 2). These defence reponses include such diverse mechanisms as the synthesis and accumulation of low molecular weight anti-microbial compounds called phytoalexins (3–6), the production of glycosyl hydrolases capable of attacking microbial surface polymers (7), the synthesis of proteins that inhibit microbial degradative enzymes (8, 9), and the modification of plant cell walls by deposition of callose (10, 11), hydroxyproline-rich glycoproteins (12, 13), and/or lignin (14).

Biochemical analysis of the induction of plant defence responses has been facilitated by the recognition that cell-free extracts of microbial and plant origin are capable of inducing defence responses when applied to plant tissues. The active components in the extracts are commonly referred to as 'elicitors'. The term 'elicitor' was originally used to refer to molecules and other stimuli that induce phytoalexin synthesis in plant cells (15), but the term is now used to describe molecules that elicit any of the observed plant defence responses.

Crude preparations of biotic elicitors have been prepared from a variety of sources (reviewed in references 16–18). Elicitors that have been purified to apparent homogeneity include oligosaccharide fragments of fungal (19–22) and plant (23–25) cell wall polysaccharides, and fungal cell surface lipids (26). A number of elicitor-active polypeptides of fungal and bacterial origin have also been purified and partially characterized. Several of these have been shown to be glycosylhydrolases that release elicitor-active oligosaccharides from plant cell walls (27–29). Other elicitor-active polypeptides have not (yet) been shown to have enzymatic activities (30–32). Oligosaccharide elicitors that derive from plant and fungal cell wall polysaccharides are the best-characterized elicitors. Their purification and characterization are described in detail in this chapter.

2. Chemical and biological assays

2.1 Carbohydrate assays

Colorimetric assays which are widely used for quantitating the amounts of neutral sugars, uronic acids, and reducing sugars present in the carbohydrate elicitor preparations are described. The response (i.e. the amount of colour produced) of these colorimetric assays is dependent on the glycosyl-residue composition of the carbohydrate being assayed. For example, the molar response of galactose is about 50% less than that of glucose in the anthrone assay for neutral sugars. Thus, if quantitative data are required, appropriate standards must be used to account for the composition (see Section 2.2) of the elicitor preparation being analysed. All test tubes used for the colorimetric assays should be baked in a muffle furnace for 4 h at 400°C to remove residual carbohydrate and reduce background interference.

Protocol 1. Colorimetric assays

A. *Anthrone assay for neutral sugars* (33)

1. Prepare a 0.2% (w/v) solution of anthrone in concentrated H_2SO_4 (**caution!**) and transfer into a glass repipettor. The anthrone reagent should have a bright yellow colour, and can usually be stored for two weeks at room temperature. The reagent should be discarded when its colour begins to turn darker (yellow–brown).

2. To a 12 × 75 mm test tube containing 2–20 µg of neutral sugar (glucose) in 200 µl of water, **carefully** add 400 µl of the anthrone reagent.

3. Mix thoroughly using a vortex mixer (**CAUTION!** Reaction is exothermic) and heat samples for 5 min in a boiling water-bath.

4. Allow samples to cool to room temperature.

5. Read the absorbance of the samples at 620 nm within 1 hour.

B. *Meta-hydroxybiphenyl assay for uronic acids* (34)

1. Prepare a 12.5 mM solution of sodium tetraborate (4.77 g of $Na_2B_4O_7 \cdot 10\,H_2O$ per litre) in concentrated H_2SO_4 (this will usually require stirring overnight). Once the borate has dissolved, transfer the solution into a glass repipettor, and refrigerate at 4°C.

2. Prepare a 0.15% (w/v) solution of *m*-hydroxybiphenyl in 0.5% (w/v) NaOH. The solution should be stored at 4°C in a bottle wrapped in aluminium foil.

3. To a 12 × 75 mm test tube containing 1–10 µg of uronic acid in 100 µl of water, **carefully** add 0.6 ml of the ice-cold borate solution.

4. Mix thoroughly using a vortex mixer (**CAUTION!** Reaction is exothermic) and heat samples for 5 min in a boiling water-bath.

5. Immediately cool in an ice-water bath.

6. Add 10 µl of the *m*-hydroxybiphenyl solution and mix thoroughly.

7. Allow colour to develop for 5 min at room temperature and read absorbance at 520 nm within 1 hour.

C. *Para-hydroxybenzoic acid hydrazide (PAHBAH) assay for reducing sugars (35)*

1. Prepare PAHBAH reagent by mixing four volumes of 0.5 M NaOH with one volume of 5% (w/v) PAHBAH (Sigma) in 0.5 M HCl.

2. Add 1.5 ml of fresh PAHBAH reagent to a 13 × 100 mm test tube containing 10–100 nmol of reducing sugar in 500 µl of water.

3. Mix thoroughly using a vortex mixer, and heat for 10 min in a boiling water-bath.

4. Allow samples to cool to room temperature.

5. Read the absorbance of the samples at 410 nm within 30 min.

2.2 Glycosyl-residue and glycosyl-linkage composition analyses

Determining the structure of oligosaccharides requires knowledge of the sugars present in the oligosaccharides (glycosyl-residue composition) and knowledge of the way in which the various types of sugars are glycosidically linked within the oligosaccharide (glycosyl-linkage composition). Methods for determining the glycosyl-residue and glycosyl-linkage compositions of oligosaccharides are given in the following sections. Methods for determining the primary structure of oligosaccharides (glycosyl sequence) are beyond the scope of this chapter. These methods include analyses using HPLC, GLC, mass spectroscopy, NMR spectroscopy, and/or a combination of these techniques. A number of reviews summarize the methods used to determine the primary sequences of complex carbohydrates (36–39).

2.2.1 Glycosyl-residue composition analysis

Two methods for determining the glycosyl-residue composition of oligosaccharides will be described. The first method (*Protocol 2* A) involves the formation of alditol acetate derivatives of the glycosyl residues of the oligosaccharide which can then be analysed by GLC (40). Glycosyluronic acid residues in oligosaccharides will not be detected by this method unless the carboxyl groups of the glycosyluronic acids have been reduced prior to derivatization (41), a relatively difficult, normally non-quantitative procedure. The alditol acetate method gives a single derivative for each glycosyl residue in the oligo- or polysaccharide and relatively simple chromatograms are obtained that are easily interpreted and quantitated. The second method (*Protocol 2* B) involves the formation of trimethylsilyl ethers of the methyl glycosides and methyl esters, where appropriate, from each glycosyl residue. These derivatives can then be analysed by GLC (42). This method gives several (usually two for neutral glycosyl residues, four for glycosyluronic acid residues) derivatives for each glycosyl component. However, glycosyluronic acid residues and neutral glycosyl residues can be quantitated simultaneously using the second method.

Protocol 2. Glycosyl-residue composition analyses

A. *Method 1. Formation of alditol acetates* (40)[a]

1. Transfer the sample to be analysed to a 13 × 100 mm tube fitted with a Teflon-lined screwcap. 100 μg is a convenient amount of sample, although as little as 5 μg can be analysed by micromethods (43). In order to calculate detector response factors that take the recovery of each sugar into account, it is important to include, in each analysis, a tube containing known amounts (for example 100 μg) of each of the neutral sugars likely to be present in the sample being analysed. For example, in the analysis of a plant cell wall component, this tube would contain rhamnose, fucose, arabinose, xylose, mannose, galactose, and glucose; listed in the order that their alditol acetate derivatives are eluted from a Supelco SP 2330 capillary GLC column (see *Figure 1*). Stock solutions of mixtures of known sugars can be prepared in advance and stored at −20°C.

2. Add 250 μl of 2 M trifluoroacetic acid (TFA) containing 25 μg of myo-inositol as an internal standard to each sample to be analysed (including a sample with a mixture of known sugars). Seal the tubes and place them in a heating block at 121°C for 1 h to hydrolyse the glycosidic linkages of the samples to form mixtures of free glycoses.

3. Cool the samples and evaporate the 2 M TFA at 40°C under a stream of filtered air. This is best done using a manifold to direct the filtered air to the bottom of each tube (for example Reacti-Vap Evaporator, Pierce; N-EVAP Model 111, Organomation Assoc. Inc.). All subsequent evaporations are performed in the same manner.

4. When the tubes appear dry, add 300 μl of isopropyl alcohol to each sample and evaporate the solvent to dryness. The evaporation of isopropyl alcohol results in more complete removal of the TFA.

5. Reduce the glycoses to the corresponding alditols by the addition to each sample of 250 μl of a solution of 1 M ammonia containing 10 mg/ml of

sodium borohydride. Allow the reaction to stand for 1 h at room temperature, after which any remaining sodium borohydride is converted into borate by the addition of 50 μl aliquots of glacial acetic acid until effervescence ceases.

6. Evaporate the borate as its trimethylester by the addition of 250 μl of acetic acid–methanol, 1:9 (v/v), to each tube and evaporate the solvent at 40°C until dry. Perform three additional evaporations with 250 μl of acetic acid–methanol (1:9), followed by four evaporations with 250 μl of methanol.

7. Per-O-acetylate the resulting alditols by adding 50 μl of acetic anhydride and 50 μl of pyridine to each sample, sealing the tubes, and heating at 121°C for 20 min.

8. Add approx. 200 μl of toluene to the resulting solution and evaporate the solvent at room temperature until dry. The toluene forms an azeotrope with acetic anhydride facilitating evaporation. Add a further 200 μl of toluene and evaporate the solvent at room temperature until dry.

9. Add 500 μl of methylene chloride and 500 μl of water to each sample and mix the solutions using a vortex mixer. Separate the organic and water phases by centrifugation at low speed. Transfer the methylene chloride layer (lower layer) with a Pasteur pipette to a clean tube and evaporate the methylene chloride with a gentle stream of air at room temperature.

10. The resulting per-O-acetylated alditols are dissolved in acetone (approx. 1 μl/μg of carbohydrate in the original sample) and then analysed by GLC. A variety of gas chromatography columns are suitable for these analyses (40, 44, 45). The method described here uses a 15-metre SP 2330 (Supelco, Inc.) fused-silica capillary column (0.25 mm i.d.). The per-O-acetylated alditols are injected using the split mode (ratio of amount injected to amount loaded on to column of 50:1) to prevent overloading the column. The oven temperature is 235°C (isothermal). The per-O-acetylated alditols are detected with a flame-ionization detector as they elute from the column. A typical chromatogram of a sample containing derivatives of neutral sugar standards is shown in *Figure 1*. The response factor of each alditol acetate derivative is determined empirically by injecting a known volume of a derivatized standard mixture containing known amounts of each sugar and myo-inositol, the internal standard, and determining the peak areas for each sugar derivative.

---

[a] The accurate quantitation of the glycosyl-residue composition of an oligo- or polysaccharide using the alditol acetate method can be hindered by several factors. The glycosidic linkages of some sugars, for example uronic acids and N-acetyl glucosamine, are difficult to hydrolyse, which leads to an underestimation of glycosyl residues glycosidically linked to these sugars. In addition, the yields of the alditol acetate derivatives of some plant cell wall glycosyl residues, including apiose and aceric acid are low (46, 47). Furthermore, 3-deoxy-*manno*-octulosonic acid and 3-deoxy-*lyxo*-heptulosaric acid cannot be detected by the conventional alditol acetate procedure (48, 49).

Protocol 2. *Continued*

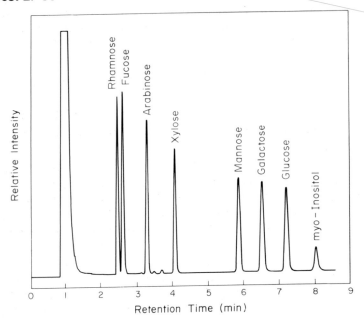

Figure 1. GLC of per-*O*-acetylated alditol derivatives of seven neutral sugars and an internal standard (myo-inositol). Derivatives were eluted from a fused silica SP 2330 (Supelco) 15 metre by 0.25 mm i.d. capillary column using helium as the carrier gas. The oven temperature programme is described in the text.

B. *Method 2. Formation of trimethylsilyl ethers of methyl glycosides* (42)

1. Separately place the samples to be analysed and a mixture of known amounts of the monosaccharides (including galactosyl- and glucosyluronic acids) in 13 × 100 mm test tubes that can be sealed with Teflon-lined screw caps.

2. Add 250 µl of 1 M HCl in methanol, seal the tubes, and heat at 80°C for 15 h. [The 1 M HCl in methanol can be prepared by slowly adding acetyl chloride (Alltech) to methanol (**CAUTION**: as this is a *very* exothermic reaction). Alternatively, HCl gas can be bubbled into methanol. Determine the HCl concentration by titration, and then dilute the HCl to 1 M by the addition of methanol.] Methanolysis converts the oligo- or polysaccharide into a mixture of methyl glycosides of the neutral glycosyl residues and methyl ester, methyl glycosides of the glycosyluronic acid residues.

3. Remove the methanolic HCl by adding 100 µl of *t*-butyl alcohol (50) and then evaporating the solvent to dryness at room temperature with a stream of filtered air. Perform two additional evaporations with 100 µl of *t*-butyl alcohol.

4. Silylate the hydroxyl groups of the methyl glycosides and methyl ester methyl glycosides using a mixture of 50 μl of pyridine, 100 μl of hexamethyldisilazane, and 50 μl of trimethylchlorosilane (51), which can be purchased conveniently in these proportions as Tri-Sil (Pierce Chemical Company). Heat the samples at 80°C for 20 min.

5. Cool the samples and carefully evaporate the silylating reagent at room temperature.

6. Dissolve the derivatives in hexane (1 ml) and allow insoluble salts to settle.

7. Transfer the supernatant to a clean test tube and carefully evaporate the solvent just to dryness.

8. Dissolve the residue in 100 μl of hexane, and analyse 1 μl of this solution by GLC. The GLC analyses are best performed on capillary columns; a fused-silica DB1 (J & W Scientific) 30 metre capillary column (0.25 mm internal diameter) is used in our laboratory. Injections are made using the split mode (split ratio 50:1). The following oven temperature programme is used: an initial temperature of 160°C that is held for 3 min, followed by a linear temperature increase to 200°C at a rate of 2°C/min, and then an increase to 250°C at 10°C/min. Condition the column for the next injection by increasing the oven temperature at a rate of 30°C/min to 275°C and maintain this temperature for 10 min. A typical chromatogram of a sample containing standard sugar derivatives is shown in *Figure 2*. As each sugar yields several derivatives, the peak areas of the major derivatives of each sugar must be added together before response factors can be calculated and glycosyl compositions determined.

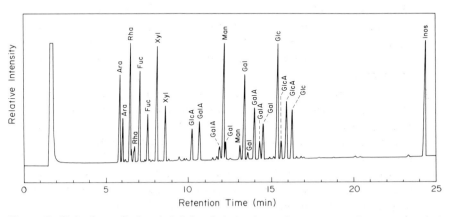

Figure 2. GLC of per-*O*-trimethylsilylated derivatives of seven neutral sugars (methyl glycosides) and two acidic sugars (methyl ester methyl glycosides) and an internal standard (myo-inositol). Derivatives were eluted from a fused-silica DB1 (J & W Scientific) 30 metre by 0.25 mm i.d. capillary column using helium as the carrier gas. The oven temperature programme is described in the text.

2.2.2 Glycosyl-linkage composition analysis

The glycosyl-linkage composition of an oligo- or polysaccharide is determined by methylation analysis. It is worth noting several points that will improve the success of this procedure for some types of oligo- and polysaccharides. Reduction of the hemi-acetal at the reducing end of the oligo- or polysaccharide to the corresponding alcohol (see steps 5 and 6, *Protocol 2* A) will prevent base-catalysed degradation during the methylation reaction, particularly with 3-linked oligo- and polyaccharides.

The presence of glycosyluronic acid residues in an oligo- or polysaccharide complicates the successful methylation analysis of such carbohydrates. Oligo- or polysaccharides containing 4-linked methyl-esterified glycosyluronic acid residues will undergo base-catalysed elimination reactions and degrade under the strongly basic conditions of the methylation procedure. De-esterification of methyl esters of glycosyluronic acid residues can be accomplished by treatment with dilute NaOH at 2°C (52), conditions that do not cause detectable degradation of the carbohydrate. The partially *O*-acetylated, partially *O*-methylated derivatives of the glycosyluronic acid residues are not stable and will not be detected by the standard methylation protocol. Thus, the glycosyluronic acid residues must be reduced to the corresponding 6,6-dideuterioglycosyl residues at some point during the methylation protocol. In one method, the glycosyluronic acid residues are reduced prior to methylation. This can be accomplished by reduction, using sodium borodeuteride, of the derivatives formed by reaction of the glycosyluronic acid residues with a water-soluble carbodiimide (41). Use of this method prevents β-elimination reactions during methylation. However, it is difficult to achieve complete reduction of the glycosyluronic acid residues by this method, and substantial losses of material (~50%) can occur due to the necessity of repeating the carbodiimide reduction several times. An alternative method, described in the following sections, is to reduce the glycosyluronic acid residues following a single methylation reaction. This method is more convenient and can be done with a smaller amount of material than the carbodiimide method. However, it is sometimes difficult to obtain complete methylation of an oligo- or polysaccharide in a single methylation reaction, and degradative base-catalysed elimination reactions can occur with some polysaccharides (52).

Protocol 3. Glycosyl-linkage composition analysis

A. *Methylation of oligo- or polysaccharides*

Oligo- and polysaccharides are most successfully methylated using modifications of the Hakomori procedure (53, 54).

1. Transfer the carbohydrate (0.1 mg) to be methylated and a small magnetic stir-bar to a test tube (13 × 100 mm) fitted with a Teflon-lined screw cap. Dry the carbohydrate in a vacuum oven at 40°C for at least 12 h.

2. Add 500 μl of dry dimethyl sulphoxide to the test tube, flush with nitrogen gas for approx. 1 min, cap the tube, and stir the mixture magnetically until the oligo- or polysaccharide is dissolved. A period of 2–4 h is usually sufficient, although some carbohydrates must be stirred overnight. A bath sonicator and/or heating the sample to 50°C can also be used to assist in the solubilization of the carbohydrate.

3. Add 500 μl of 4 M sodium dimethylsulphinyl anion solution (prepared as described in *Protocol 4*) to the sample using a dry glass syringe. Flush the tube with nitrogen, cap, and stir the mixture for at least 2 h at room temperature. Rinse the syringe with ethanol, then with water, and dry.

4. Dimethylsulphinyl anion must be maintained in excess throughout the reaction. To determine whether dimethylsulphinyl anion is still present at the end of the reaction period, test an aliquot (for example 5 μl) of the reaction mixture with a crystal of triphenylmethane. The triphenylmethane turns red in the presence of sodium dimethylsulphinyl anion. If the triphenylmethane solution does not turn red, add another 500 μl aliquot of the anion solution to the methylation reaction, and stir the mixture for an additional 2 h.

5. Place the test tube in an ice water-bath until the sample solution solidifies. Using a dry glass syringe, carefully (drop-wise) add a volume of methyl iodide equal to that of the added dimethylsulphinyl anion (0.5 or 1 ml) to the reaction mixture. (**CAUTION**: Methyl iodide is a strong mutagen. This chemical should only be used in a fume hood. Wear gloves while handling solutions containing methyl iodide, and discard gloves and wash hands immediately if any spillage occurs.) Cap the tubes immediately, place them in a water-bath (20–25°C) and stir for at least 2 h (usually overnight).

B. *Purification of per-O-methylated oligo- or polysaccharides*

The per-O-methylated oligo- or polysaccharide is recovered and purified by reverse-phase chromatography (43, 55). If several samples are being analysed at the same time, the procedures described below can be carried out using a vacuum manifold (for example Visiprep solid phase extraction manifold, Supelco).

1. Pre-flush a SEP-PAK C-18 (Water Associates, Inc.) cartridge with 20 ml of 100% ethanol to remove contaminants from the cartridge and to increase the recovery of the per-O-methylated carbohydrate.

2. Pre-condition the cartridge by passing 2 ml of 100% acetonitrile and then 10 ml of water through the cartridge.

3. Dilute the methylation-reaction mixture containing the per-O-methylated oligo- or polysaccharide with at least an equal volume of water to produce a biphasic solution. Carefully vortex the solution. Gently bubble

Protocol 3. *Continued*

nitrogen gas through the solution to evaporate the methyl iodide (**CAU-TION**: Perform in a fume hood, see step A5 above and step B4 below). The solution will become clear when most of the methyl iodide is gone.

4. Load this solution carefully (one to two drops per second, using the syringe plunger to control the flow) on to the SEP-PAK cartridge until the liquid level is just above the resin bed. (**CAUTION**: Disconnect the syringe from the cartridge before withdrawing the plunger from the syringe. Wear gloves during this procedure to prevent absorption of dimethyl sulphoxide and methyl iodide through the skin. Discard gloves and wash hands immediately if any spillage occurs. Quench all waste solutions with concentrated NH_4OH to destroy residual methyl iodide.)

5. Elute the more polar contaminants in the methylation reaction mixture, including the dimethyl sulphoxide and the sodium iodide, from the cartridge containing the methylated material with 8 ml of water, pushed completely through the cartridge with air.

6. Elute the less polar contaminants in the methylation reaction mixture from the cartridge with an 8 ml flush of aqueous 20% (v/v) acetonitrile, pushed completely through the cartridge with air.

7. Elute the per-*O*-methylated oligo- or polysaccharide from the SEP-PAK C-18 cartridge with a 2 ml flush of 100% acetonitrile, followed by a 4 ml flush of 100% ethanol. *Note*: Some methylated polysaccharides bind very strongly to the SEP-Pak C-18 cartridges and chloroform–methanol (1:1, v/v) is required to elute them (55) (don't throw away the cartridge until the GLC analyses have been completed!). Combine the 100% acetonitrile and 100% ethanol flushes, containing the per-*O*-methylated carbohydrate, in a test tube and evaporate the solvent to dryness under a stream of filtered air at room temperature. The per-*O*-methylated oligo- or polysaccharide can then be hydrolysed, reduced, and *O*-acetylated to form partially *O*-acetylated, partially *O*-methylated alditols as described below.

C. *Formation of partially O-acetylated, partially O-methylated alditols*

A partially *O*-methylated oligo- or polysaccharide containing only neutral glycosyl residues is converted into its corresponding partially *O*-acetylated, partially *O*-methylated alditols using the procedure described below. If the oligo- or polysaccharide contains any glycosyluronic acid residues, these must first be reduced to their corresponding dideuterated glycosyl residues using the procedure described in *Protocol 5*.

1. Add 250 µl of 2 M TFA, containing 25 µg of myo-inositol as an internal standard, to the per-*O*-methylated carbohydrate in a test tube (13 × 100 mm) fitted with a Teflon-lined screw cap.

2. Seal the test tube and heat for 1 h at 121 °C to form a mixture of *O*-methylated aldoses.

3. Cool the samples and evaporate the TFA with filtered air at room temperature.

4. Add 250 μl of isopropanol to the tube and evaporate at room temperature to remove residual TFA. Repeat the evaporation with an additional 250 μl of isopropanol.

5. Reduce the resulting partially *O*-methylated aldoses to the corresponding partially *O*-methylated alditols by dissolving them in 220 μl of 95% ethanol and adding 200 μl of aqueous NaBD$_4$ (Aldrich Chemical Co.) solution (10 mg/ml in 1 M NH$_4$OH). Lightly cap (do not seal) the test tube and react for 1 h at room temperature.

6. Convert the excess borodeuteride into borate by adding 50 μl aliquots of glacial acetic acid to the tube until effervescence ceases.

7. To evaporate the borate as its trimethyl ester, add 200 μl of acetic acid–methanol (1:9, v/v) to the tube, mix the contents, and evaporate the solvents with filtered air at room temperature. Perform three additional evaporations with 200 μl of 1:9 (v/v) acetic acid–methanol and follow with two evaporations with 200 μl of methanol.

8. To form the partially methylated, partially acetylated alditols, add 50 μl of acetic anhydride to the test tube containing the partially *O*-methylated alditols, seal the tube, and heat for 3 h at 121 °C.

9. Allow the tube to cool to room temperature and add 500 μl water.

10. Convert any remaining acetic anhydride to sodium acetate by adding solid Na$_2$CO$_3$, a small amount (25 mg) at a time, until effervescence ceases. If all of the Na$_2$CO$_3$ does not dissolve, more water can be added.

11. Add 500 μl of dichloromethane to the tube, and mix the contents. Separate the organic and water phases by centrifugation at low speed.

12. Transfer the methylene chloride phase (lower layer) to a clean tube, and evaporate the solvent just to dryness. Great care (i.e. use as little evaporation as possible) must be taken in evaporating the methylene chloride to prevent the loss of some of the more volatile partially *O*-acetylated, partially *O*-methylated alditols.

13. Determine the positions of *O*-acetyl and *O*-methyl groups on the partially *O*-acetylated, partially *O*-methylated alditols by combined gas chromatography–electron impact mass spectrometry (GLC–MS). The GLC–MS analysis is performed with a 30 m fused-silica SP 2330 (Supelco, Inc.) capillary column (0.25 mm i.d.) in the splitless mode using the following oven temperature programme. Two minutes at an initial temperature of 80 °C, increased to 170 °C at 30 °C/min, then to 240 °C at 4 °C/min, and hold for 5 min at 240 °C. The identity of each

Protocol 3. *Continued*

partially *O*-acetylated, partially *O*-methylated alditol is determined by its retention time in GLC and its electron-impact fragmentation patterns in the mass spectrometer. The fragmentation patterns of partially *O*-acetylated, partially *O*-methylated alditols are well known, and guidelines for interpreting these spectra are given elsewhere (44, 52).

The positions of *O*-acetyl groups in each partially *O*-acetylated, partially *O*-methylated derivative denote the hydroxyl groups involved in glycosidic linkages or involved in the formation of the sugar ring in the original oligo- or polysaccharide.

Protocol 4. Preparation of potassium dimethylsulphinyl anion

1. The following procedure for the preparation of potassium dimethylsulphinyl anion should be carried out in a fume hood behind an explosion shield. All glassware needed for the preparation of the reagent should be thoroughly cleaned and oven-dried. **CAUTION**: potassium hydride **reacts violently** with water and extreme care should be taken when handling this chemical. Dimethyl sulphoxide is readily absorbed through the skin, so gloves should be worn throughout the procedure and disposed of appropriately. The gloves should be changed immediately if any dimethyl sulphoxide is spilled on them, and hands should be washed before putting on a fresh set of gloves.

2. Place 4.8 g of thoroughly mixed 35% potassium hydride in mineral oil (Aldrich Chemical Co.) in a 50 ml, three-necked flask with a magnetic stirbar. Fit the flask with a pressure equalization funnel having a Teflon stopcock and an entrance and exit port for nitrogen or argon gas. Flush nitrogen or argon gas (argon is preferable because it is heavier than nitrogen) through the flask during the entire procedure.

3. Carefully add 25 ml of hexane and stir the suspension for 1–2 min. Stop stirring and allow the potassium hydride to settle. Using a pipette, carefully transfer the hexanes containing the mineral oil into a beaker containing absolute ethanol (which quenches the small amounts of potassium hydride that are left in suspension, thus reducing the possibility of explosion). Wash the potassium hydride with 25 ml of hexane three more times in the same manner to completely remove the mineral oil.

4. Magnetically stir the washed potassium hydride (free from mineral oil) while flushing the flask with nitrogen or argon gas until a dry powder is obtained. **CAUTION**: dry potassium hydride is extremely reactive and should be handled with great care (see step 1 above).

5. Using the pressure equalization funnel, carefully add 10 ml of anhydrous

dimethyl sulphoxide (DMSO, 99%, stored over molecular sieves) drop-wise with constant stirring. **CAUTION**: if the DMSO is added too rapidly, the reaction may bubble vigorously and the solution may flow out of the flask.

6. After the addition of DMSO is complete (10–15 min), continue stirring the solution until no further bubbling from the reaction is observed (60–90 min). The potassium dimethylsulphinyl anion solution should be clear and have a grey–green colour; if the solution is black, discard the anion.

7. Transfer 500 µl aliquots of the dimethylsulphinyl anion to 1 ml serum vials and seal them with Teflon-lined caps. The anion can be stored at −20°C for several months.

8. Determine the concentration of the potassium dimethylsulphinyl anion by titration as follows: using a dry 0.5 ml glass syringe with an 18-gauge needle, transfer 100 µl of the anion to a flask containing approximately 5 ml of water and a drop of phenolphthalein solution (0.1% in ethanol). Titrate this solution with 0.1 M HCl, and calculate the concentration of anion from the amount of HCl consumed. The concentration of the pot-assium dimethylsulphinyl anion should be about 4 M.

Protocol 5. Reduction of esterified glycosyl uronic acid residues in per-*O*-methylated oligo- or polysaccharides

1. The following procedure will successfully reduce *esterified* glycosyl uronic acid residues to their corresponding neutral glycosyl residues. This pro-cedure will not reduce non-esterified glycosyl uronic acid residues. Thus, an oligo- or polysaccharide containing glycosyluronic acid residues must first be methylated as described in *Protocol 3*, parts A and B.

2. Using a dry glass syringe, add 250 µl of lithium triethylborodeuteride dissolved in tetrahydrofuran (Superdeuteride, Aldrich Chemical Co.). **CAUTION**: lithium triethylborodeuteride reacts violently with water. Allow the reaction to stand for 2 h at room temperature. Deuteride (as opposed to hydride) reduction yields 6,6-dideuteriohexosyl residues. Thus, the glycosyl residues originating from glycosyluronic acid residues can be distinguished from neutral glycosyl residues present in the native oligo- or polysaccharide by their fragmentation patterns in electron-impact mass spectrometry (44, 52).

3. Destroy excess borodeuteride by adding 50 µl aliquots of glacial acetic acid until effervescence ceases.

4. Evaporate the solvents with a stream of filtered air at room temperature until dry.

Protocol 5. *Continued*

5. Prepare a 2 ml column of Dowex 50W-X12 (H^+ form) and pre-wash the column with at least 5 ml of aqueous 50% (v/v) ethanol. Dissolve the reduced, partially methylated oligo- or polysaccharide in 1 ml of 50% ethanol and apply to the column. Elute the column with an additional 2 ml of 50% ethanol. Collect everything eluting from the column and evaporate the solvents with a stream of filtered air, or nitrogen, until dry.

6. The oligo- or polysaccharide containing the reduced glycosyluronic acid residues can be further derivatized in one of two ways. The oligo- or polysaccharide can be immediately per-*O*-acetylated as described in *Protocol 3*, part C, in which case the glycosyluronic acid derivatives will all be *O*-acetylated at C-6, in addition to being labelled with two deuterium atoms at the same carbon. Alternatively, the entire methylation procedure (*Protocol 3*, A and B) can be repeated prior to performing the acetylation (*Protocol 3* C) to yield glycosyluronic acid derivatives that are *O*-methylated and dideuterated at C-6. The latter alternative can also yield a more fully methylated oligo- or polysaccharide.

2.3 Fast-atom-bombardment mass spectrometry for determining the mol. wt of oligosaccharide elicitors

2.3.1 Sample preparation

Oligosaccharides to be analysed by fast-atom-bombardment mass spectrometry (FAB-ms) must be free of buffer salts, and oligogalacturonides should be converted to their ammonium salt forms for optimum FAB-ms analysis. Oligoglucosides (see Section 4) and their tyramine conjugates (see Section 6.1), prepared as described later in this chapter, can be analysed successfully without further sample preparation. Oligogalacturonides (see Section 5) and their tyrosine conjugates (see Section 6.2) must first be desalted and converted into their ammonium salt forms as described in *Protocol 6*.

Protocol 6. Sample preparation for fast-atom-bombardment mass spectrometry of oligogalacturonides

1. Desalt the oligogalacturonides by chromatography on a Sephadex G-10 column (3 × 18 cm) equilibrated in deionized water, or by dialysis using 1000 dalton cut-off membranes (Spectrapor-7, Spectrum Medical).

2. Prepare a Dowex 50–X-200 (NH_4^+ form) column by washing a column (column volume approx. 1 ml) of Dowex 50–X-200 (H^+ form) with 5 ml of 1 M NH_4OH solution followed by 10 column volumes of deionized water. The size of the column (column volume) is dependent on the amount of

oligouronide being applied. The capacity of the column should exceed the amount applied (in milliequivalents) by at least ten-fold.

3. Load the desalted oligogalacturonides on to the Dowex column and elute with approx. 1 ml of deionized water. Lyophilize the eluate if necessary to concentrate.

2.3.2 Mass spectrometry

We record fast-atom-bombardment mass spectra of oligosaccharides using a VG ZAB-SE mass spectrometer operating at low resolution (1:1000) in the negative ion mode with an accelerating voltage of 8 kV. The mass spectrometer is calibrated with CsI. Aliquots (approx. 1 µl; 2–10 µg/µl) of the underivatized oligosaccharides are applied to the probe tip of the mass spectrometer together with 1–2 µl of 1-amino-2,3-dihydroxypropane (for oligoglucosides) or 1–2 µl of thioglycerol acidified with HCl (for oligogalacturonides). Isotopomeric ions of high-mass ion clusters are not resolved under the low resolution conditions used. Instead, signals corresponding to the average mass of the ion clusters are detected. The m/z values can be converted into the nominal masses of the isotopomers containing only ^{12}C, ^{1}H, and ^{16}O isotopes (reported here) using the CARBOMASS mass spectrometry software developed by and available from W. S. York (Univ. of Georgia) (56).

Fast-atom-bombardment mass spectrometry provides information about the molecular weight of oligosaccharide elicitors. Valuable structural information about oligosaccharides can also be obtained by examining the fragmentation pattern of derivatized oligosaccharides (for example fully alkylated oligosaccharides) in the FAB-ms (57) and combining this information with glycosyl-residue and glycosyl-linkage composition data (see Sections 2.2.1 and 2.2.2 above). Underivatized neutral oligosaccharides consisting of up to 20 sugars have been successfully analysed by FAB-ms in our laboratory. The molecular weights of native oligogalacturonides of DP ≤ 15 have also been obtained by the procedures described above.

Success in obtaining molecular weight data from FAB-ms is usually dependent on the purity of the sample, particularly for oligogalacturonides. Minor constituents of a mixture that are more readily volatilized in the spectrometer can dominate the observed spectrum. Contaminants particularly salts, can also suppress ionization, thus impairing the ability to obtain a spectrum. Mixtures of cations (for example Na^+ and K^+) lead to large numbers of different salt forms of the oligogalacturonides, which greatly reduce sensitivity and complicate interpretation of the spectra.

2.4 Soybean cotyledon bioassay for elicitors of phytoalexin accumulation

The type of bioassay used to monitor the biological activity of an elicitor is the critical factor in determining how rapidly the elicitor can be purified. The bioassay should be as rapid, quantitative, and as reproducible as possible. The soybean cotyledon bioassay (58) described below is one example of a bioassay that meets these requirements, and has been used in our laboratories for over a decade to monitor elicitation of phytoalexin accumulation by cell wall-derived elicitors.

Protocol 7. Bioassay for elicitors of phytoalexin accumulation

1. Detach cotyledons from young soybean plants whose primary leaves have just expanded (usually eight or nine days after seed planting). As with any biological assay, the plant used as source material must be healthy and grown under reproducible conditions (for example light, temperature, moisture, etc.) in order to obtain reproducible bioassay results. The cotyledons should be turgid, uncurled, and free of blemishes or other obvious signs of infection or stress.

2. Surface-sterilize the cotyledons for 5 min in 0.5% hypochlorite (for example 10% Chlorox), and rinse extensively with deionized water.

3. Perform steps 3 and 4 in a sterile hood to minimize microbial contamination. Cut a section of tissue, 7×5 mm and approx. 1 mm thick, from the convex abaxial (bottom) surface of each cotyledon using an ethanol-sterilized, single-edged razor blade (Weck & Co. Inc.). Discard the tissue excised. The wound in the cotyledon should not extend into the vascular system, should be free of discolorations, and should have a flat, smooth, wet surface. Arrange the cotyledons, cut surface up, on wet filter papers (7.5 cm, grade No. 615, VWR Scientific) in sterile 10 cm Petri dishes (10 cotyledons/dish).

4. The samples to be bioassayed are dissolved either in sterile deionized water, for oligogalacturonides and oligogalacturonide/oligoglucoside mixtures (59), or in a sterile aqueous solution of 4 mM sodium acetate and 3 mM sodium bicarbonate, for oligo-β-glucosides (19, 59). Apply 90 μl of sample solution to each cotyledon and cover the dish. Incubate the dishes of cotyledons in a dark culture-chamber maintained at approx. 70% relative humidity for 20 h at 26°C. Generally, two dishes of 10 cotyledons each are used for each sample. Control samples, one with the buffer and a second with a known amount of an elicitor (see discussion below), are included in each essay.

5. After the incubation, transfer the 10 cotyledons, with their wound droplets, from the Petri dish to 20 ml of water. A portion of this wound-droplet solution is diluted 10-fold and its absorbance at 285 nm (A_{285}) is used as a measure of elicitor activity. The A_{285} is linearly correlated with the phytoalexin content of the wound-droplet solutions (23, 58).

The response of soybean cotyledons to increasing amounts of an elicitor is not linear. In particular, the response of the assay (i.e. the amount of phytoalexins accumulated) saturates at high elicitor concentrations (see *Figure 3A*). To accurately evaluate the activity of an elicitor, the elicitor must be assayed at several concentrations in order to determine the range over which a concentration dependence of the elicitor activity can be observed. The activities of different elicitors are most accurately compared by determining the elicitor concentrations that give half-maximum activity ($A/A_{std} = 0.5$; see below).

The soybean cotyledon bioassay exhibits some day-to-day variability that complicates the comparison of elicitor activities of oligosaccharides assayed on different days. Both the sensitivity of cotyledons to elicitor and the maximum amount of phytoalexin accumulated in response to elicitor vary. Two methods for minimizing the effects of these variations can be used (60), depending on the degree of quantitation desired. One correction that minimizes the effects of daily fluctuations in the maximal amount of phytoalexins induced in the cotyledons is routinely applied as follows. A sample containing a standard amount of an elicitor known to elicit a maximum response from the cotyledons is included in each assay. We generally use the mixture of

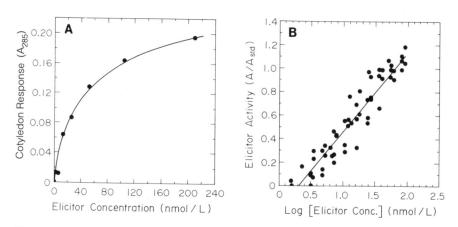

Figure 3. Response of soybean cotyledons to synthetic hepta-β-glucoside elicitor (73, 74). Increasing amounts of the elicitor were applied to sets of ten wounded soybean cotyledons as described in the text. **A.** The soybean response was measured as the absorbance at 286 nm of the ten-fold diluted wound-droplet solution. The response of the cotyledons to buffer blank has been subtracted from each datum. The data were obtained from a single bioassay and each datum is the average of two sets of 10 cotyledons. **B.** The response of soybean cotyledons (*Elicitor Activity*) to the elicitor measured on different days. The elicitor activity is expressed as A/A_{std} (see text). A_{std} is the response of the cotyledons to the 'Void glucan' elicitor (see *Figure 5*) applied at a concentration of 1 μg/ml. The response of the cotyledons to a buffer blank has been subtracted from each datum. Each datum is the average of two sets of 10 cotyledons. The data represent the results of seven independent bioassays. The line through the data was calculated by least squares analysis of the data ($r^2 = 0.86$).

glucans ('Void glucan elicitor') solubilized from the mycelial walls of *Phyto-phthora megasperma* f. sp. *glycinea* by partial acid hydrolysis and eluting in the void volume of a low-resolution P-2 column (see Section 4.2.1; *Figure 5*). The *Elicitor Activities* of the samples of interest are then expressed as a fraction of this maximum response of the cotyledons, that is A/A_{std} where A is the absorbance at 285 nm of the wound-droplet solution obtained from 10 cotyledons treated with the sample of interest, and A_{std} is the absorbance at 285 nm of the wound-droplet solution obtained from 10 cotyledons treated with the standard elicitor (for example 1 μg/ml of Void glucan elicitor). Standard curves of the *Elicitor Activity* for a given elicitor assayed on different days and plotted as A/A_{std} are fairly reproducible (±10%) (see *Figure 3B*).

The application of a second correction method minimizes the effect of day-to-day variations in the sensitivity of the cotyledons to elicitor. A complete concentration curve for the activity of a standard elicitor is included in each assay. The *Elicitor Activities* of molecules of interest are then reported with reference to the curve obtained for the standard elicitor. The *relative elicitor activity* of a molecule is defined as the concentration of that molecule required to achieve half-maximum induction of phytoalexin accumulation ($A/A_{std} = 0.5$) in the bioassay in relation to the concentration of the standard elicitor required to give $A/A_{std} = 0.5$. This procedure permits the most quantitative comparisons of the activities of different elicitors.

3. Isolation of cell walls

The procedures described in this chapter for the preparation of cell walls from cell homogenates of plants and fungi are based on the assumption that cell walls are insoluble in aqueous buffers and non-polar solvents. These procedures may remove some molecules present in the intact wall that are elicitor-active in a particular plant. However, as a source of material for the two classes of elicitors described in detail in this chapter, cell walls will be defined as the material purified by the procedures described below.

3.1 Plant cell walls

Plant cell walls can be obtained from isolated plant tissues and from suspension-cultured cells. Suspension-cultured plant cells are a relatively homogeneous source of primary plant cell walls. A large quantity of cells can be cultured and harvested under well-defined conditions (approx. 20 g fresh weight/litre of culture), leading to the production of several grams of cell walls. The composition of such walls is reproducible from batch to batch. The procedure described below was developed for the isolation of primary cell walls from suspension-cultured sycamore maple cells (52).

Protocol 8. Isolation of plant cell walls[a]

All solutions, equipment, and operations in the purification protocol prior to the organic solvent washes (step 6) are kept at 4°C.

1. Maintain suspension cultures of sycamore maple (*Acer pseudoplatanus*) cells on M-6 medium (61). Grow cultures intended for use as a cell wall source on a modified M-6 medium (62), in which mannan-free yeast extract is used to minimize contamination of the plant cell wall preparation with yeast polysaccharides (52). Inoculate sycamore maple cells into fresh medium every 7–10 days. Grow cells through at least three transfers on modified M-6 medium (10% inoculum into 100 ml of medium/ 500 ml flask) before starting large scale cultures for cell wall preparation. Initiate large scale sycamore maple cell suspension cultures by transferring 50 ml of a 7–10-day-old culture into 1 l of medium in a 2.8 l Fernbach flask. Incubate cell cultures on gyratory platform shakers (80–100 r.p.m. for small flasks; 60–80 r.p.m. for large flasks) in the dark at 23°C. Harvest sycamore maple cells for cell wall preparation during the late log phase (7–10 days after inoculation).

2. Pass the plant cell suspension culture through a coarse sintered-glass funnel. Retain the cells and use the culture fluid to isolate sycamore maple extracellular polysaccharides (63) if desired.

3. The volume of washes given in steps 3–8 are based on a cell wall preparation starting with 30 litres of suspension culture (~600 g wet weight of cells.) Add ~2.5 l of cold 100 mM potassium phosphate, pH 7, to the mass of cells in the sintered-glass funnel and stir to resuspend the cells. Apply gentle vacuum to the funnel in order to remove buffer and solutes. Repeat this procedure five times with the 100 mM phosphate buffer, and then four times with cold 500 mM potassium phosphate buffer, pH 7.

4. Resuspend the cells in a volume of cold 500 mM potassium phosphate buffer, pH 7, approximately equal to the volume of packed cells, and place in a pressure bomb (Parr Instrument Co.), bring to high pressure (at least 1000 p.s.i.) with nitrogen gas (**CAUTION**: use an explosion shield). After 15 min, release the cell suspension from the pressure bomb via a small orifice. The sudden change in pressure incurred during release causes virtually every cell to break open.

5. Centrifuge the suspension of broken cells at 2000 g for 10 min. The cell wall suspension should not be subjected to forces that exceed 2000 g, as this could cause other plant cell organelles to sediment with the walls.

6. Decant the supernatant solution and wash the pellet twice by suspending it in 2 l of the 500 mM phosphate buffer and centrifuging as described above. Repeat this washing procedure four more times with distilled water.

Protocol 8. *Continued*

7. Suspend the washed cell walls by vigorous stirring in 2 l of chloroform–methanol (1:1, v/v) and transfer the suspension to a coarse sintered-glass funnel. Remove the organic solvent by applying gentle vacuum to the funnel. Repeat the chloroform–methanol wash one or two more times.

8. Suspend the cell walls by vigorous stirring in 2 l of acetone. Remove the acetone by vacuum as before, and air-dry the cell walls. Crumble the mass of cell walls periodically to facilitate drying.

9. At this point the cell wall preparation is contaminated with starch granules. Remove these starch granules by suspending the dried cell walls (10 g/l) in potassium phosphate buffer (100 mM, pH 7.0) containing α-amylase from porcine pancreas (50 000 units/litre; Sigma type I-A, phenylmethylsulphonyl fluoride treated, 2 × recrystallized) and 0.01% (w/v) thimerosol as an antibiotic. This α-amylase is not demonstrably contaminated with β-glucanase(s) which can degrade 1,3/1,4-linked β-glucans present in some cell walls (i.e. walls of cereals). An alternative α-amylase which is often employed to remove starch from cell wall preparations is prepared from *Bacillus* sp. (Sigma Type II-A, 4 × recrystallized). This amylase should not be used to remove starch from cell walls of cereals since it contains a β-glucanase that degrades mixed-linkage glucans present in those cell walls. The amylase/wall suspension is stirred for 48 h at 25 °C. The amylase-treated cell walls are collected by centrifugation at 10 000 g and washed at least twice with distilled water (~1 l per 10 g of starting material). The cell walls are then washed with the organic solvents listed above (steps 17 and 18) and dried.

[a] Plant cell walls prepared by this procedure typically yield about 1 g of cell walls per litre of cell suspension culture.

3.2 Fungal cell walls

Fungal cell walls can be obtained from fungal cultures grown *in vitro*, preferably on a chemically-defined medium in liquid culture. The use of chemically-defined media avoids possible contamination of the cell wall preparations with elicitor-active constituents originating from complex media components, for example yeast extract (64). The following procedure was developed for the isolation of mycelial walls from *Phytophthora megasperma* f. sp. *glycinea* (19, 60).

Protocol 9. Isolation of fungal cell walls[a]

All solutions, equipment, and operations in the purification protocol up to the organic solvent washes (step 7) are kept at 4 °C.

1. Maintain cultures of *P. megasperma* on potato dextrose agar (Difco)

slants. Prepare inocula for large scale cultures by transferring the fungus to V8 agar plates (25% (v/v) clarified V8 juice, 1.5% (w/v) agar). Grow large cultures of the fungus in 2.8 l Fernbach flasks containing 500 ml of asparagine medium (65). Inoculate the liquid cultures with several 4 mm diameter mycelial agar plugs cut from the periphery of radial mycelial colonies on V8 agar plates. Grow the cultures at 24°C as standing cultures. On the second and fourth days after inoculation, vigorously swirl the cultures by hand for ~15 sec to disperse the mycelia on the surface of the culture. Harvest mycelia for cell wall preparation during the late log phase, when cultures are confluent (usually 2–3 weeks after inoculation).

2. Collect the mycelia on a nylon screen (37 μm pore size; #3-400-37, Tetko Inc.) supported in a Buchner funnel. Wash the mycelia extensively with distilled water and then with 0.5 M NaCl.

3. Homogenize the mycelial mat in 5 volumes (ml/g mycelia) of 0.5 M NaCl in a commercial Waring Blender at full speed. Collect the homogenized hyphae by filtration on a nylon screen as before.

4. Repeat homogenization and filtration in 0.5 M NaCl at least three more times.

5. Further homogenize the mycelial walls in 5 volumes (ml/g mycelia) of 20 mM Tris–HCl, pH 7.5, containing 50 mM EDTA, and collect the walls by filtration on the nylon screen as before. Repeat homogenization procedure until examination by light microscopy shows hyphal ghosts with no cellular contents.

6. Stir the walls for 3 h in the Tris–EDTA buffer, collect with suction on a sintered-glass funnel, and wash with 15 volumes (ml/g of mycelia) of Tris–EDTA buffer, followed by extensive washing with at least 10 volumes of cold distilled water.

7. Suspend the washed mycelial walls by vigorous stirring in 5 volumes (ml/g mycelia) of chloroform–methanol (1:1, v/v) and place in a coarse sintered-glass funnel. Remove the organic solvent by applying gentle vacuum to the funnel.

8. Resuspend the mycelial walls in 5 volumes (ml/g mycelia) of acetone. Remove the acetone by vacuum as before, and air-dry the walls under suction.

9. Resuspend the walls in water (100 ml/g walls) and dialyse six times against 80 volumes of 0.5 M acetic acid to remove inorganic salts. Repeat the dialysis an additional six times against 80 volumes of distilled water.

10. Lyophilize the walls.

[a] Mycelial walls prepared by this procedure typically account for ~3% of the wet weight of the harvested mycelia. The composition of walls isolated from *P. megasperma* and prepared as described here contain approximately 93% carbohydrate and 7% protein (60).

4. Preparation of fungal cell wall hepta-β-glucoside elicitor

Oligosaccharides that are derived from three types of polysaccharides present in the mycelial walls of most fungi have been shown to induce various plant defence responses. Linear oligomers of (1 → 4)-linked β-D-glucosamine, originating from chitosan, have been identified as elicitors of phytoalexin accumulation in pea pods (66), protease inhibitor accumulation in tomato plants (67–69), and callose synthesis in suspension-cultured soybean cells (70, 71). Linear oligomers of (1 → 4)-linked *N*-acetyl-β-D-glucosamine, derived from chitin, elicit chitinase in melon (72) and lignification in wheat (22). The third and best characterized elicitor isolated from fungal walls is a branched hepta-β-D-glucoside (19, 20) (*Figure 4*) derived from branched mycelial wall β-glucans. This elicitor was purified to homogeneity from partial acid hydrolysates of isolated mycelial walls of the fungal phytopathogen *Phytophthora megasperma* f. sp. *glycinea* (19) by the procedures described below. The structure of the elicitor-active hepta-β-glucoside has been determined (20) and confirmed by chemical synthesis (73–75).

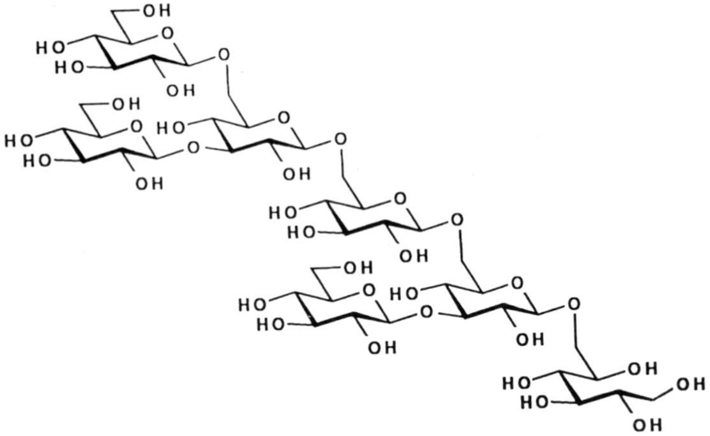

Figure 4. Structure of the hexa-β-glucosyl glucitol elicitor purified from partially hydrolysed mycelial walls of *P. megasperma* f. sp. *glycinea* (20).

4.1 Release of elicitor-active oligosaccharides from fungal walls

Elicitor-active mycelial wall components were originally released from *P. megasperma* mycelial walls by aqueous extraction at 121 °C, which solubilizes about 5% of the carbohydrate present in the walls (76). More efficient release of the elicitor-active mycelial wall carbohydrates is achieved by partial acid hydrolysis (19). The partial acid hydrolysis conditions described below have been optimized for hydrolysis of *P. megasperma* walls and cannot a priori be applied to the release of elicitor-active oligosaccharides from the mycelial walls of other fungi. Parameters such as the concentration of acid, hydrolysis time, and temperature must be experimentally determined for the optimal release of elicitors from mycelial walls of different fungi.

Protocol 10. Release of active oligosaccharides from fungal walls[a]

1. Suspend 1 g of lyophilized *P. megasperma* mycelial walls in 100 ml of 2 M trifluoroacetic acid (TFA) in a 160 ml KIMAX milk dilution bottle fitted with a rubber-lined screwcap (VWR Scientific). A typical preparation starts with 10 bottles.

2. Heat the mycelial wall suspensions for 2.5 h in an 85°C water-bath and briefly shake the bottles every half-hour to mix the contents.

3. Cool the bottles rapidly by placing them in an ice water-bath. Remove the insoluble residues from the solubilized material by centrifugation at 20 000 g for 20 min and passage of the supernatant through glass-fibre filter paper (GF/C, Whatman).

4. Pool the filtrates of 10 bottles and concentrate to a syrup by rotary evaporation at 35–40°C under reduced pressure in a silanized round-bottomed 2-litre flask. Silanization of glassware improves the recovery of poly- and oligosaccharides.

5. Add approx. 400 ml of methanol to the resulting syrup and repeat the rotary evaporation. Repeat the methanol evaporation three or four times to remove residual TFA.

6. Dissolve the resulting syrup, containing a mixture of various-sized fungal cell wall oligosaccharides, in 100 ml of water, adjust to pH 7.0 with 1 M NaOH, and bring to a final volume of 300 ml.

[a] Partial acid hydrolysis of *P. megasperma* mycelial walls solubilizes ~50% of the carbohydrate present in the walls (19). The released material is composed of approximately 3% polypeptide and 97% carbohydrate, of which 4% is mannose and 96% is glucose (19).

4.2 Gel-permeation chromatography of acid-released fungal cell wall oligosaccharides

The hepta-β-glucoside elicitor is purified from the mixture of oligo-β-glucosides released by partial acid hydrolysis of the fungal mycelial walls by using several chromatography techniques, including gel-permeation and normal- and reversed-phase column chromatographies. The β-glucan oligo-saccharides behave anomalously on gel-permeation columns when compared with the elution volumes of oligosaccharide standards such as malto-oligosaccharides. As a consequence, calibrations of gel-permeation columns with commercially available malto-oligosaccharides will not give reliable information about the size of the mycelial wall-derived oligo-β-glucosides. Thus, the degree of polymerization (DP) of the components present in individual chromatography peaks must be determined experimentally. Fast-atom-bombardment mass spectrometry (*Protocol 6*) is the method of choice. Measuring the molar ratio of reducing sugar (PAHBAH assay, *Protocol 1 C*) to total hexose (anthrone assay, *Protocol 1 A*) can give an estimate of the DP that is sufficiently accurate to calibrate the column. Glycosyl-residue and glycosyl-linkage compositions of the oligosaccharide-containing fractions eluting from the different chromatography columns are determined by methods described above (*Protocols 2* and *3*).

4.2.1 Low-resolution P-2 gel-permeation chromatography of acid-solubilized mycelial wall material

Low-resolution gel-permeation chromatography of 700 mg of carbohydrate solubilized from mycelial walls is performed on a 1 litre Bio-Gel P-2 column (4 cm × 1.2 m, 100–200 mesh; Bio-Rad) using distilled water at a linear flow rate of 7 ml/cm^2/h. Collect 10 ml fractions and assay aliquots of each fraction colorimetrically for hexose content as described earlier (*Protocol 1*) and for elicitor activity (*Protocol 7*). The fractions containing oligosaccharides having six or more glycosyl residues possess elicitor activity. Pool fractions containing oligosaccharides having DPs of 5–9 (*Figure 5*) and lyophilize. (*Note*: The oligosaccharides eluting at the void volume of the low-resolution P-2 column can be pooled (*Figure 5*) and used as a standard elicitor ('Void glucan' elicitor) in the cotyledon bioassay (*Protocol 7*).)

4.2.2 High-resolution P-2 gel-permeation chromatography of the low-resolution gel-permeation-purified oligosaccharides

High-resolution gel-permeation chromatography of 10 mg of low-resolution P-2-purified oligosaccharides is performed on two Bio-Gel P-2 columns (1.6 cm × 97 cm, minus 400 mesh) connected in series, equilibrated with distilled water at 45 °C; the linear flow rate is 10 ml/cm^2/h. The glass columns should be silanized prior to packing the columns to minimize adsorption of carbohydrate to the glass. Collect 3 ml fractions in silanized test tubes. Determine the elution profile of carbohydrate of the columns with the anthrone assay (*Protocol 1*), and the DP of the components eluting in individual peaks by FAB-ms (*Protocol 6*). Pool fractions containing oligosaccharides of DP = 7 (*Figure 6*) and lyophilize. These pooled oligosaccharides should contain less than 1% peptide and should be composed of 100% glucose (19).

4.3 NaBH$_4$ reduction of oligoglucosides

The separation of oligosaccharides on reversed-phase HPLC columns with water (77) can be complicated by the partial separation of the α- and β-anomers of each oligosaccharide. Thus, the heptaglucosides purified by gel-

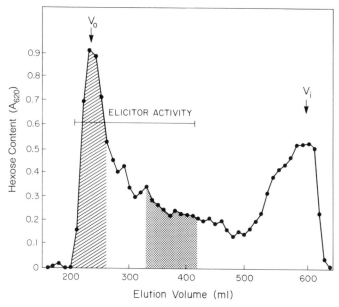

Figure 5. Low-resolution P-2 gel permeation chromatography of 700 mg of acid-solubilized oligosaccharides from *P. megasperma* f.sp. *glycinea* mycelial walls. Chromatography conditions are described in the text. The amount of neutral sugar in each fraction was determined by the anthrone assay (see *Protocol 1* A) and is represented by the A_{620} value. The shaded region (elution volume = 330–420 ml) represents the elicitor-active oligosaccharide-containing fractions that were pooled for further purification. The cross-hatched region (elution volume = 200–260 ml) can be pooled and used as a standard elicitor ('Void glucan' elicitor) in the soybean cotyledon assay (see *Protocol 7*). (Redrawn from reference 19 with permission, copyright The American Society for Biochemistry and Molecular Biology.)

permeation chromatography are converted to hexaglucosyl-glucitols prior to HPLC purification to simplify their elution profiles and subsequent analyses. Reduction of the heptaglucoside pool does not significantly alter the elicitor activity of these oligoglucosides (19).

Protocol 11. NaBH₄ reduction of oligoglucosides

1. Dissolve the heptaglucosides (50–100 mg), previously purified on the high-resolution Bio-Gel P-2 column, in 20 ml of 1 M NH₄OH containing 10 mg/ml NaBH₄ in a silanized 100 ml round-bottomed flask, and incubate for 4 h at room temperature.

2. Neutralize the mixture with glacial acetic acid, with stirring, until all the remaining NaBH₄ has reacted (i.e. no further effervescence), and add 25 ml of freshly prepared 10% (v/v) acetic acid in Spectranalyzed grade methanol. (The use of high quality methanol reduces the introduction of

Protocol 11. *Continued*

impurities into the sample.) Remove the solvent by rotary evaporation, and follow with three additional rotary evaporations of 25 ml of fresh 10% acetic acid in methanol.

3. Repeat the rotary evaporations four additional times with 25 ml of 100% Spectranalyzed grade methanol. Steps 2 and 3 evaporate the borate as its trimethyl ester from the hexaglucosyl glucitol samples.

4. Dissolve the hexaglucosyl-glucitols in ~5 ml of water. Desalt the mixture by elution with water through a Dowex 50W–X12-200 cation-exchange column (H$^+$ form; 10 ml bed volume). Wash the column with at least 10 ml of deionized water. Collect and lyophilize the eluant.

4.4 High-performance liquid chromatography (HPLC)

HPLC of the hexaglucosyl-glucitols purified by low-resolution gel-permeation chromatography is carried out using both normal- and reverse-phase columns. Guard columns containing either polar amino-cyano (PAC) or octadecyl silyl (ODS) functional groups corresponding to the functional groups on the chroma-

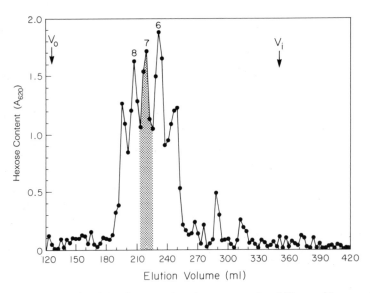

Figure 6. High-resolution P-2 gel permeation chromatography of 10 mg of low-resolution P-2 purified elicitor-active oligosaccharides. Chromatography conditions are described in the text. The amount of neutral sugar in each fraction was determined by the anthrone assay (*Protocol 1* A) and is represented by the A_{620} value. Each peak is labelled according to the degree of polymerization of the major components in that peak as determined by FAB-ms (see Section 2.3, *Protocol 6*). The heptaglucoside fraction that was pooled for further purification is shown in the shaded region.

tography columns should be used for all HPLC to prevent degradation of the columns. In preference to colorimetric assays which consume valuable sample, the column effluent is monitored for absorbance at 206 nm, or by using a refractive index detector whose temperature is maintained at 30.00°C ± 0.01°C with an appropriately precise circulating water-bath and insulated tubing connections to the detector. All HPLC solvents are filtered before use through a 47 mm diameter nylon-66 membrane filter (0.2 μm pore size, Rainin) in an all-glass filter apparatus (47 mm diameter, Millipore). Effluent fractions are collected in silanized test tubes. Column fractions are assayed for elicitor activity after removal of the HPLC solvents. This is done by placing aliquots of each fraction in a silanized test tube, concentrating to dryness under a stream of dry nitrogen, and dissolving the residue in sterile water. Appropriate aliquots of these solutions are then assayed for elicitor activity in the soybean cotyledon assay (*Protocol 7*).

4.4.1 Normal phase HPLC

Normal-phase partition chromatography of 12 mg of hexaglucosyl-glucitols is accomplished on a semi-preparative Magnum-9 PAC carbohydrate column (9.4 mm × 25 cm; Alltech Assoc. Inc.), equilibrated at 2.0 ml/min in 65% acetonitrile in water (v/v) containing 0.01% ammonium hydroxide (pH 8–9).

Protocol 12. Normal phase HPLC

1. Dissolve lyophilized samples first in 225 μl of HPLC-grade water and then slowly add 225 μl acetonitrile.
2. Inject samples on to the column (see above) via a 1 ml injection loop.
3. Collect 2 ml fractions and screen for elicitor activity as described above.
4. Pool fractions containing elicitor-active hexaglucosyl-glucitols (*Figure 7*), and reduce in volume to approximately 1 ml under a stream of nitrogen.
5. Filter the pooled hexaglucosyl-glucitols by centrifuging through a nylon-66 filter (0.2 μm pore size) in a microcentrifuge tube (Microfilterfuge Tube, Rainin Instrument Co.) at full speed in a bench microfuge to remove precipitated silica from the column bleed, and further reduce the volume under a stream of nitrogen at room temperature to between 50 and 90 μl volumes per 3 mg of hexaglucosyl-glucitols.
6. Store at −20°C.

4.4.2 Reverse-phase HPLC

Reversed-phase liquid chromatography of PAC-purified 3 mg of hexaglucosyl-glucitols is performed on two Spherisorb-5 ODS analytical columns (4.6 mm × 25 cm; Alltech) connected in series.

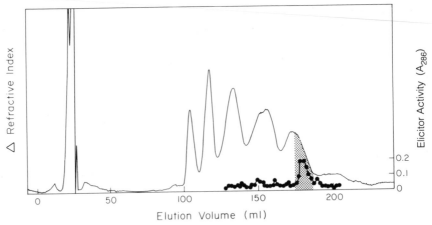

Figure 7. Normal-phase partition chromatography of 12 mg of high-resolution P-2 purified hexaglucosyl-glucitols on a Magnum-9 PAC column. Chromatography conditions are described in the text. The refractive index detector was set at an attenuation of 8×. The elicitor activity (●, A_{286}) of one-100th of each fraction was determined using the soybean cotyledon bioassay (see Section 2.4, *Protocol 7*). The first two eluting peaks contained no measurable elicitor activity (data not shown). The fractions in the shaded region were pooled for further purification. (Reproduced from reference 19 with permission, copyright *The American Society for Biochemistry and Molecular Biology*.)

Protocol 13. Reverse-phase HPLC

A. *Isolation of hexaglucasyl-glucitol fractions*

1. Pre-condition the columns (see above) with 100% acetonitrile and then equilibrate with aqueous 1.75% (v/v) acetonotrile at 0.3 ml/min and 800 p.s.i.

2. Ensure reproducibility in solvent strength by first degassing the appropriate volume of filtered HPLC-grade water and then adding the appropriate volume of degassed, filtered, HPLC-grade acetonitrile. (This procedure prevents evaporative losses of acetonitrile during degassing that would result in small changes in the acetonitrile concentration of the final solvent mixture, which can dramatically affect column performance (see below).)

3. Inject a sample of hexaglucosyl-glucitols (3 mg in 50–90 μl) on to the column via a 200 μl loop.

4. Collect 0.5 ml fractions and screen for elicitor activity as described earlier.

5. Wash the columns, after each chromatography run, with aqueous 15% (v/v) acetonitrile followed by 100% acetonitrile.

6. Pool fractions containing elicitor-active hexaglucosyl-glucitols (*Figure 8*) and reduce in volume to between 50 and 100 μl under a stream of nitrogen.

7. Store at −20°C.

B. *Further purification*

Hexaglucosyl-glucitol fractions isolated by chromatography on the Spherisorb ODS columns are further purified on two Ultrasphere ODS analytical columns (4.6 mm × 26 cm, Beckman) connected in series.

1. Pre-condition the Ultrasphere columns with 100% acetonotrile, and equilibrate in aqueous 2.0% (v/v) acetonitrile at 0.3 ml/min and 900 p.s.i.

2. Inject a sample of 50–90 μl on to the column via a 200 μl injection loop.

3. Collect 0.5 ml fractions and screen for elicitor activity as described.

4. Pool fractions containing the elicitor-active hexaglucosyl-glucitol (*Figure 9*) and reduce in volume to between 50 and 100 μl under a stream of nitrogen.

5. Store at −20°C.

The purification protocol described above yields about 200 μg of the pure elicitor-active hexa-β-glucosyl glucitol from 10 g of fungal mycelia (19). The purified hexa-β-glucosyl glucitol elicitor is, on a molar basis, the most active elicitor of photoalexin accumulation yet isolated, showing half maximum activity at a concentration of 10^{-8} M (19, 78) (see *Figure 3B*). The elicitor-active hexa-β-glycosyl glucitol is the only active hexaglucosyl glucitol that was observed at any stage of the purification in the mixture of hexaglucosyl glucitols released from *P. megasperma* walls (19). Judging by the numbers of peaks observed in the various chromatographic steps, we estimate that there

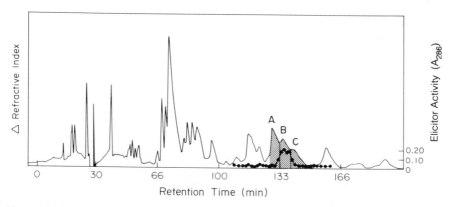

Figure 8. Reversed-phase chromatography of 3 mg of PAC-purified hexaglucosyl glucitols on the Spherisorb-5 ODS columns. Chromatography conditions are described in the text. The refractive index detector was set at an attenuation of 4×. The elicitor activity (●, A_{286}) of one-2000th of each fraction was determined using the soybean cotyledon bioassay (see Section 2.4, *Protocol 7*). All other fractions had no detectable elicitor activity (data not shown). the fractions in the shaded regions were pooled as indicated (A, B, C) for further purification. (Redrawn from reference 19 with permission, copyright The American Society for Biochemistry and Molecular Biology.)

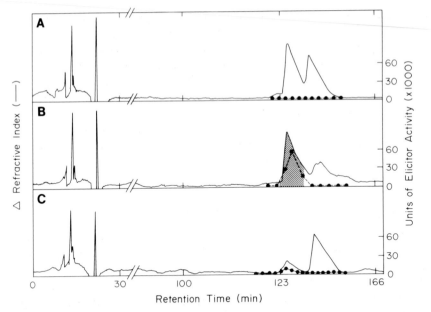

Figure 9. Reversed-phase liquid chromatography of the Spherisorb-ODS-purified hexa-glucosyl glucitols fractions **A, B,** and **C** on the Ultrasphere ODS columns. Chromatography conditions are described in the text. The refractive index detector was set at an attenuation of 2×. Retention times were calculated with respect to internal standards to correct for pump and solvent variations. The elicitor activity (●) of each fraction was determined using the soybean cotyledon bioassay (see Section 2.4, *Protocol 7*). The shaded region in **B** indicates fractions containing the elicitor-active hexa-β-glucosyl glucitol (see *Figure 4*). (Adapted from reference 19, with permission, copyright The American Society for Biochemistry and Molecular Biology.)

could be as many as 300 structurally distinct elicitor-inactive hexaglucosyl glucitols present in the crude mixture of heptaglucosides. The tremendous heterogeneity of the mycelial wall fragment oligoglucosides greatly increases the difficulty of purifying the elicitor.

It is important to recognize a few chromatographic parameters that are important for high-resolution reversed-phase liquid chromatography of oligosaccharides and how these parameters can effect such separations. HPLC is a rapidly changing and improving technique, with new types of columns being developed and old ones being improved. This may lead to new columns giving greater resolution and yields compared to the columns described here. Furthermore, the properties of the column packing materials made by different manufacturers can vary, which may lead to differences in column resolution. For example, hexa(β-D-glucopyranosyl)-D-glucitols that co-elute on one manufacturer's column separate on another supposedly identical column sold by another manufacturer (compare *Figures 8* and *9*). Thus, elution conditions for optimal separation of the mycelial wall fragments must be determined experimentally for every column, and it may prove advantageous to try reversed-phase columns made by different manufacturers.

The composition of the elution solvent is another critical parameter for successful separation of oligosaccharides. Increases in the acetonitrile concentration of the chromatography solvent by as little as 0.2% result in significant decreases in the elution volume of the hexa(β-D-glucopyranosyl)-D-glucitols on a particular column. On the other hand, structurally diverse oligosaccharide-glucitols require very different solvent conditions for separation. For instance, 4-linked oligo-α-glucosides, which separate well in water on ODS columns (77), are not retained under the solvent conditions used to separate hexa(β-D-glucopyranosyl)-D-glucitols.

5. Preparation of oligogalacturonide elicitor

Elicitor-active oligosaccharide fragments of plant origin were first identified in the material released by partial acid hydrolysis of plant cell walls (23). It was later shown that the same, or very similar, oligosaccharides could be released by treatment of plant cell walls with polygalacturonic acid-degrading enzymes purified from micro-organisms (27, 28, 79). The elicitor-active plant cell wall fragments appear to be linear polymers of $(1 \rightarrow 4)$-linked α-D-galacturonic acid for example $GalA\alpha(1 \rightarrow [4)GalA\alpha(1 \rightarrow]_{9-11}4)GalA$ and $\Delta4,5GalA\alpha(1 \rightarrow [4)GalA\alpha(1 \rightarrow]_{8}4)GalA$ (24, 25, 79, 80). Treatment of several plant species with micromolar concentrations of oligo-α-galacturonide elicitors induce, in various plants or plant tissues, the synthesis and accumulation of phytoalexins (23–25, 67), the accumulation of proteinase inhibitors (81, 82), and the lignification of plant cell walls (28, 80). The highest phytoalexin- and lignin-inducing activity is obtained with oligo-α-D-galacturonides that have a DP greater than 10 (24, 25, 28, 79, 80). Shorter oligogalacturonides as small as the disaccharide induce proteinase inhibitor accumulation (81, 82).

5.1 Solubilization of oligogalacturonides from pectin or polygalacturonic acid

Oligogalacturonide preparations are most conveniently prepared from commercial sources of citrus pectin or polygalacturonic acid (partially hydrolysed, methyl-de-esterified pectin). These oligogalacturonides appear to be similar to those isolated from plant cell walls (24). Two methods for releasing elicitor-active oligogalacturonides from the larger polymers are described below. Partial acid hydrolysis offers the advantage that enzyme purification (*Protocol 14* B) is not required, while partial enzymatic digestion provides an approximately 10-fold higher yield of the larger elicitor-active oligogalacturonides. Furthermore, oligogalacturonides obtained by partial enzymatic hydrolysis have not been exposed to the harsh conditions necessary for chemical hydrolysis, conditions that may lead to alterations in the structures of some of the oligogalacturonides.

Protocol 14. Solubilization of oligogalacturonides from pectin or polygalacturonic acid

A. *Partial acid hydrolysis of citrus pectin*

1. Suspend citrus pectin (Sunkist Growers) (1 g/100 ml) in 2 M TFA acid in a 160 ml KIMAX milk dilution bottle fitted with a rubber-lined screwcap (VWR Scientific). A typical preparation starts with 10 bottles.

2. Heat the pectin suspensions for 4 h in an 85°C water-bath and shake the bottles every 30 minutes to mix the contents.

3. Cool the suspensions to room temperature and filter through Whatman GF/A paper. Combine and concentrate the filtrates to a paste by rotary evaporation under reduced pressure at 35°C.

4. Suspend the paste in 10 ml of methanol and concentrate by rotary evaporation as before. Repeat four times.

5. Suspend the residue in 100 ml of deionized water, titrate to pH 7 with 1 M imidazole–HCl, and stir the suspension overnight at 4°C.

6. Pass the suspension progressively through Whatman GF/A, Millipore type HA (0.45 μm), and Millipore type GS (0.22 μm) filters. The final filtrate is lyophilized. The expected yield is ~3.5 g of crude oligogalacturonides from 10 g of pectin.

B. *Purification of endopolygalacturonase*

Endopolygalacturonase can be purified from a commercially available preparation of pectic-degrading enzymes secreted by *Aspergillus niger*. All operations described below are carried out at 4°C.

1. Pectinase solution (100 ml; Sigma P-5146) in 40% glycerol is ultraconcentrated and dialysed using a PM-10 membrane (Amicon) against 100 mM sodium acetate, pH 5.0, to a final volume of 10–15 ml and a final glycerol concentration of 4–5%.

2. Load the concentrated and dialysed pectinase on a 2.5 × 70 cm gel permeation column packed with Bio-Gel P-10 (100–200 mesh; Bio-Rad) pre-equilibrated in 10 mM sodium acetate, pH 5.0. Elute the column with the same buffer at a linear flow rate of 25 ml/cm^2/h and collect 20 ml fractions.

3. Screen the fractions for protein by the Bradford assay (83) and uronic acid content by the *m*-hydroxybiphenyl assay (*Protocol 1* B). The endopolygalacturonase elutes in the quantitatively predominant protein peak eluting at or near the column void volume, while the majority of uronic acid-containing carbohydrate contaminants elute closer to the included volume of the column.

4. Pool the fractions containing the bulk of the protein and load on to a 50 ml (2.5 × 10 cm) column packed with CM–Trisacryl (IBF Biotechnics) pre-equilibrated in 10 mM sodium acetate, pH 5.0.

5. Wash the column with 100 ml of the 10 mM sodium acetate buffer, and elute the endopolygalacturonase with a linear gradient made from 100 ml of 10 mM sodium acetate, pH 5.0, and 100 ml of 10 mM sodium acetate, pH 5.0, containing 0.5 M NaCl. Collect 2–3 ml fractions and screen for protein by the Bradford assay (83), and endopolygalacturonase activity by measuring the production of reducing groups with the PAHBAH assay (*Protocol 1* C) using a 1% (w/v) solution of polygalacturonic acid (Sigma P-1879) as substrate (9). The endopolygalacturonase elutes at a salt concentration between 0.2 and 0.25 M NaCl.

6. Pool fractions containing the endopolygalacturonase activity and ultraconcentrate to a final volume of 5 ml using a PM-10 membrane (Amicon).

7. Further purify the concentrated endopolygalacturonase by loading 0.5 ml aliquots on a 1.5 × 100 cm column packed with Sephacryl S-200 (Pharmacia) pre-equilibrated in 100 mM sodium acetate, pH 5.0. Elute the column at a linear flow rate of 45 ml/cm^2/h and collect 2 ml fractions. Screen the fractions for protein by A_{280} or by the Bradford assay (83). The endopolygalacturonase elutes as the major protein peak, and is flanked by two smaller peaks of contaminating proteins. Pool the fractions containing endopolygalacturonase activity. The purified endopolygalacturonase, which represents approximately 90–95% of the protein loaded on the Sephacryl column, gives a single band on 10% SDS–PAGE gels and has no other detectable glycosidase activities.

[a] This protocol will yield ~1–2 mg of purified endopolygalacturonase having a specific activity of 2400 ± 100 RGU/mg of protein. One RGU (reducing group unit) is defined as the amount of enzyme producing 1 microequivalent of reducing groups per minute at 30°C in 50 mM sodium acetate, pH 5.0, using a 1% (w/v) solution of polygalacturonic acid as substrate (84).

C. *Partial endopolygalacturonase digestion of polygalacturonic acid*

1. Dissolve 2 g of polygalacturonic acid (Na$^+$ salt, Sigma P-1879) in 1 l of 20 mM sodium acetate buffer, pH 5, containing 2 mg of bovine serum albumin to stabilize the endopolygalacturonase.

2. Add 12–15 RGUs of purified fungal endopolygalacturonase (*Protocol 14* B).

3. Digest the polygalacturonic acid for 8 h at room temperature with gentle stirring.

4. Terminate the digestion by subdividing the solution into four 250 ml batches, and autoclaving for 12 min at 121°C.

5. Cool the solutions to room temperature and adjust the pH to 7 with 2 M imidazole free base.

5.2 Low-resolution ion-exchange chromatography of oligogalacturonides

---

Protocol 15. Low-resolution ion-exchange chromatography

1. Equilibrate a QAE-Sephadex (Sigma, Q-25-120) anion-exchange column (3 × 12 cm) with 0.125 M imidazole–HCl, pH 7.0.

2. Ensure that the crude oligogalacturonide mixtures (either acid- or enzyme-released) hve a pH ≈ 7 and a conductivity of less than 6.7 mΩ^{-1}. If necessary, adjust the pH by the addition of 5 M imidazole–HCl, pH 7.0, and the conductivity by dilution with water.

3. Load the crude oligogalacturonide mixture on to the QAE–Sephadex anion-exchange column. Wash the column with 400 ml of 0.125 M imidazole–HCl, pH 7.0. Elute the bound oligogalacturonides stepwise with 400 ml of 0.55 M imidazole–HCl, pH 7.0, and then 200 ml of 0.75 M imidazole–HCl, pH 7.0, collecting each wash separately. The 0.55 M imidazole wash contains oligogalacturonides having a DP of 5–9; the 0.75 M imidazole wash contains oligogalacturonides having a DP of 6–20.

4. If necessary, concentrate the mixtures of oligogalacturonides obtained from the low resolution anion exchange chromatography by rotary evaporation under reduced pressure at 35°C.

5. Desalt the mixtures of oligogalacturonides either by dialysis, using 1000 dalton cut-off dialysis membranes (Spectrapor-7, Spectrum Medical), or by chromatography on a Sephadex G-10 column (3 × 18 cm) eluted with deionized water.

5.3 High-resolution ion-exchange chromatography of larger oligogalacturonides

---

Protocol 16. High-resolution ion-exchange chromatography

1. Equilibrate a Q-Sepharose Fast Flow (Pharmacia) anion-exchange column (3.5 × 40 cm) with 0.4 M imidazole–HCl, pH 7.0.

2. Ensure that the conductivity of the oligogalacturonides (0.75 M imidazole wash from low-resolution column described above) is less than 15 mΩ^{-1}; if necessary, adjust by adding deionized water.

3. Load about 130 mg of the oligogalacturonide mixture on to the Q-Sepharose Fast Flow column. Elute the column, at a linear flow rate of 3 ml/cm^2/min, with the linear gradient prepared from equal volumes of 0.4 M imidazole–HCl, pH 7.0 (3.5 l), and 0.75 M imidazole–HCl, pH 7.0

(3.5 l). Collect 10 ml fractions and quantitate the galacturonic acid content of each using the *m*-hydroxybiphenyl assay (*Protocol 1* B). A typical elution profile is shown in *Figure 10*.

4. Further purify individual peaks eluting from the Q-Sepharose column by re-chromatography on an anion exchange column, or by gel-permeation chromatography on two Sephadex G-25 (Pharmacia) columns (1.6 × 95 cm) connected in series and equilibrated in aqueous 1% (v/v) acetic acid with a linear flow rate of 4.5 ml/cm²/h.

5. Convert the purified oligogalacturonides to their sodium salt form prior to assay for elicitor activity, as the presence of buffer salts, particularly imidazole, in the oligogalacturonides can affect the elicitor bioassay. Dialyse the oligogalacturonides three times, using 1000 dalton cut-off dialysis tubing (SpectraPor 7, Spectrum Medical) against 20 volumes of distilled water overnight at 4°C. Repeat the dialysis once against 100 mM NaCl followed by three more dialyses against water. Lyophilize the oligogalacturonides.

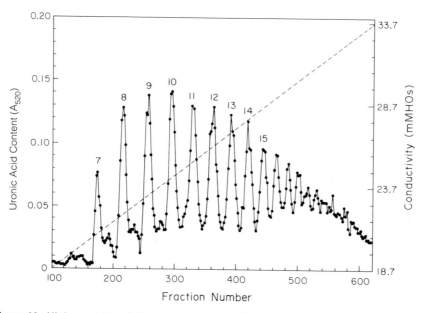

Figure 10. High-resolution Q-Sepharose anion-exchange chromatography of ~130 mg of low-resolution anion exchange-purified oligogalacturonides obtained by partial enzyme digestion of polygalacturonic acid. Chromatography conditions are described in the text. The amount of uronic acid in each fraction was determined by the *m*-hydroxybiphenyl assay (see *Protocol 1* B) and is represented by the A_{520} value. Each peak is labelled according to the degree of polymerization of the major components in that peak as determined by FAB-ms (see Section 2.3, *Protocol 6*).

The individual oligomer fractions (for example DP ≤ 14) prepared by the methods described in this section are composed primarily (> 90%) of galacto-syluronic acid, but they still contain minor amounts of other neutral sugars, particularly rhamnose. The possibility remains that a minor component of these oligogalacturonide preparations is the actual elicitor-active component, although the elicitor-active component is sensitive to digestion with purified α-1,4-endopolygalacturonase. Improved chromatographic separations, such as high-pH anion-exchange chromatography with pulsed amperometric detection (85, 86), will undoubtedly lead to the identification of the structure of the active species. Such proof would best be obtained by an unambiguous chemical synthesis of the active unit, for example oligogalacturonide, but the amounts of oligogalacturonides that have been chemically synthesized to date have been too small (87–90) to rigorously bioassay the products.

6. Radiolabelling of elicitors

Elicitors have been shown to induce a variety of plant defence responses. However, the mechanism by which elicitors regulate the pathway leading to activation of plant defence responses remains unknown. Labelled homogeneous elicitors are necessary for elucidating the mechanism of action of elicitors. For example, labelled elicitors could be used in assays to detect elicitor binding to receptors. Reaction sites at which labels can be incorporated into oligosaccharides include the hemi-acetal of the reducing terminal glycosyl residue of oligosaccharides, the carboxyl groups in uronic acid-containing oligosaccharides, and amino groups in amino sugar-containing oligosaccharides. Examples of labels that permit sensitive detection of biologically active oligosaccharides include radiolabelling (see below), biotinylation (91), and fluorescent tagging (92–94). The methods available for labelling oligosaccharides significantly modify their structures. Thus, it is necessary to demonstrate that labelling has not altered the biological activity of the oligosaccharide elicitors.

One method for radiolabelling oligosaccharides is to react their reducing terminal glycosyl residue with sodium borotritiide, thereby introducing one tritium atom into the labelled oligosaccharide with limited structural alterations. However, the specific radioactivity of oligosaccharides labelled in this manner is often too low (~10 Ci/mmol) for studies requiring sensitive detection of the labelled oligosaccharide, for example in receptor–ligand binding studies (95). Radio-iodination yields oligosaccharides with much higher specific radioactivities (> 100 Ci/mmol). However this requires the attachment to the oligosaccharide of a group that can be iodinated, for example phenoxyl groups. Two methods for incorporating a phenoxyl group at the reducing end of oligosaccharide elicitors are described in the following sections. One method couples tyramine to the reducing terminal glycosyl residue of the oligosaccharide by reductive amination, the other forms a hydrazone of the reducing glycosyl residue. Radiolabelling of these oligosaccharide derivatives with ^{125}I is described in the subsequent section (*Protocol 19*).

6.1 Preparation of a tyraminylated derivative of the hepta-β-glucoside elicitor

Coupling of tyramine to oligosaccharides having a reducing terminal glycosyl residue is accomplished by reductive amination (*Figure 11*). The protocol described below was originally devised for coupling aromatic amines to oligosaccharides (96), and has been used in our laboratory to derivatize chemically synthesized hepta-β-glucoside elicitor obtained from P. Garegg (University of Stockholm, Sweden).

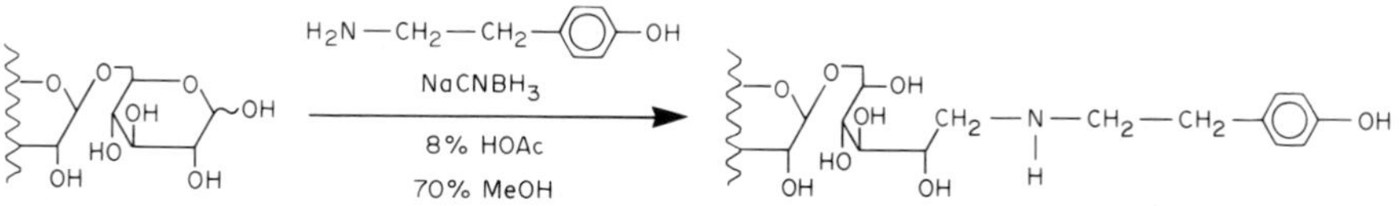

Figure 11. Formation of the tyramine conjugate of an oligoglucoside by reductive amination (96).

Protocol 17. Preparation of tyraminylated hepta-β-glucoside elicitor

1. Dissolve 1 mg of hepta-β-glucoside in 5 μl of deionized water in a 12 × 75 mm glass test tube fitted with a Teflon-lined screwcap.

2. In a separate 12 × 75 mm glass test tube prepare a reaction mixture consisting of:
   - 137 mg of tyramine (free base)
   - 35 mg sodium cyanoborohydride
   - 350 μl of methanol
   - 50 μl of deionized water
   - 41 μl of glacial acetic acid

   If necessary, briefly heat the reaction mixture in an 80°C heating block to completely dissolve the reagents.

3. Add 40 μl of the reaction mixture to the sample of hepta-β-glucoside, seal the tube, and heat the reaction mixture for 1 h at 80°C in a heating block.

4. Cool the reaction to room temperature and evaporate the methanol under a stream of air in a fume hood. Dissolve the residue in 200 μl of deionized water, heating briefly in the heating block if necessary to completely dissolve the sample.

5. Prepare a small AG 1-X8 (Bio-Rad; 200–400 mesh, Cl⁻ form) anion

Protocol 17. *Continued*

exchange column (0.5 ml bed volume) in a 3 ml disposable plastic syringe and wash the resin with 5 ml of deionized water. Apply the reaction mixture to the column, elute with 2 ml of deionized water, and lyophilize the eluate.

6. Purify the tyramine conjugate by chromatography on a Bio-Gel P-2 column (1 × 110 cm; 200–400 mesh) equilibrated in 0.1 M NH_4HCO_3. Collect 1 ml fractions and assay each fraction colorimetrically for hexose (see *Protocol 1 A*) and for absorbance at 274 nm (absorption maximum for tyramine; molar extinction coefficient = 1420 M^{-1} cm^{-1}). The tyraminylated hepta-β-glucoside will elute as a coincident peak of hexose and A_{274}. The molar ratio of tyramine to hexose content of the conjugate should be 1 to 6 (the derivatized terminal glucosyl residue will not react in the anthrone assay).[a]

---

[a] The coupling of tyramine by the reductive amination procedure described above is quantitative for neutral oligosaccharides that are soluble in 70% methanol. We have not detected significant amounts of side-products using this derivatization procedure. Coupling of tyramine to the hepta-β-glucoside elicitor had no significant effect on the ability of the elicitor to induce phytoalexin accumulation in soybean cotyledons (78).

6.2 Preparation of tyrosine hydrazones of oligogalacturonide elicitors

Reaction of oligosaccharides with hydrazides to form hydrazones (97) has been used (94) to form the tyrosine hydrazone of oligogalacturonides (*Figure 12*). This procedure is described below (*Protocol 18*).

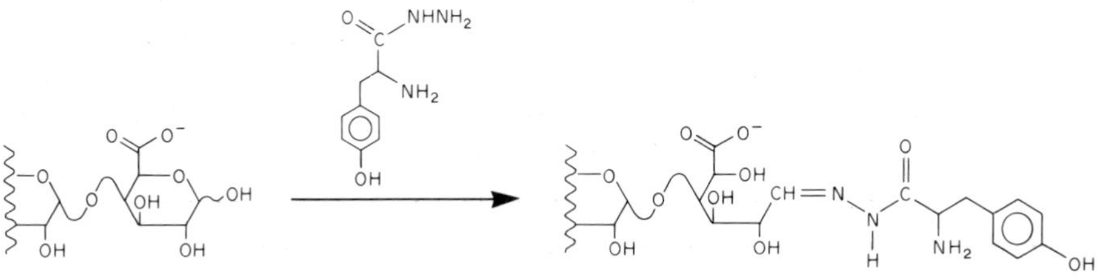

Figure 12. Formation of the tyrosine hydrazone of an oligogalacturonide (94).

Protocol 18. Preparation of tyrosine hydrazones of oligogalacturonide elicitors

1. Prepare a 1 mg/ml solution of oligogalacturonide elicitor in 0.125 M imidazole–HCl, pH 7.0.

2. Add a ten-fold molar excess of tyrosine hydrazide (Aldrich Chem. Co.) to this solution and incubate under an inert (N_2) atmosphere for 24 h at room temperature.

3. Separate the oligogalacturonide tyrosine hydrazone from unreacted tyrosine hydrazide by chromatography on a disposable PD-10 (Pharmacia) gel permeation column pre-packed with Sephadex G-25 equilibrated in 0.125 M imidazole–HCl, pH 7.0.

4. Collect fractions of 1 ml and assay each fraction for uronic acid content by the *m*-hydroxybiphenyl assay (*Protocol 1* B) and absorbance at 274 nm (absorption maximum for tyrosine; molar extinction coefficient = $1420 \, M^{-1} \, cm^{-1}$). The oligogalacturonide tyrosine hydrazone will elute as a coincident peak of uronic acid and A_{274}. The molar ratio of tyrosine to uronic acid content should be 1 to 11 for the conjugate of the dodecagalacturonide[a] (the derivatized terminal galacturonosyl residue will not react in the *m*-hydroxybiphenyl assay) if complete derivatization is obtained.

[a] The coupling of tyrosine hydrazide to dodecagalacturonide has been reported to be ~80% efficient (P. Low, Purdue Univ., personal communication). This derivatization of the oligogalacturonide had no effect on its activity in an oxidative burst bioassay (94).

6.3 Radio-iodination of tyramine– and tyrosine–oligosaccharide derivatives

The tyramine and tyrosine elicitor derivatives both contain the iodinatable phenoxyl group, and thus can be iodinated by methods developed for the iodination of proteins. Chloramine T, a reagent commonly used to iodinate proteins (98), was found to be unsuitable for iodinating oligosaccharides because this oxidizing reagent causes extensive structural modification of the oligosaccharide derivatives (unpublished results of the authors). The radio-iodination protocol below is described for the hepta-β-glucoside–tyramine conjugate. Alternative procedures are given for the purification of the iodinated oligogalacturonide (*Protocol 19*, steps 5–7).

Protocol 19. Radio-iodination of tyramine[a]– and tyrosine–oligosaccharide derivatives

1. Dissolve 2 mg of iodogen (Pierce Chemical Co.) in 25 ml of chloroform. Prepare a ten-fold dilution of this stock solution with chloroform and add 50 µl of the diluted stock to the bottoms of 12 × 75 mm glass test tubes. Evaporate the chloroform with a stream of nitrogen, or allow the solvent to evaporate in a fume hood. Store the tubes desiccated and in the dark.

2. Wash an iodogen-coated test tube with 0.5 ml of 250 mM sodium phosphate, pH 7.5

Protocol 19. *Continued*

3. Prepare a stock solution of hepta-β-glucoside–tyramine conjugate at a concentration of 26 pmol/μl (30 ng/μl) in 250 mM sodium phosphate, pH 7.5. Add ~5 μl of the conjugate to the bottom of the rinsed iodogen-coated test tube.

4. Add ~0.5 mCi of Na[^{125}I] (carrier free; specific radioactivity = 13.4 Ci/μg) to the solution in the bottom of the test tube and allow to stand at room temperature for 15 min.

5. Prepare a small AG 1-X8 (Bio-Rad; 200–400 mesh, Cl$^-$ form) anion exchange column (0.5 ml bed volume) in a disposable plastic syringe. Wash the column with ~5 ml of deionized water. Load 10 mg of maltoheptaose dissolved in 1 ml of deionized water on to the column, and wash column with an additional 5 ml of deionized water. The maltoheptaose serves to block any binding sites for oligoglucosides on the plastic syringe or on the resin, thus improving the recovery of labelled oligoglucoside derivative.

   Note: Since the oligogalacturonide conjugate is itself charged, a gel permeation column (for example Sephadex G-25, 1 × 30 cm) must be used to separate the iodinated product from unreacted iodine.

6. Add 30 μl of deionized water to the iodination reaction, and apply the total reaction mixture to the anion exchange column. Wash the reaction tube with an additional 50 μl of deionized water and apply the wash solution to the anion exchange column.

7. Elute the column with 4 ml of deionized water, collecting 0.5 ml fractions in 1.5 ml Eppendorf tubes.

8. Count 10 μl of each fraction in a gamma counter.

9. Pool the fractions containing radioactivity (usually the first three fractions will contain ~90% of the radiolabelled oligosaccharide). Store the radiolabelled oligosaccharide in a lead container at room temperature.

[a] Typical iodinations of the tyraminylated hepta-β-glucoside elicitor are ~70% efficient. The ratios of reactants given here will typically yield radio-iodinated oligosaccharides with specific radioactivities of ~100 Ci/mmol. Adjustment of the molar ratio of iodine to tyramine conjugate will yield higher or lower specific activities of the radiolabelled product as desired.

7. Concluding remarks

The purification of the two cell wall-derived oligosaccharide elicitors described in this chapter illustrates the complexity of purifying active elicitors from the crude preparations that are most commonly used in studies of elicitor-induced plant responses (19, 23, 24). The elicitor-active hepta-β-glucoside was the only active elicitor detected in the mixture of heptaglucosides released from *P. megasperma* mycelial walls. A conservative estimate of the number of different oligoglucosides present in the crude heptaglucoside preparation is at least 150, and could be as high as 300, based on the number of peaks observed in the various column chromatography elution profiles (19). It is interesting to speculate that some of the other oligosaccharides present in the crude fraction, while not having any phytoalexin elicitor activity, may have the ability to stimulate or inhibit other plant responses.

Studies utilizing elicitors have expanded our knowledge of the biochemistry and molecular biology of plant defence responses. However, the fact that most studies utilizing elicitors have been carried out with unpurified elicitor preparations limits the conclusions that can be drawn from these studies. Crude elicitor preparations have been used extensively to activate plant defence genes, enabling these genes and their products to be studied in great detail (5, 6, 99, 100). However, studies using impure elicitor preparations aimed at elucidating mechanisms of gene regulation or signal transduction must be interpreted with caution, inasmuch as one can never be sure which molecule(s) in these heterogeneous mixtures are responsible for triggering the molecular event being studied. For example, several studies have demonstrated that the level of expression of a large number of genes is altered after treatment of plant tissues with heterogeneous elicitor preparations (101–103 and unpublished results of the authors). It remains unclear whether the plethora of changes observed is caused by a single elicitor molecule, or whether observed changes in gene expression represent a sum of the activities of a number of different elicitors. Thus, although it is difficult and time-consuming to purify oligosaccharide elicitors to homogeneity, this effort is essential if the elicitor-activated signal transduction pathways are to be fully elucidated.

Acknowledgements

The authors are grateful to P. S. Low for communicating details on the preparation of tyrosine hydrazones of oligogalacturonides. We also thank M. O'Neill and D. Mohnen for their advice and critical comments on this manuscript, K. Howard for typing the manuscript, and C. L. Gubbins Hahn for drawing the figures. Research in the authors' laboratories is supported by grants from the National Science Foundation (DCB-8904574 to MGH) and the Department of Energy (DE-FG09-85ER13426 to AD and DE-FG09-85ER13425 to PA). Also supported in part by the USDA/DOE/NSF Plant Science Centres Programme; through funding by the Department of Energy grant DE-FG09-87ER13810.

References

1. Bell, A. A. (1981). *Annu. Rev. Plant Physiol.*, **32**, 21.
2. Hahn, M. G., Bucheli, P., Cervone, F., Doares, S. H., O'Neill, R. A., Darvill, A., and Albersheim, P. (1989). In *Plant–microbe interactions. Molecular and genetic*

perspectives, Vol. 3 (ed. T. Kosuge and E. W. Nester), p. 131. McGraw Hill Publishing Co., New York, NY.

3. Grisebach, H. and Ebel, J. (1978). *Angew. Chem.*, **90**, 668.
4. Dixon, R. A., Dey, P. M., and Lamb, C. J. (1983). *Adv. Enzymol.*, **55**, 1.
5. Dixon, R. A. (1986). *Biol. Rev.*, **61**, 239.
6. Ebel, J. (1986). *Annu. Rev. Phytopathol.*, **24**, 235.
7. Boller, T. (1987). In *Plant–microbe interactions. Molecular and genetic perspectives*, Vol. 2 (ed. T. Kosuge and E. W. Nester), p. 385. Macmillan Publishing Co., New York, NY.
8. Ryan, C. A. (1978). *Trends Biochem. Sci.*, **3**, 148.
9. Cervone, F., Hahn, M. G., De Lorenzo, G., Darvill, A., and Albersheim, P. (1989). *Plant Physiol.*, **90**, 542.
10. Aist, J. R. (1976). *Annu. Rev. Phytopathol.*, **14**, 145.
11. Kauss, H. (1987). *Naturwissenschaften*, **74**, 275.
12. Cooper, J. B., Chen, J. A., Van Holst, G.-J., and Varner, J. E. (1987). *Trends Biochem. Sci.*, **12**, 24.
13. Showalter, A. M. and Varner, J. E. (1989). In *The biochemistry of plants: a comprehensive treatise*, Vol. 15, *Molecular Biology* (ed. A. Marcus), p. 485. Academic Press, Inc., New York, NY.
14. Ride, J. P. (1983). In *Biochemical plant pathology* (ed. J. A. Callow), p. 215. John Wiley & Sons, New York.
15. Keen, N. T., Partridge, J. E., and Zaki, A. I. (1972). *Phytopathology*, **62**, 768.
16. West, C. A. (1981). *Naturwissenschaften*, **68**, 447.
17. Darvill, A. G. and Albersheim, P. (1984). *Annu. Rev. Plant Physiol.*, **35**, 243.
18. Anderson, A. J. (1989). In *Plant–microbe interactions. Molecular and Genetic Perspectives*, Vol. 3 (ed. T. Kosuge and E. Nester), p. 87. McGraw Hill Inc., New York, NY.
19. Sharp, J. K., Valent, B., and Albersheim, P. (1984). *J. Biol. Chem.*, **259**, 11312.
20. Sharp, J. K., McNeil, M., and Albersheim, P. (1984). *J. Biol. Chem.*, **259**, 11321.
21. Kendra, D. F. and Hadwiger, L. A. (1984). *Exp. Mycol.*, **8**, 276.
22. Barber, M. S., Bertram, R. E., and Ride, J. P. (1989). *Physiol. Mol. Plant Pathol.*, **34**, 3.
23. Hahn, M. G., Darvill, A. G., and Albersheim, P. (1981). *Plant Physiol.*, **68**, 1161.
24. Nothnagel, E. A., McNeil, M., Albersheim, P., and Dell, A. (1983). *Plant Physiol.*, **71**, 916.
25. Jin, D. F. and West, C. A. (1984). *Plant Physiol.*, **74**, 989.
26. Bostock, R. M., Kuc, J. A., and Laine, R. A. (1981). *Science*, **212**, 67.
27. Bruce, R. J. and West, C. A. (1982). *Plant Physiol.*, **69**, 1181.
28. Robertsen, B. (1987). *Physiol. Mol. Plant Pathol.*, **31**, 361.
29. Davis, K. R., Lyon, G. D., Darvill, A. G., and Albersheim, P. (1984). *Plant Physiol.*, **74**, 52.
30. Cruickshank, I. A. M. and Perrin, D. R. (1968). *Life Sci.*, **7**, 449.
31. Farmer, E. E. and Helgeson, J. P. (1987). *Plant Physiol.*, **85**, 733.
32. Parker, J. E., Hahlbrock, K., and Scheel, D. (1988). *Planta*, **176**, 75.
33. Dische, Z. (1962). In *Methods in carbohydrate chemistry*, Vol. 1 (ed. R. L. Whistler and M. L. Wolfrom), p. 478. Academic Press Inc., New York, NY.

34. Blumenkrantz, N. J. and Asboe-Hansen, B. (1973). *Anal. Biochem.*, **54**, 484.
35. Lever, M. (1972). *Anal. Biochem.*, **47**, 273.
36. McNeil, M., Darvill, A. G., Åman, P., Franzén, L.-E., and Albersheim, P. (1982). *Meth. Enzymol.*, **83**, 3.
37. Dabrowski, J. (1989). *Meth. Enzymol.*, **179**, 122.
38. van Halbeek, H. (1990). In *Frontiers of NMR in molecular biology* (ed. D. Live, I. Armitage, and D. Patel), p. 195. Alan R. Liss Inc., New York, NY.
39. Dell, A., Carman, N. H., Tiller, P. R., and Thomas-Oates, J. (1988). *Biomed. Environ. Mass Spectrom.*, **16**, 19.
40. Albersheim, P., Nevins, D. J., English, P. D., and Karr, A. (1967). *Carbohydr. Res.*, **5**, 340.
41. Taylor, R. L. and Conrad, H. E. (1972). *Biochemistry*, **11**, 1383.
42. Chambers, R. E. and Clamp, J. R. (1971). *Biochem. J.*, **124**, 1009.
43. Waeghe, T. J., Darvill, A. G., McNeil, M., and Albersheim, P. (1983). *Carbohydr. Res.*, **123**, 281.
44. Björndal, H., Hellerqvist, C. G., Lindberg, B., and Svensson, S. (1970). *Angew. Chem., Int. Ed. Engl.*, **9**, 610.
45. Darvill, A. G., Roberts, D. P., and Hall, M. A. (1975). *J. Chromatogr.*, **115**, 319.
46. Spellman, M. W., McNeil, M., Darvill, A. G., Albersheim, P., and Dell, A. (1983). *Carbohydr. Res.*, **122**, 131.
47. Spellman, M. W., McNeil, M., Darvill, A. G., Albersheim, P., and Henrick, K. (1983). *Carbohydr. Res.*, **122**, 115.
48. York, W. S., Darvill, A. G., McNeil, M., and Albersheim, P. (1985). *Carbohydr. Res.*, **138**, 109.
49. Stevenson, T. T., Darvill, A. G., and Albersheim, P. (1988). *Carbohydr. Res.*, **179**, 269.
50. Chaplin, M. F. (1982). *Anal. Biochem.*, **123**, 336.
51. Sweeley, C. C., Bentley, R., Makita, M., and Wells, W. W. (1963). *J. Am. Chem. Soc.*, **85**, 2495.
52. York, W. S., Darvill, A. G., McNeil, M., Stevenson, T. T., and Albersheim, P. (1985). *Meth. Enzymol.*, **118**, 3.
53. Hakomori, S. (1964). *J. Biochem.*, **55**, 205.
54. Sandford, P. A. and Conrad, H. E. (1966). *Biochemistry*, **5**, 1508.
55. Mort, A. J., Parker, S., and Kuo, M.-S. (1983). *Anal. Biochem.*, **133**, 380.
56. York, W. S., Doubet, R. S., Darvill, A., and Albersheim, P. (1988). In *XIVth International Carbohydrate Symposium, Stockholm, Sweden, August 14–19* (Abstract).
57. Dell, A. (1987). *Adv. Carb. Chem. Biochem.*, **45**, 19.
58. Ayers, A. R., Ebel, J., Finelli, F., Berger, N., and Albersheim, P. (1976). *Plant Physiol.*, **57**, 751.
59. Davis, K. R., Darvill, A. G., and Albersheim, P. (1986). *Plant Mol. Biol.*, **6**, 23.
60. Cline, K. and Albersheim, P. (1981). *Plant Physiol.*, **68**, 221.
61. Torrey, J. G. and Shigemura, Y. (1957). *Am. J. Bot.*, **44**, 334.
62. Bauer, W. D., Talmadge, K. W., Keegstra, K., and Albersheim, P. (1973). *Plant Physiol.*, **51**, 174.
63. Stevenson, T. T., McNeil, M., Darvill, A. G., and Albersheim, P. (1986). *Plant Physiol.*, **80**, 1012.

64. Hahn, M. G. and Albersheim, P. (1978). *Plant Physiol.*, **62**, 107.
65. Keen, N. T. (1975). *Science*, **187**, 74.
66. Hadwiger, L. A. and Beckman, J. M. (1980). *Plant Physiol.*, **66**, 205.
67. Walker-Simmons, M., Hadwiger, L., and Ryan, C. A. (1983). *Biochem. Biophys. Res. Commun.*, **110**, 194.
68. Walker-Simmons, M., Jin, D., West, C. A., Hadwiger, L., and Ryan, C. A. (1984). *Plant Physiol.*, **76**, 833.
69. Walker-Simmons, M. and Ryan, C. A. (1984). *Plant Physiol.*, **76**, 787.
70. Köhle, H., Young, D. H., and Kauss, H. (1984). *Plant Sci. Lett.*, **33**, 221.
71. Köhle, H., Jeblick, W., Poten, F., Blasjek, W., and Kauss, H. (1985). *Plant Physiol.*, **77**, 544.
72. Roby, D., Gadelle, A., and Toppan, A. (1987). *Biochem. Biophys. Res. Commun.*, **143**, 885.
73. Ossowski, P., Pilotti, Å., Garegg, P. J., and Lindberg, B. (1984). *J. Biol. Chem.*, **259**, 11337.
74. Fügedi, P., Birberg, W., Garegg, P. J., and Pilotti, Å. (1987). *Carbohydr. Res.*, **164**, 297.
75. Sharp, J. K., Albersheim, P., Ossowski, P., Pilotti, Å., Garegg, P. J., and Lindberg, B. (1984). *J. Biol. Chem.*, **259**, 11341.
76. Ayers, A. R., Ebel, J., Valent, B., and Albersheim, P. (1976). *Plant Physiol.*, **57**, 760.
77. Heyraud, A. and Rinaudo, M. (1980). *J. Liquid Chromatog.*, **3**, 721.
78. Cheong, J.-J., Birberg, W., Fügedi, P., Pilotti, Å., Garegg, P., Hong, N., Ogawa, T., and Hahn, M. G. (1991). *The Plant Cell*, **3**, 127.
79. Davis, K. R., Darvill, A. G., Albersheim, P., and Dell, A. (1986). *Z. Naturforsch.*, **41c**, 39.
80. Robertsen, B. (1986). *Physiol. Mol. Plant Pathol.*, **28**, 137.
81. Bishop, P. D., Makus, D. J., Pearce, G., and Ryan, C. A. (1981). *Proc. Natl. Acad. Sci. (USA)*, **78**, 3536.
82. Bishop, P. D., Pearce, G., Bryant, J. E., and Ryan, C. A. (1984). *J. Biol. Chem.*, **259**, 13172.
83. Bradford, M. (1976). *Anal. Biochem.*, **72**, 248.
84. Cervone, F., De Lorenzo, G., Degrà, L., and Salvi, G. (1987). *Plant Physiol.*, **85**, 626.
85. Townsend, R. R., Hardy, M. R., and Lee, Y. C. (1989). *Meth. Enzymol.*, **179**, 65.
86. Hotchkiss, A. T., Jr. and Hicks, K. B. (1990). *Anal. Biochem.*, **184**, 200.
87. Nakahara, Y. and Ogawa, T. (1987). *Tetrahedr. Lett.*, **28**, 2731.
88. Nakahara, Y. and Ogawa, T. (1987). *Carbohydr. Res.*, **167**, c1.
89. Nakahara, Y. and Ogawa, T. (1989). *Tetrahedr. Lett.*, **30**, 87.
90. Nakahara, Y. and Ogawa, T. (1989). *Carbohydr. Res.*, **194**, 95.
91. Debbage, P. L., Lange, W., Hellmann, T., and Gabius, H.-J. (1988). *J. Histochem. Cytochem.*, **36**, 1097.
92. Glabe, C. G., Harty, P. K., and Rosen, S. D. (1983). *Anal. Biochem.*, **120**, 287.
93. Ogamo, A., Matsuzaki, K., Uchiyama, H., and Nagasawa, K. (1982). *Carbohydr. Res.*, **105**, 69.
94. Horn, M. A., Heinstein, P. F., and Low, P. S. (1989). *Plant Cell.*, **1**, 1003.
95. Schmidt, W. E. and Ebel, J. (1987). *Proc. Natl. Acad. Sci. (USA)*, **84**, 4117.

96. Wang, W. T., LeDonne, N. C., Jr., Ackerman, B., and Sweeley, C. C. (1984). *Anal. Biochem.*, **141,** 366.

97. Mester, L. and El Khadem, H. S. (1980). In *The Carbohydrates*, Vol. IB, *Chemistry and Biochemistry* (ed. W. Pigman, D. Horton, and J. D. Wander), p. 929. Academic Press, New York.

98. Greenwood, F. C., Hunter, W. M., and Glover, J. S. (1963). *Biochem J.*, **89,** 114.

99. Hahlbrock, K. and Scheel, D. (1989). *Annu. Rev. Plant Physiol. Plant Mol. Biol.*, **40,** 347.

100. Lamb, C. J., Lawton, M. A., Dron, M., and Dixon, R. A. (1989). *Cell*, **56,** 215.

101. Cramer, C. L., Ryder, T. B., Bell, J. N., and Lamb, C. J. (1985). *Science*, **227,** 1240.

102. Hamdan, M. A. M. S. and Dixon, R. A. (1986). *Physiol. Mol. Plant Pathol.*, **28,** 329.

103. Hamdan, M. A. M. S. and Dixon, R. A. (1987). *Physiol. Mol. Plant Pathol.*, **31,** 105.

9

Xyloglucan oligosaccharides

1. Introduction

Xyloglucan (XG) is a polysaccharide found in the primary cell walls of higher plants. Its main role is architectural, probably limiting cell expansion (1–3). In addition, xyloglucan oligosaccharides (XGOs), enzymically released from XG, are potent plant growth regulators at nanomolar to micromolar concentrations, i.e. they are 'oligosaccharins' (4). They are not elicitors.

XGOs are formed from XG by the action of cellulase (EC 3.2.1.4), which hydrolyses the polysaccharide backbone at Glc residues that lack side-chains. The major penta- to decasaccharides produced are composed of Glc + Xyl ± Gal ± Fuc (1) (*Figure 1, Table 1*).

The first XGO found to be biologically active was XG9: at 1–10 nM it antagonizes the growth-promoting effect of 1 µM 2,4-dichlorophenoxyacetic acid (2,4-D) (an auxin) on pea stem segments (5, 6). The effect of nanomolar

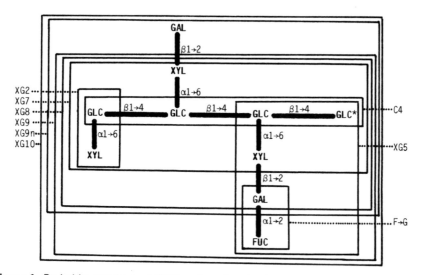

Figure 1. Probable structure of XGOs referred to in the text. C4, cellotetraose; F→G, fucosylgalactose (H-disaccharide); *, reducing terminus.

Table 1. Data on xyloglucan oligosaccharides

Oligosaccharide (see *Figure 1*)	Sugar composition (Glc:Xyl:Gal:Fuc)	Most convenient source[a,b] (method of purification)	Biological activity		
			Anti-auxin (nM)	Anti-auxin (μM)	Auxin mimic (μM)
XG10	4:3:2:1	1° walls; cellulase; GPC	−		
XG9	4:3:1:1	1° walls; cellulase; GPC	+	−	+
XG9n	4:3:2:0	Nasturtium; cellulase; GPC	−	−	+
XG8	4:3:1:0	Nasturtium; cellulase; GPC	−	−	+
XG7	4:3:0:0	1° walls; cellulase; GPC	−	−	+
XG5	2:1:1:1	1° walls; cellulase; GPC	+	+	
XG2	1:1:0:0	Nasturtium; Driselase; PC			
Cellobiose	2:0:0:0	Available commercially			
Cellotriose	3:0:0:0	Available commercially			
Fuc→Gal	0:0:1:1	Chemical synthesis (21)			
Fucosyl-lactose	1:0:1:1	Human milk and available commercially	+	+	

[a] Starting materials: 1° walls = XG isolated from primary cell walls of cell cultures or legume seedlings; Nasturtium = XG isolated from nasturtium seeds.
[b] Methods of fragmentation (of XG into XGOs): Cellulase = *Trichoderma* cellulase; Driselase = a mixture of enzymes from *Irpex lacteus*.

concentrations of XG9 is mimicked by XG5 and by 2′-fucosyl-lactose at similar concentrations (7), but XGOs that lack an α-fucose residue are not anti-auxins (8). XG9 also antagonizes H^+-induced growth in pea stem segments (Lorences and Fry, in preparation).

At 0.1–1 μM, XG9 loses its anti-auxin effect (5, 6), possibly because it gains an auxin-like effect (9). XG5 and fucosyl-lactose do not do this (7). XG7, XG8, XG9n and XG9 show auxin-like activity at 1 μM (9)—clearly a fucose-independent effect. The auxin-mimicking effect of XGOs may be related to their ability (9, 10) to 'activate' cellulase *in vitro*. Complex feedback control loops can be envisaged.

This chapter presents methods for the preparation, purification, characterization, radioactive labelling, and bioassay of XGOs. Further details of many of the procedures are given elsewhere (11).

2. Isolation of xyloglucan

Protocol 1. Preparation of fucose-rich XG from primary cell walls

A. *To remove starch*

1. Collect suspension-cultured Dicot cells (for example *Rosa* or *Spinacia*) on muslin, and transfer 20 g fresh weight into 180 ml of dimethylsulphoxide (DMSO).

2. Sonicate vigorously, for example with an MSE 'Soniprep' fitted with a titanium probe operated at 23 kHz and 24 μm amplitude.

3. Stir at ~25°C overnight to solubilize starch.

B. *To remove protein*

1. Filter the de-starched cells through sintered glass and wash with:
- 100 ml 90% DMSO
- 100 ml water
- 100 ml of PAW (a mixture of 100 ml acetic acid, 250 ml 80% (v/v) aq. phenol, and 50 ml water)

2. Resuspend the residue in a further 200 ml of PAW and stir overnight at ~25°C to solubilize proteins.

3. Filter on sintered glass and wash the residue (cell walls) with 100 ml PAW followed by 1 litre H_2O.

C. *To solubilize XG*

1. Suspend the walls in 25 ml of 24% NaOH/1% $NaBH_4$ and stir at ~25°C for 1–2 days.

2. Centrifuge.[a]

3. Adjust the supernatant to pH 5 with acetic acid.

4. Repeat the centrifugation and dialyse the supernatant (which contains XG) against several changes of water (each approx. 1 l).

5. Freeze-dry.

[a] 2500 g for 10 min at room temperature is adequate for all applications described in this chapter.

Protocol 2. Preparation of XG (fucose-less) (12) from nasturtium seed

1. Vigorously homogenize 200 dry nasturtium seeds (*Tropaeolum majus* L.) in 100 ml water in a food-mixer and add the homogenate to 900 ml 8% NaOH/0.05% $NaBH_4$.

2. Incubate at 100°C for 1 h.

3. Centrifuge, 2500 g for 10 min.

4. Add 2 volumes of ethanol–acetic acid (10:1) to the supernatant and stand for 30 min, then centrifuge as above.

5. Re-dissolve the pellet in 800 ml water + 10 ml Fehling's solution B (35% potassium sodium tartrate tetrahydrate in 25% KOH) by heating at 100°C for 30 min.

Protocol 2. *Continued*

6. Cool and pellet any insoluble matter. Add 100 ml of 7.5% $CuSO_4 \cdot 5H_2O$ to the solution and pellet the blue complex.

7. Re-dissolve the pellet in the minimum volume of acetic acid–water (1:4) and then repeat step (3).[a]

8. Wash the final precipitate several times with 50% ethanol followed by absolute ethanol.

9. Air-dry.

[a] Repetition of step 7 yields a white (largely copper-free) XG sample. If the blue colour proves difficult to eliminate, wash the precipitate with a saturated, neutral solution of EDTA in 50% ethanol.

---

3. Preparation of xyloglucan oligosaccharides (XGOs)

Protocol 3. Cellulase-catalysed hydrolysis of XG to yield XGOs (6, 11)

1. Dissolve 100 mg of XG in 10 ml of 0.1% NaOH and adjust the pH to 4.7 with acetic acid.

2. Add 10 mg of *Trichoderma viride* cellulase (Sigma Chemical Co., Type V, C 2274) and incubate at ~25°C for 3–4 h.[a]

3. Pellet any insoluble matter, and store the supernatant frozen at −20°C (if not to be used at once).[b]

[a] Primary wall XG is hydrolysed to XG7, XG9, and a little XG5 and XG10; nasturtium XG yields XG8 (two isomers), XG9n and some XG7.

[b] The cellulase preparation contains slight α-fucosidase activity, but <10% of the fucose residues of XG are lost under the conditions described.

If the substrate XG has not been subjected to alkali (for example if it is not extracted from the cell walls), the digestion-products may include naturally-acetylated derivatives of the XGOs mentioned above (13, 14). Acetylated derivatives behave like the corresponding simple XGOs on gel-permeation chromatography (see Section 4.1), but have higher R_f-values on paper chromatography (PC) (see Section 4.2).

Protocol 4. Driselase-catalysed hydrolysis of XG to isoprimeverose (11)

The diagnostic disaccharide XG2 (isoprimeverose) is released in high yield by digestion as described opposite.

1. Digest 0.1 g XG in 5 ml buffer (acetic acid–C_5H_5N–H_2O, 1:1:100 (v/v/v), pH 4.7) containing 25 mg of crude Driselase[a] (Sigma, D9515).

2. Add 0.05% chlorbutol to suppress microbial growth.

3. Incubate at 25–37°C overnight.

---

[a] Driselase lacks α-xylosidase while possessing all the other enzyme activities needed to digest XG completely.

---

4. Separation of xyloglucan oligosaccharides

Xyloglucan oligosaccharides can be separated by gel-permeation chromatography (GPC) (*Protocol 5*); paper chromatography (PC) (*Protocol 6*); or by high-pressure liquid chromatography (HPLC) (Section 4.3).

4.1 Gel-permeation chromatography (GPC) of XGOs

Protocol 5. Separation of XGOs by GPC

1. Pass 5 ml of the cellulase-digestion products through a column (100×1.5 cm) of Bio-Gel P-2 equilibrated with buffer (acetic acid–C_5H_5N–H_2O, 1:1:100 (v/v/v), pH 4.7).

2. Run the column slowly (< 10 ml/h); resolution may be improved by jacketing the column at 70°C.

3. Collect 50 4 ml fractions; assay 10 μl of each for hexose (15).

4. Pool the fractions of interest and freeze-dry to remove the volatile buffer.[a]

---

[a] Approximate k_{av} values [= elution-volumes relative to dextran ($k_{av} = 0$) and glucose ($k_{av} = 1$)] are: XG5, 0.65; XG7, 0.52; XG8, 0.46; XG9n 0.40; XG10, 0.36. A useful marker is maltoheptaose (k_{av}, 0.54).

4.2 Paper chromatography (PC) of XGOs

Small oligosaccharides, for example XG2,[a] are better purified by preparative PC[b] than by GPC.

---

[a] A useful marker is sucrose, which approximately co-chromatographs with XG2.

[b] PC can also be used for the separation of some of the larger XGOs. Solvents include butan-1-ol–C_5H_5N–H_2O (4:3:4) and ethylacetate–acetic acid–H_2O (10:5:6), run for 24 h by the descending method. Staining is with $AgNO_3$ (11).

Protocol 6. Separation of XGOs by PC

1. Load a Driselase-digest on to Whatman 3MM paper at 1 ml/20 cm of paper.
2. Develop the chromatogram for 40 h by the descending method, in ethyl acetate–C_5H_5N/H_2O (8:2:1) (11).
3. Stain with aniline hydrogen phthalate.

Table 2. Paper chromatographic data for Driselase digestion products and sucrose in ethyl acetate–$C_5H_5N–H_2O$
(8:2:1)

Compound	R_{Glc}[a]	Colour reaction with[b]	
		AHP	AgNO$_3$
Cellobiose[c]	0.26	brown	brown
XG2	0.50	brown	brown
Sucrose	0.50	none	faint
Xylobiose[d]	0.62	red	brown
D-Galactose	0.84	brown	brown
D-Glucose[d]	1.00	brown	brown
D-Mannose[c]	1.31	brown	faint
L-Arabinose	1.56	red	brown
L-Fucose	1.94	brown	brown
D-Xylose[d]	2.00	red	brown

[a] R_{Glc} = chromatographic mobility relative to glucose.
[b] AHP = aniline hydrogen phthalate; for details of this and the AgNO$_3$ stain, see reference 11.
[c] May be produced in small amounts by autolysis of Driselase.
[d] May be produced (by partial hydrolysis of paper) while the Driselase solution is being dried on to chromatography paper; can be prevented by inactivation of the Driselase (e.g. boiling for 1 min or addition of cold formic acid to 2 M) prior to sample application.

4.3 High-pressure liquid chromatography (HPLC) of XGOs

Two HPLC systems have proved particularly valuable. System *i* gives excellent resolution of XGOs according to size, for example separating XG7, XG8, and XG9 by wide margins. System *ii* barely separates XG7 from XG9, but resolves XG8 into two isomers and readily discriminates between XG9 and XG9n (*Table 3*).

Table 3. HPLC data for cellulase digestion products

Compound	Retention time relative to that of XG7 [a] on	
	Amino Spheri-5	CarboPac PA1
XG7	1.00	1.00
XG8	1.26	1.13 + 1.21 [b]
XG9n	1.57	1.34
XG9	1.33	1.04
XG10	1.66	1.14

[a] Approximate retention time of XG7 = 30.5 min on Amino Spheri-5 and 12.8 min on CarboPac PA1. For chromatographic conditions, see text.
[b] Two isomers resolved.

i. Amino column

HPLC on a column (220 × 4.6 mm) of NH_2-substituted silica (Brownlee Amino Spheri-5); isocratic elution using 55% CH_3CN at 0.5 ml/min; refractive index monitoring

ii. Pellicular column

HPLC on a column (250 × 4 mm) of Dionex CarboPac-PA1, a pellicular anion-exchange resin; gradient elution using 100 mM NaOH/50 mM sodium acetate → 100 mM NaOH/100 mM sodium acetate in 20 min at 1 ml/min; monitoring with a pulsed amperometric detector (gold electrode). If the eluted oligosaccharides are to be investigated further, it is important to remove the NaOH from the eluate quickly because oligosaccharides are unstable at high pH.

5. Characterization of XGOs

Monosaccharide composition is determined by chromatography (11) after hydrolysis in 2 M trifluoroacetic acid at 120°C for 1 h. Molecular weight is estimated by GPC (see *Protocol 5*) on Bio-Gel P-2 calibrated with malto-oligosaccharides. A rigorous method for the determination *ab initio* of the molecular structures is available (16). Two simpler methods, suitable for defining particular key attributes of XGOs, are described below.

5.1 Non-reducing terminus

Xylosylated and non-xylosylated Glc residues are released from XGOs by Driselase as XG2 and Glc, respectively (see *Protocol 4*). The XG2:Glc ratio, determined by PC (*Protocol 6*) indicates the proportion of Glc residues in the XGO that were xylosylated. (The production of a small amount of Glc by autolysis of Driselase must be taken into account.) Furthermore, an unusual β-glucosidase ('xyloglucosidase') can, if pure, be used to determine the sequence of substituents along the backbone of the XGO: the enzyme attacks the oligosaccharide, starting from the non-reducing end, and liberates XG2 or Glc units sequentially until it meet a Glc residue which carries a longer side-chain than Xyl→, for example Gal→Xyl→ or Fuc→Gal→Xyl→, where the enzyme stops (17).

5.2 Reducing terminus

The reducing end of an XGO is usually Glc, linked only through its O-4 position; however, at least one naturally-occurring XGO has a xylosylated Glc unit as its reducing end (18). The presence of a Xyl residue attached to the reducing terminal Glc unit can be tested for by Driselase-digestion after $NaBH_4$-reduction—the expected product is Xyl→glucitol instead of glucitol; these can be separated from each other and from Glc and XG2 by a two-step scheme (18) involving:

- PC in butan-1-ol–acetic acid–water (12:3:5) to give two zones:

XG2	$R_F = 0.17$	zone 1
Xyl→glucitol	$R_F \simeq 0.17$	
Glc	$R_F = 0.28$	zone 2
Glucitol	$R_F = 0.30$	

followed by:

- paper electrophoresis in 2% $Na_2MoO_4 \cdot 2H_2O$ (pH adjusted to 5.0 with H_2SO_4) with bromophenol blue present as an internal marker:

XG2	$m_{\text{bromophenyl blue}} \simeq 0.0$
Xyl → glucitol	$m_{\text{bromophenyl blue}} \simeq 1.3$
Glc	$m_{\text{bromophenyl blue}} \simeq 0.0$
Glucitol	$m_{\text{bromophenyl blue}} \simeq 1.7$

6. Radioactive labelling of XGOs

[³H]- and [¹⁴C]oligosaccharides are valuable in studies of the transport (19), binding (20), and metabolic fate (14) of oligosaccharins. One excellent method of labelling any reducing oligosaccharide, including XGOs, is tritiation by catalytic exchange with ³H₂ (11). The structure of the XGO is not altered by the process, and the ³H is incorporated at position 1 of the reducing terminus.

Labelling can also be conveniently carried out *in vivo* by aseptic feeding of radioactive sugars to cultured plant cells (11). The period of labelling is typically 2–7 days. Useful precursors and the sugar residues of XG labelled by them include:

D-[U-^{14}C]glucose	Glc, Xyl, Gal, Fuc (and any Ara)
L-[1-^{3}H]arabinose	Xyl (and any Ara)
L-[1-^{3}H]fucose	Fuc

XG is extracted from the labelled cells and digested with cellulase to yield XGOs in the normal way (see Section 4).

7. Bioassay of XGOs for anti-auxin and auxin-like activity

Protocol 7. Bioassay for anti-auxin and auxin-like activity (6, 9)

1. Soak pea seeds (*Pisum sativum* L., cv. Alaska) in running tap water for 6–8 h.

2. Sow in 5 cm deep moist-vermiculite and incubate in the dark at 25 °C. Conduct the following operations in dim red-light at 25 °C.

3. After 7–8 days, when the third internode is 1–3 cm long, cut a straight 6 mm segment from each third internode, starting 2 mm below the hook.

4. Shake the segments gently in water for 30 min, and then in bioassay medium (1% sucrose, 5 mM KH_2PO_4, 0.02% K^+ benzyl penicillin, pH adjusted to 6.1 with NaOH) for a further 90 min.

5. Transfer eight segments into each 5 cm Petri dish containing 5 ml of the bioassay medium and add the XGOs, 1 µM 2,4-D, both, or neither.[a]

6. Incubate in the dark with gentle shaking for 18 h.

7. Measure the lengths of the segments (working in the light); this is facilitated by projecting their image on to a screen.

[a] A range of XGO concentrations centred on 1 nM is suitable for demonstration of anti-auxin activity and a range centred on 1 µM for demonstration of auxin-like activity.

Acknowledgements

I am very grateful to Dr G. J. McDougall for valuable discussions and to Mrs J. G. Miller for technical assistance. Previously unpublished work was supported by the Agricultural and Food Research Council.

References[a]

1. Fry, S. C. (1989). *J. Exp. Bot.*, **40**, 1.
2. Hayashi, T. (1989). *Annu. Rev. Plant Physiol. Plant Mol. Biol.*, **40**, 139.
3. Fry, S. C. (1989). *Physiol. Plant.*, **75**, 532.
4. Darvill, A. G. and Albersheim, P. (1985). *Sci. Am.*, **Sept 1985**, 44.
5. York, W. S., Darvill, A. G., and Albersheim, P. (1984). *Plant Physiol.*, **75**, 295.
6. McDougall, G. J. and Fry, S. C. (1988). *Planta*, **175**, 412.
7. McDougall, G. J. and Fry, S. C. (1989). *J. Exp. Bot.*, **40**, 233.
8. McDougall, G. J. and Fry, S. C. (1989). *Plant Physiol.*, **89**, 883.
9. McDougall, G. J. and Fry, S. C. (1990). *Plant Physiol.*, **93**, 1042.
10. Farkas, V. and Maclachlan, G. (1988). *Carbohydr. Res.*, **184**, 213.
11. Fry, S. C. (1988). *The growing plant cell wall: chemical and metabolic analysis.* Longman, London; and Wiley, NY.
12. Edwards, M., Dea, I. C. M., Bulpin, P. V., and Reid, J. S. G. (1985). *Planta*, **163**, 133.
13. Kiefer, L. L., York, W. S., Darvill, A. G., and Albersheim, P. (1989). *Phytochemistry*, **28**, 2105.
14. Baydoun, E. A.-H. and Fry, S. C. (1989). *J. Plant Physiol.*, **134**, 453.
15. Dische, Z. (1962). In *Methods in carbohydrate chemistry* (ed. R. L. Whistler and M. L. Wolfrom), Vol. 1, p. 475. Academic Press, NY.
16. Valent, B. S., Darvill, A. G., McNeil, M., Robertsen, B. K., and Albersheim, P. (1980). *Carbohydr. Res.*, **79**, 165.
17. Matsushita, J., Kato, Y., and Matsuda, K. (1985). *Agric. Biol. Chem.*, **49**, 1533.
18. Fry, S. C. (1986). *Planta*, **169**, 443.
19. Baydoun, E. A.-H. and Fry, S. C. (1985). *Planta*, **165**, 269.
20. Schmidt, W. E. and Ebel, J. (1987). *Proc. Natl Acad. Sci. (USA)*, **84**, 4117.
21. Wegmann, B. and Schmidt, R. R. (1988). *Carbohydr. Res.*, **184**, 254.

[a] Literature search completed November 1989.

10

Analysis of calcium involvement in host–pathogen interactions

BJØRN K. DRØBAK, DOUGLAS S. BUSH, RUSSELL L. JONES,
ALAN P. DAWSON, and IAN B. FERGUSON

1. Introduction

Calcium has often been associated with pathogenic infections in a variety of plant tissues; but the processes involved are still poorly understood. However, present evidence suggests that Ca^{2+} is likely to be involved in three major areas of the host–pathogen interaction (see Sections 1.1–1.3).

1.1 Cell wall physiology

Maintenance of relatively high concentrations of Ca^{2+} in the plant cell wall may reduce cell wall breakdown induced by fungal attack. For instance, in bulky tissues such as potato tubers and carrot roots, increased concentrations of Ca^{2+} have been associated with reduced levels of infection of pathogens such as *Sclerotium rolfsii* (1). The soft-rot symptoms found in many storage organs are associated with pectinases, whose activity may be modified by the Ca^{2+} status of the cell walls. The role of Ca^{2+} generally in cell wall breakdown has still not been fully elucidated (2, 3), and there are difficulties in measuring fractions of Ca^{2+} in the cell wall which have any physiological meaning.

1.2 Ca^{2+}-dependence of enzymes

The pectic lyases of *Erwinia* sp. are Ca^{2+} dependent (4). Phospholipases A, C, and D, which are all involved in the degradation of membrane lipids, all have degrees of Ca^{2+}-dependence, which may influence the effectiveness of such pathogens as the blast fungus. The other major groups of enzymes which show Ca^{2+}-dependence are the Ca^{2+}-ATPases and protein kinases, which in many cases are additionally stimulated by the Ca^{2+}-binding protein calmodulin.

1.3 Signal transduction

A role for Ca^{2+} as a second messenger is well-established in animal cells, and is beginning to be characterized in plant cells (5, 6). As in animal cells, cytoplasmic Ca^{2+} concentrations are very low (i.e. nanomolar) in unstimulated cells but increase rapidly in response to stimuli. The regulatory activity of Ca^{2+} may have importance in the mechanism of elicitor action. Elicitors such as chitinase activate β-1,2-glucan synthase resulting in callose production which is important in resistance of plants to fungal attack (7). This process appears to be dependent on an influx of Ca^{2+} into the cell. Phytoalexin production in carrot cells (8), soybean cells (9), and potato tubers (10) has been shown to be dependent on the maintenance of an external Ca^{2+} supply. Alteration of transmembrane fluxes of ions, such as Ca^{2+} resulting from wounding, pathogen attack, and stress, has led to suggestions of specific transmembrane proteins combining the properties of receptor proteins and ion channels (11). Calcium has long been ascribed a role in maintaining membrane structure and function (2). The performance of plant cell membranes under the stress of pathogen attack may depend on the Ca^{2+} status of the membranes and the availability of Ca^{2+} inside and outside the cell.

1.4 Ca^{2+}-associated physiological disorders

These mostly result from localized deficiencies arising from transport problems and the supply of Ca^{2+} to specific organs, such as the fruit. Where organ growth is largely dependent on a phloem supply, even for water, Ca^{2+} concentrations may become relatively low and result in, for example, bitter pit in apple fruit (12), where localized cell collapse occurs in fruit flesh. As fruit ripening proceeds (usually post-harvest), sites low in Ca^{2+} may senesce at a greater rate, resulting in the characteristic tissue pitting. These problems and other disorders, such as blossom-end rot in tomatoes, highlight a need for both measurement of tissue Ca^{2+} and of transport within the plant. An appreciation of both these aspects is necessary for establishing a role for Ca^{2+} in pathogen attack.

Measurement of Ca^{2+} in plant tissues in relation to plant pathogen attack must, therefore, be approached at several levels. Whole tissue analysis is necessary for determining the mineral status of the tissue. However, the unequal distribution of Ca^{2+} in tissues means that analysis of the separate parts of the tissue are required if contents are to be correlated with resistance to pathogen attack. It is useful to estimate extracellular (both soluble and cell wall-bound) versus intracellular Ca^{2+}, and the quantities transported. At the cellular level, concentrations in the cytosol and in organelles should be assessed since these are critical to signal transduction. Furthermore, fluxes across membranes and between compartments are essential to a response system and need to be monitored. Selected techniques involved in such measurements in plant tissues are outlined in this chapter.

2. Problems associated with investigations of the physiological roles of Ca^{2+}

Some of the problems associated with cellular Ca^{2+} research are caused by the ubiquity of Ca^{2+} and the specific characteristics of its distribution within cells.

As the Ca^{2+} concentrations in various cellular compartments show great variation, from nanomolar in the cytosol to millimolar in the vacuole, it is essential that media used for investigations of specific Ca^{2+} related/induced reactions closely mimic the *in vivo* concentrations. Thus, studies of reactions likely to occur in the cytosol, must be performed in a very low Ca^{2+} environment. Many sources of Ca^{2+} contamination exist in the average laboratory, so even when no exogenous Ca^{2+} is added to a buffer solution the final concentration of Ca^{2+} is, in most instances, at least in the micromolar region.

Several steps can be taken to reduce Ca^{2+} levels in buffers. The three most common sources of Ca^{2+} contamination are found in chemicals, glassware, and water. The chemicals used for low Ca^{2+} buffers should be of high purity—most suppliers give information about the Ca^{2+} content of high-grade inorganic chemicals. Glassware should, as far as possible, be avoided in low-Ca^{2+} studies, as large amounts of exchangeable Ca^{2+} are present in many types of glass, and whenever possible plastic containers should be used. If it is impossible to avoid the use of glass the problem can be reduced somewhat by rinsing several times in strong ethyleneglycol(aminoethyl ether)tetraacetic acid (EGTA) solutions (see below) before use.

Water is, in many cases, the main source of Ca^{2+} contamination, so the highest quality water available must be used. Before starting any experiment it is advisable to determine the Ca^{2+} levels in the water. High Ca^{2+} concentrations in water can be removed by ion-exchange resins such as Chelex-100.

Even when adequate precautions are taken to avoid Ca^{2+} contamination from external sources, it is virtually impossible to adjust Ca^{2+} concentrations (or more specifically: *activities*) in the low/sub-micromolar range accurately without the use of Ca^{2+} buffers. The most commonly used Ca^{2+} buffers are the tetracarboxylic relatives of EDTA, namely BAPTA, EGTA, and Quin-2. It is important to realize that the apparent affinity of such buffers for Ca^{2+} is strongly pH-dependent. Although BAPTA has the advantage that its apparent affinity for Ca^{2+} is relatively constant above pH 7.0, the affinity of the other tetracarboxylic chelators for Ca^{2+} varies greatly in the physiological pH range, due to the fact that at near neutral pH, they exist as several chemical species that differ in their degree of protonation. In the case of EGTA, for example, the predominant species at neutral pH will be doubly protonated, i.e. H_2EGTA, but $HEGTA$, H_3EGTA and H_4EGTA also exist. Only the unprotonated species or the singly protonated species complex Ca^{2+} to any significant extent. Solution pH influences the relative amounts of these Ca^{2+}-binding chelator species, thereby influencing the apparent affinity of the overall chelator solution for Ca^{2+}. Conversely, Ca^{2+} also influences the concentration of the protonated species, with the net effect that the addition of Ca^{2+} decreases the total concentration of protonated species. It is, therefore, important when using these compounds that strong pH buffers are included to absorb the protons released by the Ca^{2+} chelation. It should also be remembered that certain solutes of biological interest, such as ATP, chelate

Ca^{2+} and thus may influence the Ca^{2+} concentration. The preparation of solutions containing low and/or fixed amounts of Ca^{2+} is described in *Table 1*.

When standard Ca^{2+} solutions are prepared it is important to accurately determine the concentration of the Ca^{2+} chelator. Because the purity and hydration of chelators vary, the stock chelator solution should be titrated against standard Ca^{2+} solutions, as described in *Protocol 1*.

Protocol 1. Determination of chelator concentration using the method of Bers (14)

1. Prepare a 1 M calcium solution from oven-dried $CaCO_3$. Dissolve dried $CaCO_3$ by adding 6 M HCl to a final pH of 7.
2. Prepare a series of standard Ca^{2+} solutions in the range where the response of the electrode is Nernstian (usually about 0.1–10 mM), and determine the response of the Ca^{2+} electrode in the absence of added chelator, for example BAPTA (*Figure 1*).
3. Prepare a solution with a desired nominal chelator concentration (for example 1 mM BAPTA, *Figure 1*). Titratre this solution with the Ca^{2+} standard and record the electrode response.
4. Calculate the free Ca^{2+} concentration from the electrode response after each addition of Ca^{2+}, using the relationship between electrode response and $[Ca^{2+}]$ determined in step 2. The amount of chelated Ca is calculated as the difference between total Ca and Ca^{2+} (*Figure 1*).
5. Plot the ratio of chelated Ca/free Ca^{2+} as a function of the chelated Ca concentration. The *x*-axis intercept gives the chelator concentration and the slope the apparent binding constant (*Figure 1*).

In complex solutions, 'setting' the level of Ca^{2+} is greatly aided by the use of computer programs (for example that shown in Section 9) that allow the free Ca^{2+} concentration to be calculated.

Computer programs can obviously only be used to calculate the Ca^{2+} concentration in solutions if the exact composition of the solutions is known. The program (Section 9) was written in its original form by Westall (15) and has been modified for use on IBM PC-type computers with the IBM version 3.0 of BASIC. This program requires three types of information input as data statements on lines 9000 to 9999. An example of the output from the program is shown in Section 9 of this chapter, which also contains lists for each species, the stoichiometry of the components and their stability constant (both input by the user), and the calculated concentration of each component of the solution expressed in molarity and log molarity. Also listed are the differences between the input concentrations and the computed concentrations of each species, which provides an approximation of the error in computation.

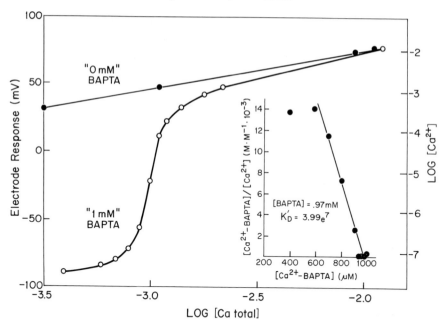

Figure 1. Potentiometric titration of a nominal 1 mM BAPTA solution and Scatchard plot (inset) showing the actual BAPTA concentration and its apparent binding constant using the method of Bers (14). The electrode response was measured over a limited range of total Ca in the absence of BAPTA (0 mM BAPTA, closed circle) and in the presence of 1 mM BAPTA (1 mM BAPTA, open circle). The free Ca^{2+} concentration in the presence of BAPTA was calculated from the electrode response in its absence (right-hand ordinate). Ca^{2+}-BAPTA was calculated as the difference between total Ca (abscissa) and the calculated free Ca^{2+} concentrations. The Scatchard plot (inset) uses the calculated values of bound (Ca^{2+}-BAPTA) and free Ca^{2+}, (Ca^{2+}) plotted as a function of the bound Ca^{2+} (Ca^{2+}-BAPTA) to determine the actual concentration of BAPTA (intercept of the linear portion of the curve) and the apparent binding constant (slope of the linear portion of the curve).

Finally, the commonly used Ca^{2+} isotope, $^{45}Ca^{2+}$, has, from an experimental viewpoint, several virtues, such as a convenient half-life (164 days) and a β-energy (max) of 257 keV, which is useful for scintillation counting and autoradiography, but it is highly toxic in humans.

3. Analysis of Ca^{2+} in plant tissues

The preferred method for analysing Ca^{2+} in whole plant tissue is by wet digestion and atomic absorption spectrometry (17). The basic method is outlined in *Protocol 2*. A number of points are highlighted in Sections 3.1 to 3.4.

Protocol 2. Analysis of Ca^{2+} in plant tissue[a]

1. Place 1–2 g fresh weight (100–200 mg dry weight) of tissue in a 30 ml Kjeldahl digestion flask, together with a few boiling chips and 2 ml conc. HNO_3.

2. Heat slowly with increasing temperature until all visible organic material has been digested.

3. Add 0.7 ml 70% (v/v) $HClO_4$ and increase the heat, **CAUTION**, see Section 3.2. The HNO_3 will be driven off as brown fumes of NO_2, and the volume of the liquid will be reduced. Digestion will be completed when the solution boils vigorously with the emission of dense white fumes. Turn the temperature down a little to allow the solution to reflux for 20–30 min.

4. Make up the digestate (approx. 0.5 ml) to a given volume (usually 20 ml) in a volumetric flask with $LaCl_3$ (2000 p.p.m. La, 5.34 g/l).

5. Aspirate digestate into an atomic absorption spectrophotometer for measurement of Ca^{2+}.

[a] See Section 3.2.

3.1 Sampling

Larger quantities of tissue can be used (up to 5 g fresh weight) by increasing the volume of HNO_3 and allowing a longer pre-digestion. Sub-samples will need to be taken if more than 5 g are used. This can be done by drying the tissue and milling into a powder. However, oven-drying fruit tissue poses problems because of the high sugar content, and it is better to grind tissue in liquid N_2 and freeze-dry the powders to obtain homogeneous samples. Homogeneity is particularly important in fruit tissues where skin, flesh, and seeds are included, since all these differ markedly in Ca^{2+} content. To relate finite tissue Ca^{2+}-content to specific sites of infection or disorder incidence, it is best to take small plugs of tissue in the outer cortex of the flesh (18). Segments of fruit will allow a better assessment of whole fruit content, as will homogenates of fruit blended in known volumes of water (19).

3.2 Digestion

CAUTION: perchloric acid is a **dangerous** oxidant. If tissue is heated in $HClO_4$ alone, it may explode violently. Pre-digestion with nitric acid is **essential** for safe digestion. Sulphuric acid can be used but may result in calcium sulphate precipitates. If the pre-digestion is inadequate, the digestate will turn brown or black after the first part of the perchloric stage when the nitric acid has been driven off. This colour will disappear on refluxing but it should be avoided where possible, since if too much organic matter remains at the time of $HClO_4$ addition, the digestate may carbonize, dry, and burn. It is also essential to have very good venting of fume hoods where $HClO_4$ is used, since absorption into woodwork, or accumulation in unvented sites raises the risk of explosion. Despite the safety requirements, $HClO_4$ is by far the most effective oxidant, and is the acid solvent with least interference for atomic absorption spectrophotometry.

Dry ashing has been used for Ca^{2+} analysis (17), involving acid digestion and ashing in a furnace. It is less convenient for large sample numbers, and usually requires filtration to provide a clean digestate for analysis. This raises the risk of loss of Ca^{2+} and is best avoided.

Boiling chips should be acid-washed, since they often provide high background for Ca^{2+}. Blanks should always be run through the digestion procedure. Strontium and caesium chlorides can also be used in making the digestate up to volume, since, they, along with $LaCl_3$, provide protection from the formation of refractory calcium oxides. At the end of the digestion, $KClO_4$ crystals may form if the K content of the tissue is high. These will dissolve in $LaCl_3$ as the digestate is made up to volume.

3.3 Analysis

The digestate prepared as above can be used directly for atomic absorption spectrophotometry. For fruit tissues, the quantities given in *Protocol 2* provide a Ca concentration in the digestate usually between 0.5–10 p.p.m. This is within the working range for most instruments, operating at a wavelength of 422.7 nm. Digestates of leaf material may need diluting. Magnesium and potassium can be measured in the same digestate, usually with appropriate dilutions (using the $LaCl_3$ solution). Calcium can be measured with an air/acetylene flame, although fewer interferences and a greater sensitivity can be achieved with nitrous oxide/acetylene. Standards should contain the same concentrations of acid and La as tissue samples.

Atomic absorption spectrophotometry is usually sufficiently sensitive to allow analysis of most plant tissues. Most instruments have sensitivities in the range of 20–50 ng/ml for Ca. With appropriate adjustment of tissue weights and solution volumes in the digestion procedure, there should be few problems. Use of a graphite furnace allows smaller samples to be analysed with greater sensitivity.

Analyses of Ca^{2+} in plant tissues (particularly fruit) are best expressed on a fresh weight basis. In all tissues, however, changes in water content of tissue, particularly over a time sequence, can distort the expression of real values. (There are still instances of equivalents being used as the quantitative expression, allowing some estimate of charge. This may be useful where ion balance is being considered, but is now not common.)

3.4 Ca^{2+} fractions

Because of the uneven distribution of Ca^{2+} in plant tissues (millimolar concentrations in the extracellular fluid and vacuoles, and sub-micromolar in the cytosol) there is a need to identify Ca^{2+} pools. At the level of tissue analysis, this principally means identifying chemical fractions of Ca^{2+} and perhaps the Ca^{2+} associated with the cell wall, or water-free space.

There have been several attempts to chemically fractionate Ca^{2+}. The basic scheme is to sequentially extract tissue in water, 80% acetic acid, and 0.25 M HCl (20). This purports to provide fractions of Ca^{2+} which can be identified as soluble, that bound as pectate, that as oxalate, and a residue fraction which may contain insoluble forms of Ca^{2+}, such as silicate. However, an examination of this scheme (20) has shown that these fractions are not precise enough. Cell wall fractions in particular have proven very difficult to prepare such that they have any biochemical or physiological meaning. Where high concentrations of oxalate are found in tissue, such as in beet species, and in tissue infected with *S. rolfsii* (21), there may be advantages in extracting tissue for oxalate-bound Ca^{2+} (20). This procedure is described in *Protocol 3*, and provides an initial fraction of water-soluble salts of organic and inorganic acids, as well as exchangeable forms of Ca^{2+} that might be associated with pectate and phytate. The limited solubility of oxalate in the acetic acid means that the HCl fraction represents the oxalate-bound Ca^{2+}. Hence, if an analysis for oxalate is carried out in conjunction with this procedure, a good estimate of oxalate-bound Ca^{2+} is obtained (20, 22).

Protocol 3. Analysis of oxalate-bound Ca^{2+} in tissues

1. Add 100 mg freeze-dried tissue (homogeneous powder) to 3.5 ml 80% acetic acid containing 10 mM cysteine.

2. Shake the sample at 25°C for 30 min, and then centrifuge at 500 g for 10 min.

3. Remove the supernatant and repeat the extraction.

4. Extract the sample twice more with 0.25 M HCl containing 10 mM cysteine. Collect the supernatants again.

5. Digest the supernatants and residue in $HNO_3/HClO_4$ as described above for Ca^{2+} analysis (*Protocol 2*).

6. Measure calcium by atomic absorption spectrophotometry (AAS).

Water- and ethanol-soluble fractions have been obtained from apple fruit (23), but the physiological meaning of these fractions is unclear. There remains the large problem of identifying Ca^{2+} in cell walls, particularly that unexchangeably bound, and that available for release. The size of these pools may be important in cell wall degradation in tissue senescence and fruit ripening, and in susceptibility to fungal attack. Unfortunately, there is, as yet, no reliable method that will provide unequivocal identification of the size of these fractions.

4. Cellular Ca^{2+} fluxes

The most widely used methods for identifying the transporters involved in regulating Ca^{2+} flux have utilized isolated organelle or membrane vesicle preparations. By studying transport across membranes of intact organelles or vesicles *in vitro* it is possible to biochemically dissect the involvement of various types of pumps/carriers or channels involved in Ca^{2+} transport into or out of the organelle/vesicle. Intact organelles offer several advantages over membrane vesicles in transport studies; notably that an efflux of endogenous Ca^{2+} from internal stores can be monitored. Membrane vesicles, on the other hand, lose some of their contents before they reseal and, therefore, must be pre-loaded with Ca^{2+} before commencing efflux experiments. It is clearly advantageous to study the physiology of a membranous compartment whose contents approximate *in vivo* conditions.

In vitro studies with purified membranes have permitted the determination of some of the kinetic properties of transport systems and the identification of various potential regulators of Ca^{2+} transport. *In vitro* experiments also allow testing of the effects of inhibitors on Ca^{2+} transport. Such experiments have been instrumental in identifying the molecular basis of Ca^{2+} transport. On the other hand, the use of isolated organelles or membrane vesicles to study transport has the disadvantage that these membrane compartments may function differently *in vitro* than they do *in vivo*. Despite this problem, membraneous compartments isolated from plant cells provide a quick and reliable means of studying Ca^{2+} flux.

Several techniques are currently employed in the study of cellular Ca^{2+} fluxes. Basically these techniques can be divided into three categories:

(a) Ca^{2+} in a specific pool(s) can be 'tagged', for example by labelling with $^{45}Ca^{2+}$, enabling the rate of Ca^{2+} exchange between pools to be monitored. It is important to realize that when this method is used a change in labelling intensity is not necessarily indicative of a change in Ca^{2+} concentration.

(b) By ensuring isotopic equilibrium the $^{45}Ca^{2+}$ technique mentioned in (a) can be used for the study of changes in total calcium. This can also be achieved by direct chemical analysis of calcium, for example by atomic absorption spectrometry.

(c) Changes in Ca^{2+}-activities (concentrations) can be monitored by the use of electrodes or Ca^{2+}-sensitive dyes, allowing either a decrease of Ca^{2+} (emptying of a Ca^{2+} pool), or an increase of Ca^{2+} (filling of pools), or sometimes both to be detected. It will often be advantageous to monitor the pool where the largest relative change in Ca^{2+} levels occurs, but this can, in some instances, be technically difficult.

5. Studies of Ca^{2+} fluxes using $^{45}Ca^{2+}$

5.1 Isolation and purification of organelles and membrane vesicles for Ca^{2+} flux studies

The isolation of tightly sealed and transport-competent membranes is essential for studies of Ca^{2+} transport. Whereas several investigators have described the isolation of microsomal membranes for use in *in vitro* Ca^{2+} transport experiments, the use of intact organelles for transport studies is still limited. Our procedures (24, 25) for the isolation of transport-competent microsomal membranes from zucchini hypocotyls and barley aleurone layers are shown in *Protocol 4*. These differ from other more commonly used protocols in two important respects. We have found that methods generating high shear forces to break open cells should be avoided as they do not permit the subsequent isolation of microsomal vesicles in high yield (26). This is particularly true for cells having thick cell walls (for example aleurone tissue). By chopping with razor blades, cells in tissues with thick walls can be broken with low shear. In the preparation of zucchini hypocotyl microsomes we have found that Ca^{2+} uptake competence is very much dependent on the type of homogenizer used. Best results are obtained if a smooth mortar and pestle is used. The addition of bovine serum albumin BSA to the homogenization medium is essential to produce membrane vesicles that can accumulate Ca^{2+} (*Figure 2*). As yet, the stimulatory nature of added BSA on Ca^{2+} transport is not understood, although we speculate that BSA protects membranes from attack by lytic enzymes during chopping/homogenization.

Because microsomal membranes contain a mixture of endoplasmic reticulum (ER), Golgi apparatus, plasma membrane, and tonoplast membranes, these components must be separated from each other by, for example, sucrose density gradient centrifugation (25, 27, 28) if the contribution of each to Ca^{2+}-transport is to be understood. In the case of barley aleurone, the combination

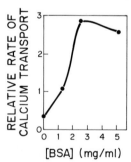

Figure 2. The effect of BSA addition during isolation of aleurone microsomal membranes on their capacity to transport Ca^{2+}. Aleurone layers were homogenized according to the procedure outlined in *Protocol 4* in a range of BSA concentrations or in its absence. Microsomes were isolated and their Ca^{2+} transport capacity measured according to the methods in *Protocol 5*.

of discontinuous and isopycnic sucrose density gradients effectively separate ER, plasma membrane, and tonoplast membranes from each other (*Protocol 4* and reference 25).

Protocol 4. Isolation and purification of membranes from zucchini hypocotyls and barley aleurone layers

A. *Zucchini hypocotyls*

Perform all steps in the extraction procedure at 2–4°C.

1. Germinate zucchini seeds in darkness in moist vermiculite at 27°C.

2. Harvest hypocotyls at a height of approx. 10 cm. Homogenize hypocotyls in a smooth mortar and pestle in isolation medium A containing:
   - 0.25 M sucrose
   - 200 mM Tris
   - 10 mM EDTA
   - 40 mM mercaptoethanol
   - 2 g/l BSA
   - 10 g/l PVP
   - pH 7.8

   Use 1 ml medium per 0.5 g tissue

3. Strain homogenate through two layers of cheesecloth and centrifuge at 1500 g for 8 min to remove debris.

4. Centrifuge supernatant from step 3 at 16 000 g for 20 min.

5. Centrifuge supernatant from step 4 at 80 000 g for 30 min.

6. Resuspend the pellet obtained from step 5 in calcium uptake medium:
   - 0.5 M sucrose
   - 8 mM Tris–MES (pH 7.2)
   - 5 mM $MgCl_2$
   - 0.1 mM $CaCl_2$
   - 0.8 g/l BSA

   Keep isolated membranes on ice until use.

B. *Barley aleurone*

Perform all steps in the extraction procedure at 2–4°C.

1. Chop aleurone layers with a motorized razor blade chopper (7) in isolation medium B containing:
   - 25 mM Hepes adjusted to pH 7.4 with BTP
   - 0.2 mM $MgSO_4$

Protocol 4. *Continued*

- 3 mM EDTA
- 1 mM dithiothreitol (DTT)
- 0.5% BSA

2. Filter homogenate through two layers of Miracloth and centrifuge at 1000 g for 10 min.

3. Centrifuge supernatant on a discontinuous sucrose density gradient consisting of a 1 ml cushion of 50% (w/w) sucrose overlaid with 5–7 ml of 13% (w/w) sucrose[a] in a 7.6 cm long, 2.5 cm diameter cellulose nitrate tube. Centrifuge for 2.5 h at 70 000 g in a Beckman SW 27.1, or equivalent, rotor.

4. Remove the microsomal membranes (turbid band), that collect at the interface between 13% and 50% sucrose, with a Pasteur pipette. Layer on to a continuous sucrose (21–45%, (w/w)) density gradient and centrifuge for 14 h at 70 000 g in the same type of tube and rotor.

5. Collect 18 1 ml fractions and store on ice. Membranes can be frozen at −40°C to −80°C for up to several months.

[a] All sucrose density gradient solutions contain 0.1 mM DTT and are buffered with 25 mM Hepes–BTP, pH 7.4.

5.2 Assay of Ca^{2+} fluxes across membranes using $^{45}Ca^{2+}$

The general procedure for experiments with isolated plant membranes has changed little since it was first described by Gross and Marme (29, 30). The protocols we have developed for measuring Ca^{2+} uptake by microsomes isolated from zucchini hypocotyls and barley aleurone layers are shown in *Protocol 5*.

Protocol 5. Measurement of $^{45}Ca^{2+}$ uptake by membrane fractions

A. *Zucchini hypocotyl membranes*

1. Adjust protein concentration of membranes in the calcium uptake medium (*Protocol 4* A, step 6) to approx. 0.75 mg/ml.

2. Add NaN_3 (3 mM final) to the calcium uptake medium.

3. Add $^{45}Ca^{2+}$ to approx. 10 kBq/ml.

4. Initiate Ca^{2+} uptake by addition of 2 mM ATP (final concentration).

5. Remove 0.05 m samples at appropriate times and filter on 0.45 μm filter discs (pre-wet filters in ice-cold uptake medium). Wash the filter with 1 ml of ice-cold uptake medium after addition of the sample.

6. Remove filters, air-dry, and measure radioactivity by liquid scintillation spectrometry.

B. *Barley aleurone membranes*

1. Mix 100 µl of microsomal membranes (approx. 100 µg protein/ml) with 400 µl of uptake medium containing (final concentration):

   - 25 mM Hepes adjusted to pH 7.4 with BTP
   - 10 mM potassium oxalate
   - 3 mM $MgSO_4 \cdot 7 H_2O$
   - 10 µM $CaCl_2$
   - 100 µM sodium azide
   - 3.7×10^4 Bq/ml $[^{45}Ca]CaCl_2$

2. Initiate Ca^{2+} uptake by adding 1 mM ATP (final concentration). Remove duplicate 200 µl samples after 20 min and filter on 0.45 µm filter discs under vacuum.

3. Wash filter disc with 3.5 ml buffer containing 250 mM sucrose, 2.5 mM Hepes–BTP (pH 7.4), and 0.2 mM $CaCl_2$.

4. Air-dry filters and measure radioactivity by liquid scintillation counting.

Figure 3 shows a typical uptake curve obtained using zucchini hypocotyl microsomes prepared as described in *Protocol 4*. The effect of inositol(1,4,5)trisphosphate on Ca^{2+} flux is also illustrated. A procedure for the assay of Ca^{2+} fluxes across the plasma membrane of protoplasts is described in *Protocol 6*.

Protocol 6. Assay of Ca^{2+} fluxes across the plasma membrane of plant protoplasts[a]

The procedure below is described in further detail by Klein and Ferguson (31).

1. Produce protoplasts.

2. Incubate 2 ml samples of protoplasts in 10 ml vials (for details on incubation medium consult reference 31).

3. Add $^{45}Ca^{2+}$ in the incubation medium to approx. 3 kBq/ml.[b]

4. Add the following components to 1.5 ml microcentrifuge tubes:

   - 20 µl $HClO_4$ (6% v/v)
   - 200 µl dense silicone oil (e.g. Wacker, AR200)
   - 50 µl low-Ca^{2+} water (on top of the oil layer).

5. At appropriate times remove 100 µl samples of protoplasts from the incubation medium. Place samples on top of the oil layer.

6. Immediately centrifuge tube(s) at 12 000 g for 10 sec.

Protocol 6. *Continued*

7. Freeze the tubes quickly *or* remove the oil and residual medium by aspiration.

8. Cut off the centrifuge tube tips and place in scintillation vials.

9. Determine radioactivity by liquid scintillation spectrometry.

[a] This method is designed to study Ca^{2+} uptake, but by pre-loading the protoplasts with $^{45}Ca^{2+}$ efflux can be monitored.

[b] If $^{3}H_2O$ is incorporated in the incubation medium a quantitative measure of protoplasts can be obtained if ^{3}H is determined simultaneously with $^{45}Ca^{2+}$. It is important if oils other than the type described in the *Protocol* are used: (a) to ascertain that no (or only little) ^{3}H is transferred to the bottom of the tube in the absence of protoplasts, and (b) to test the density of the oil to ensure that protoplasts and incubation solution can be separated. If problems occur with oil density it may be necessary to mix oils having different densities.

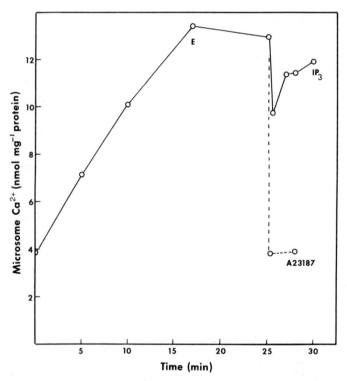

Figure 3. Uptake (and release) of Ca^{2+} by zucchini hypocotyl microsomes. The experimental protocol is described in *Protocols 4* and *5*, ATP (2 mM) was added at time zero. EGTA addition (final concentration 4×10^{-4}) is indicated by E. 20 μM Inositol(1,4,5) trisphosphate and A.23187 (final concentration 10^{-4} mM) were added to separate samples at 25 min. (Reproduced with permission from Drøbak and Ferguson, reference 24.)

6. Ca^{2+}-sensitive electrodes

6.1 Introduction

With the development of new, improved ionophores (32) Ca^{2+}-sensitive electrodes have now become a very convenient means of following Ca^{2+} sequestration and release from cells and organelles. One of the advantages of modern Ca^{2+}-sensitive electrodes is that they are useful over a wide range of Ca^{2+} concentrations (pCa 3–8). Additionally, modern electrodes show very good response times (seconds) even at 10^{-7} M free [Ca^{2+}], and also an excellent discrimination against Mg^{2+} (as well as K^+, Na^+, and H^+). Many commercially available electrodes have a near-Nernstian response in the range of pCa 7–6 with a change of 29.5 mV for a 10-fold change in Ca^{2+} concentration. Electrodes must be calibrated against solutions of known Ca^{2+} concentration, and recipes for solutions having pCa from 3–8 are shown in *Table 1*. These recipes assume the use of Ca^{2+}-free water, but the Ca^{2+}

Table 1. Composition of calibration solutions [a,b]

Solution	CaCl$_2$ (mM)	Ca ligand (mM)	KCl (mM)	NaCl (mM)	MgCl$_2$ (mM)	H$^+$ buffer (10 mM)	pH
3	1	—	98	—	—	Mops	7.30
4	5	NTA, 10	90	—	—	Hepes	7.39
5	5	NTA, 10	90	—	—	Taps	8.42
6	5	HEEDTA, 10	90	—	—	Mops	7.70
7	5	EGTA, 10	90	—	—	Mops	7.29
8	5	EGTA, 10	90	—	—	Hepes	7.80
∞	0	EGTA, 10	100	—	—	Hepes	7.80
∞, EDTA	0	EDTA, 10	100	—	—	Hepes	7.80
∞, 1 mg	0	EGTA, 2.33	115	—	1.33	Hepes	7.60
∞, 5 mg	0	EGTA, 3.67	110	—	6.67	Hepes	7.60
∞, 20 mg	0	EGTA, 8.67	95	—	26.67	Hepes	7.60
∞, 10 Na	0	EGTA, 10	90	10	—	Hepes	7.80
∞, 100 Na	0	EGTA, 10	—	100	—	Hepes	7.80
∞, 200 K	0	EGTA, 10	175	—	—	Hepes	7.80
∞, pH 6.4	0	EDTA, 10	100	—	—	Pipes	6.40
7, 5 Mg	4	EGTA, 9.33	99	—	6.33	Mops	7.29

[a] The solutions are designated by their pCa ($-\log_{10}$ [free Ca^{2+}]/M) and in other than the standard series, pCa 3 to '∞', by the free concentration of the interfering ion under test. Thus ∞, 5 mg is a solution with no added Ca, EGTA, and 5 mM free Mg^{2+}. The pH is brought to the desired level by titration with KOH and the concentration of K is close to 0.125 M, except in pCa 3 solution and of course ∞, 200 K. The H$^+$ buffers were all at 10 mM. NTA, nitrilotriacetic acid; HEEDTA, *N*-(2-hydroxyethyl)-ethylenediamine *N,N'*, *N'*-triacetic acid.
[b] Reproduced with permission from Tsien and Rink (13).

buffering capacity of the solutions is sufficiently high to permit the use of water containing low concentrations (1–2 μM) of Ca^{2+}.

6.2 Principle of Ca^{2+}-sensitive mini-electrodes

The principle of operating a Ca^{2+} electrode is similar to that of an ordinary pH electrode, a membrane which is semipermeable to Ca^{2+} separates a solution containing Ca^{2+} (the electrode filling solution) from the unknown Ca^{2+} solution outside and a membrane potential is set up, the magnitude of which is determined by the ratio of free Ca^{2+} concentration inside and outside the membrane. Under ideal conditions, the electrode potential (ψ) is given by the Nernst equation (equation 1):

$$\psi = \frac{2.3\,RT}{nF} \cdot \log_{10} \frac{[C^{n+}]_0}{[C^{n+}]_i}. \tag{1}$$

The membrane potential is measured with an Ag/AgCl electrode dipping into the electrode filling solution relative to a suitable reference electrode (for example calomel, as used as a reference for pH electrodes) in contact with the unknown solution (*Figure 4*). As Ca^{2+} is divalent ($n = 2$ in equation 1) the slope of the electrode is only about 29 mV/decade as compared with 59 mV for a pH electrode.

Since the electrode measures changes in free Ca^{2+} concentrations, it is very important when designing experiments to ensure that the Ca^{2+} buffering capacity of the experimental medium is matched to the sensitivity of the measuring system and the Ca^{2+} uptake and release capacity of the experimental material. Also, changes in Ca^{2+} buffering capacity during the experiment (such as may be caused by changes in concentration of Ca^{2+}-binding species such as ATP) can give rise to rather confusing artefacts.

A full treatment of Ca^{2+} buffering and of the construction of large Ca electrodes is given in reference 33.

6.3 Construction of Ca^{2+}-sensitive mini-electrodes

Protocol 7 is based on the technique described by Clapper and Lee (34). The electrodes which we produce have an external diameter of 3 mm and, with a suitably small reference electrode, can be used to measure Ca^{2+} concentrations in incubation volumes down to about 800 μl. All the reagents for making the electrodes are available from Fluka, and the procedure for construction is detailed in *Protocol 7*.

Protocol 7. Construction of Ca^{2+}-sensitive mini-electrodes

1. Prepare the stock ionophore cocktail as follows: mix 50 mg of Ca^{2+} ionophore ETH 1001 with 450 mg O-nitrophenyloctylether, and 5 mg of sodium tetraphenylborate (the latter added as 50 μl of 100 mg/ml solution

in tetrahydrofuran (THF)). This cocktail is usable for at least 12 months if stored in aliquots at −20°C.

2. Prepare the PVC solution for forming the electrode membranes by dissolving 100 mg high molecular weight PVC powder in 1 ml tetrahydrofuran (THF). The solution must be freshly prepared for each batch of electrodes. Sprinkle the PVC powder slowly on to the surface of the THF with continuous shaking, otherwise it turns into a totally intractable insoluble mass. Take care to avoid evaporation of the THF during this process, otherwise the resulting solution will be too viscous (it should be pipettable with a positive displacement pipette such as a Gilson Microman).

3. Cut 12 pieces (each about 35 mm long) of suitable plastic tubing. The tubing used to form the barrel of the electrode should ideally also be made from PVC but must at least be soluble in THF, so that it will bond tightly to the PVC membrane. We use Portex No. 4 tubing, o.d. 3 mm, i.d. 2 mm.

4. Mix 40 μl of ionophore cocktail with 240 μl of 10% (w/v) PVC solution.

5. Dip each piece of tubing into the ionophore/PVC mixture, such that the level of the solution rises about 3 mm up the inside of the tube. Carry out the dipping procedure as rapidly as possible to minimize evaporation of THF from the mixture.

6. Suspend the treated tubes with their open ends upwards in such a way that the dipped ends can dry evenly and without contact with anything else. Using tubing of the size given above, it should be possible to make 12 electrodes from this quantity of mixture.

7. Leave the electrodes to dry in air. As the THF evaporates, a PVC membrane containing the ionophore components forms across the dipped end of the tubing. Membranes once formed are stable for at least several months if stored dry at room temperature.

6.4 Use of Ca²⁺-sensitive electrodes

The calibration and use of Ca^{2+} sensitive electrodes are more fully described in reference 33. The mini-electrode described in *Protocol 7* is filled with 10 mM $CaCl_2$ and a Ag wire electrode is inserted into the filling solution. Filling is easily carried out using a microsyringe, followed by gentle tapping of the side of the electrode tubing to remove air bubbles. The Ag electrode is connected to the pH electrode input of a pH meter via a suitably screened cable, and the calomel electrode is connected to the reference electrode input. Because Ca electrodes only give 29 mV/decade compared with 59 mV/decade for a pH electrode, the pH meter will not read in pCa ($-\log_{10}[Ca^{2+}]_{FREE}$) unless it has a microprocessor control or hard-wired modification allowing halving of its slope per decade. In the absence of such an arrangement it is more convenient to use the mV display mode and to calibrate the mV output as a function of Ca^{2+} concentration.

Having set up the electrode, it is a good idea to soak it in a solution containing 1 mM $CaCl_2$ for about 1 hour before use. After this, the calibration should remain relatively stable for several hours. The working life of the electrode is about 1 day of continuous experimental use, although if after relatively brief use the electrode is dismantled, the filling solution emptied out and the membrane left to dry, it may be re-used several times. The life of the electrode is probably limited by penetration of water into the seal between membrane and tubing. For this reason we treat them as disposable items, using them for 1 day and then discarding them.

The electrodes can be calibrated against the solutions described in *Table 1* or by the following simple calibration procedure:

- Prepare a solution of Hepes–KOH, pH 7.0, to give an ionic strength similar to that of the experimental medium. Ca^{2+}-sensitive electrodes are some-what sensitive to ionic strength, and, as mentioned earlier, the dissociation constants of certain Ca^{2+}-chelating agents are very pH-dependent so both these parameters are important.

- Add $CaCl_2$ to bring the Ca^{2+} concentration to 2×10^{-4} M, insert the electrodes into the solution, and take the reading.

- To the Ca^{2+}-containing solution, add EGTA to a final concentration of 4×10^{-4} M. The free Ca^{2+} concentration is now 2.4×10^{-7} M. Assuming Nernstian behaviour, a graph of $\log[Ca^{2+}]_{FREE}$ versus mV is a straight line.

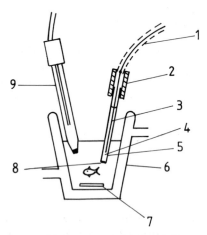

Figure 4. Apparatus for using Ca^{2+}-sensitive electrodes. 1. Screened cable to pH meter; 2. Plastic mounting collar for Ca^{2+}-electrode; 3. Ag wire; 4. 10 mM $CaCl_2$ (electrode filling solution); 5. PVC barrel of electrode; 6. Thermostatted reaction vessel; 7. Magnetic spin-bar; 8. Ca^{2+}-sensitive membrane; 9. Reference electrode.

6.5 Measurement of Ca^{2+} uptake by plant microsomal fractions using Ca^{2+}-sensitive mini-electrodes

The conditions we have found (35) to be most suitable for measurement of Ca^{2+} uptake by zucchini hypocotyl microsomes (see *Protocol 5*) are described in *Protocol 8*.

In the method described in *Protocol 8* we have found that an ATP regenerating system, in this case creatine phosphate/creatine phosphokinase, is necessary to maintain a constant Ca^{2+}-buffering capacity due to ATP. Similarly, it is better in this type of experiment to start the uptake of Ca^{2+} by the addition of microsomes to the medium, rather than by the addition of ATP, since the latter will produce a very large addition artefact due to Ca^{2+} binding by ATP.

Protocol 8. Assay of Ca^{2+} uptake in zucchini hypocotyl microsomes by Ca^{2+}-sensitive mini-electrodes

1. Prepare assay medium uptake containing:
   - 0.1 M sucrose
   - 50 mM KCl
   - 3 mM NaN_3 (to block mitochondrial Ca^{2+} uptake)
   - 1 mM dithiothreitol (DTT)
   - 5 mM $MgCl_2$
   - 5 mM ATP
   - 0.1 mg/ml creatine kinase
   - 10 mM creatine phosphate
   - 10 mM Hepes, pH 7.2
2. Prepare membrane vesicles as described in *Protocol 4 A*.
3. Start the reaction by adding microsomes to a final concentration of 1 mg protein/ml. In our hands the starting Ca^{2+} concentration is of the order of 2 μM and the microsomal vesicles should be capable of decreasing this to around 100 nM.

---

7. Ca^{2+}-sensitive dyes

7.1 Introduction

An alternative to using Ca^{2+}-sensitive electrodes for the measurement of free Ca^{2+} is the use of Ca^{2+}-sensitive dyes. Dyes have the advantage that they are generally fast in response to changes in Ca^{2+} concentration and are more sensitive to levels of Ca^{2+} in the nanomolar range. To our knowledge, the uptake of Ca^{2+} into isolated membrane fractions from plant cells at physiological concentrations of Ca^{2+} (100 nM to 1 μM) has not been measured, and only the efflux of dyes from isolated intact vacuoles has been studied using this approach. Experiments with animal cell microsomal membranes have used the fluorescent dyes Quin-2 and fura-2; the response of these dyes to altered Ca^{2+} levels falls in the range of cytosolic Ca^{2+} concentrations (36). Fura-2 is preferred over Quin-2 because its K_d for Ca^{2+} is closer to the range of cytosolic Ca^{2+} and its fluorescence is brighter. Typical reaction mixtures for monitoring changes in Ca^{2+} concentration of the ambient solution contain the same components as those used to study $^{45}Ca^{2+}$ uptake (*Protocol 5*), except that sufficient chelator (for example BAPTA, EGTA) is added to maintain Ca^{2+} in the range 100 nM to 1 μM. Dye concentrations are typically 100 μM for Quin and 0.5 μM for fura. As fluorescence from Quin and fura does not increase linearly with Ca^{2+} concentration, data should be reported in units of Ca^{2+} concentration after calibration of the dye with known additions of Ca^{2+}.

7.2 Determination of cytoplasmic free calcium in plant cells

The direct measurement of cytoplasmic Ca^{2+} in plant cells is not a simple task. The cell wall and the large central vacuole of most higher plant cells pose problems that are unique to plants. The cell wall can act as a physical barrier to the impalement of cells with microelectrodes, and it can interfere with fluorescence measurements because of the presence of phenolic compounds. As the central vacuole of plant cells frequently occupies as much as 90–95% of the volume of the cell, it also limits the use of microelectrodes and fluorescent dyes. The use of microelectrodes can be difficult when the vacuole confines the cytoplasm to a very narrow layer between the tonoplast and plasma membrane. On the other hand, the vacuole may sequester Ca^{2+}-sensitive dyes that are targetted to the cytosol thus masking the cytoplasmic signal from the probe. Despite these limitations microelectrodes and Ca^{2+}-sensitive dyes have been successfully used to measure cytosolic Ca^{2+} in plants. Recent work and methods for using microelectrodes have been reviewed by Felle (37), and we have summarized the use of intracellular probes for measuring cytosolic Ca^{2+} in Section 7.3.

7.3 Intracellular probes for measurement of cytosolic Ca^{2+} concentrations

Several types of intracellular probes have been successfully used to measure cytosolic Ca^{2+} in plant cells. Azo dyes, such as Arsenazo III, which undergo a change in absorption coefficient upon binding Ca^{2+}, photoproteins (such as aequorin which emits light upon Ca^{2+} binding), and the fluorescent probes designed by R. Y. Tsien *et al.* (36) that undergo a change in fluorescence intensity upon Ca^{2+} binding have all been used successfully with plant cells. In

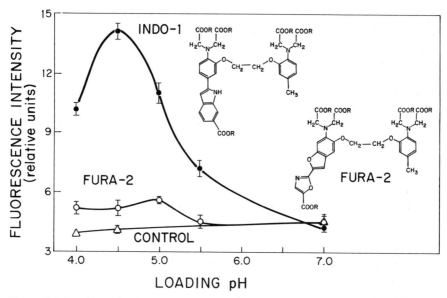

Figure 5. The effect of pH on the uptake of indo-1 and fura-2 by barley aleurone proto-plasts. Suspensions of barley aleurone protoplasts were incubated in either fura-2 (open circle), indo-1 (closed circle), or in the absence of either dye (open triangle) according to the method outlined in *Table 6*, except that the pH of the incubation medium was set between pH values of 4 and 7. Fluorescence of individual cells ($n > 50$) was measured using the apparatus described by Tsien *et al.* (40). The structures of indo-1 and fura-2 are also shown.

principle, the fluorescent probes Quin-2, fura-2, and indo-1 synthesized by Tsien and his colleagues (36) are superior to the other intracellular probes. The affinity of the fluorescent dyes for Ca^{2+}, which closely matches cytosolic Ca^{2+} levels, makes these dyes particularly useful probes. The apparent dissociation constant for indo-1, for example, is 224 nM making it ideal for measurements of Ca^{2+} between 0.1 and 1 μM. Unlike azo dyes the fluorescent dyes bind Ca^{2+} with a 1:1 stoichiometry, simplifying the calibration of the signal for Ca^{2+} quantitation. Perhaps the most valuable feature of some of the fluorescent dyes (for example fura-2 and indo-1, *Figure 5*) is that they under-go a shift in their fluorescence spectra when they bind Ca^{2+}. This feature allows Ca^{2+} levels to be determined from the ratio of fluorescence at two wavelengths. This ratio method has the advantage that Ca^{2+} estimates are more-or-less independent of fluctuations in fluorescence intensity as a result of differences in dye loading. When fluorescence is measured at only one wavelength, fluctuations in dye concentration, lamp excitation intensity, cell movement, cell volume, etc., cannot be distinguished from changes in the response of the dye to Ca^{2+}.

7.3.1 Loading fluorescent dyes

In order to use any of the intracellular Ca^{2+} probes they must be loaded into the cytoplasm and accumulate only in that compartment. The lack of rapid progress in measuring cytosolic Ca^{2+} in plant cells is due principally to the difficulties encountered in loading the fluorescent dyes into the cytoplasm. Tsien and his colleagues (36) designed their dyes as esters in order to permit non-intrusive loading into animal cells. The acetoxymethyl (AM) esters of Quin-2, fura-2, and indo-1 permeate the plasma membrane and accumulate in the cytosol as anions after cleavage of the ester bonds. The AM esters of Quin-2 and fura-2 accumulate in the cytosol of only a few plant cell types (38, 39, 40).

Uptake of the AM forms of the fluorescent dyes by animal cells is enhanced by incubating the cells for several hours in a solution containing one to several micromolar concentrations of the dye and 0.5% of the surfactant pluronic (38, 39, 40).

A potential problem in loading the AM forms of fluorescent dyes is incomplete hydrolysis of the four AM ester groups. Since the AM forms of these dyes are also fluorescent, it is impossible to tell from fluorescence alone if the dye is Ca^{2+} sensitive. To establish that AM esters have been cleaved from the dye after loading it is necessary to remove extracellular dye and lyse the loaded cells with detergents, such as digitonin or Triton X-100 at approximately 0.05%. The sensitivity of dye to added Ca^{2+} is then determined fluorimetrically.

7.3.2 Alternative dye-loading strategies

To overcome the difficulties associated with the cleavage of the AM group from the Ca^{2+}-sensitive dyes two alternative strategies have been adopted with plant cells. Cells have been loaded with Quin-2 by electroporation (41, 42) and by microinjection (43, 44). The disadvantage of these techniques is that they are intrusive and may themselves perturb cytosolic Ca^{2+} levels.

An alternative strategy that has been shown to work with a wide variety of plant cells is to load the acid form of the fluorescent dyes under mildly acidic conditions (45–47). In barley aleurone protoplasts this strategy has worked particularly well with indo-1, but fura-2 does not load well at low pH (*Figure 5*). For indo-1, loading shows a maximum at pH 4.5 with lower amounts of dye accumulating in the cell at pH values above and below 4.5 (*Figure 5*). A protocol for the loading of barley aleurone protoplasts with indo-1 is shown in *Protocol 9*.

Protocol 9. Method for loading indo-1 into barley aleurone protoplasts

1. Wash isolated barley aleurone protoplasts in WS medium containing:

- 60 mM sucrose
- 700 mM mannitol

- 100 mM glucose
- 60 mM KCl
- buffer to pH 4.5 with 10 mM dimethyl glutaric acid and KOH

2. Resuspend protoplasts in WS containing indo-1 at 25 μM and incubate at 25°C for 2 h.

3. Sediment protoplasts at 1 *g*, 30 min, and resuspend in WS without dye.

4. Repeat step 3.

5. Resuspend protoplasts in WS for fluorescence measurements.

7.3.3 Measuring fluorescence

Fluorescence measurements from dye-loaded cells can be made on single cells using an epifluorescence microscope or on populations of cells by spectro-fluorimetry. A number of commercial instruments, based on Tsien's original design (40), are manufactured for making fluorescence measurements on single cells. A simple and inexpensive apparatus can be constructed on an inverted microscope by the addition of a photomultiplier and the appropriate cut-off filters that span the emission spectrum of the dye (45), or band-pass filters that allow the peaks of fluorescence in the Ca^{2+}-bound and Ca^{2+}-free states to be measured. A xenon lamp is commonly used for its even illumination in the ultraviolet range. The intensity of the excitation beam of the microscope or fluorimeter should be kept low to minimize photobleaching of the dye. A simple and inexpensive way of making measurements on single cells has been described in detail by Bush and Jones (45) and the method is described in *Protocol 10*.

Protocol 10. Fluorescence measurements of indo-1-loaded protoplasts

1. Place 50–100 μl protoplasts in WS medium[a] on a microscope slide.

2. Measure fluorescence from a single protoplast without cut-off filter.

3. Measure fluorescence with Zeiss cut-off filters at 410, 440, 470, 500, 530, and 580 nm (45).

4. Determine fluorescence intensities in the blue and green regions of the spectrum by subtracting the value obtained with the various filters (45). For example, measure blue fluorescence by subtracting the fluorescence intensity with the 410 nm cut-off filter from the intensity of fluorescence obtained without a filter. Similarly, measure green fluorescence by subtracting fluorescence at 500 nm from the fluorescence at 470 nm. Alternatively, band pass filters at 410 and 470 nm may be used to obtain directly fluorescence intensities.

[a] See *Protocol 9*, step 1.

7.3.4 Estimating Ca²⁺ concentration

Estimates of Ca^{2+} concentration in individual protoplasts can be made from fluorescence measurements using a ratio technique. This technique quantitatively relates the Ca^{2+} concentration of a solution to the emission shift that indo-1 undergoes in response to Ca^{2+} binding:

$$Ca^{2+} = K_d\left[\frac{(R_0 - R_1)}{(R_2 - R_0)}\right]R_3 \qquad (2)$$

where K_d is the stability constant for the Ca^{2+}–indo-1 complex (250 nM); R_1 is the ratio of fluorescence intensity at two wavelengths, $\lambda 1$ (410 nm) and $\lambda 2$ (470 nm) for indo-1 in protoplasts permeabilized with 1–100 µM digitonin and immersed in 10 mM EGTA. R_2 is the ratio of fluorescence intensity at the same wavelengths for permeabilized protoplasts immersed in 10 mM $CaCl_2$ and R_0 is the ratio of fluorescence intensity for intact protoplasts. R_3 is the ratio of fluorescence intensities at $\lambda 2$ for permeabilized cells in 10 mM EGTA and 10 mM $CaCl_2$ respectively (36). The use of equation 2 requires that the background fluorescence be subtracted before ratios are calculated. This is done by subtracting the autofluorescence measured in unloaded cells from the fluorescence measured in loaded cells.

7.3.5 Measuring the Ca²⁺ concentration of subcellular compartments

Although the accumulation of Ca^{2+} probes in compartments other than the cytoplasm can interfere with the measurement of cytosolic Ca^{2+}, the sequestration of dyes in the endomembrane system can be exploited to measure the Ca^{2+} concentration of that compartment (25). Bush et al. (25) exploited the observation that indo-1 accumulated in the ER of barley aleurone cells to measure Ca^{2+} levels in this compartment. Endoplasmic reticulum was isolated from barley aleurone protoplasts loaded with indo-1 as described in *Protocol 9*.

Microsomal membranes were isolated from indo-1-loaded protoplasts as described in *Protocol 11*. To establish that the indo-1 associated with the microsomal membranes was sequestered within the lumen of the vesicles, fluorescence was measured after suspending the purified microsomal membrane fraction in 5 mM EGTA in the presence or absence of the detergent digitonin (1 µM). Using this approach Bush et al. (25) estimated the Ca^{2+} concentration of barley aleurone ER at about 3 µM. To establish that the dye was reporting Ca^{2+} associated with the lumen of the ER and not the outside surface of the membrane, they (25) showed that only when EGTA and digitonin were added together to microsomal membranes was the high Ca^{2+} signal from indo-1 quenched.

Protocol 11. Isolation of ER from indo-1-loaded aleurone protoplasts

1. Suspend dye-loaded protoplasts (10^5/ml) in a medium containing:
   - 25 mM Hepes buffer adjusted to pH 7.4 with BTP
   - 0.2 mM $MgSO_4$
   - 3 mM EDTA
   - 1 mM DTT
   - 0.5% BSA

2. Sonicate the cell suspension with five 5 sec pulses of ultrasound from a Branson Sonic Disruptor (Branson Ultrasonic Corp.).

3. Layer lysate over a discontinuous sucrose gradient consisting of 5 ml 38% (w/w) sucrose overlaid by 7 ml of 13% sucrose; all sucrose solutions contain 0.1 mM DTT, 25 mM Hepes–BTP (pH 7.4) and 0.1 mM EGTA.

4. Centrifuge gradient at 70 000 g in a Beckman SW 27 rotor with SW 27.1 tubes for 2 h at 2°C.

5. Collect microsomes enriched in ER at the 13%/38% sucrose interface.

6. Measure fluorescence, from indo-1 associated with the microsomal membranes, spectrofluorimetrically.

This experimental strategy should be applicable to the measurement of Ca^{2+} in other cellular compartments. As Ca^{2+}-sensitive dyes accumulate in compartments other than the cytosol, they can influence measurements of cytosolic Ca^{2+} levels especially if the concentration of Ca^{2+} in that compartment is high. It is possible to measure the contribution of compartmentation to the overall cytoplasmic signal by measuring the concentration of Ca^{2+} in the compartment as described above and by estimating the volume occupied by that compartment. Estimations of ER volume can be made using ESR techniques (48) or microscopic methods exploiting morphometric analysis. In the barley aleurone cell Bush *et al.* (25) have estimated that the ER occupies about 4% of the cytoplasmic volume. This value together with the determination of the Ca^{2+} concentration of the ER can be used to correct the determined value for cytosolic Ca^{2+}.

8. Concluding remarks

In this chapter we have presented a number of methods which should prove useful for investigating the role of Ca^{2+} in host–pathogen interactions. Although we have focused on the Ca^{2+} ion itself, it is important to mention that the study of signalling events involving Ca^{2+}-modulated processes can often be approached both up- and downstream of the 'Ca^{2+} signal'. Little is currently known about the upstream events which link the interactions between extracellular agonists and receptors at the cell surface to mobilization of intracellular Ca^{2+}; one candidate for fulfilling this function is the phosphoinositide system (see Chapter 11). Some events downstream of the Ca^{2+} signal (i.e. processes modulated by a rise in cytosolic Ca^{2+}) were briefly mentioned in the introduction. As only limited information exists about the likely speed and time-course of potential Ca^{2+} concentration changes in host–pathogen interactions, it may be that an initial search for downstream evidence for Ca^{2+} involvement in the specific process under study may prove more productive than if the initial point of attack is the Ca^{2+} signal itself.

9. Appendix I. A computer program in IBM BASIC for calculating the free Ca^{2+} concentration in solutions

The computer program shown in *Table A.1* has been modified by Doug Bush and Russell Jones, University of California, Berkeley, from a program called MICROQL originally written by John Westall (15). It is a general program for solving the concentrations of all specified chemical species in a solution. The computational strategy has been described by Westall *et al.* (15). A chemical species is defined as any chemical entity. Chemical species are specified by combinations of a unique set of chemical components. Chemical components are the smallest number of chemical species which in combination can give rise to all other chemical species. For example, in a solution containing EGTA, Ca, and Mg, there are four components: EGTA, Ca, Mg, and H. These components can combine to give rise to many more chemical species such as CaEGTA, MgEGTA, CaHEGTA, MgHEGTA, HEGTA, H$_2$EGTA, H$_3$EGTA, OH, and CaCO$_3$. The program requires as input in DATA statements; the total number of chemical species and components (line 9066). It then requires, for each chemical species, a separate DATA statement listing the number of moles of each component present in one mole of the species (see *Table A.2*), followed by the stability constant for the formation of the species from the components and, finally, by the name of the species. The program then requires a list of the concentration of each component in the solution (line 9138 in the example below) and a guess of the final concentration of each component at equilibrium (line 9140). Finally, the program requires the names of each component. If the concentration of one of or more of the components is fixed, as is frequently the case for H, then these components must appear at the end of the component list.

The data given in *Table A.2* have been used to calculate the free concentration of the resulting 28 chemical species at pH 7.4. The output of the program described in *Table A.1* is shown in *Table A.3*.

Table A.1. A computer program to calculate the concentration of chemical species in solution

```
1000 REM **************** CHEMSP2 ****************
1001 REM A COMPUTER PROGRAM IN IBM BASIC TO CALCLULATE THE
1002 REM CONCENTRATION OF CHEMCICAL SPECIES IN SOLUTION. REQUIRES
1003 REM DATA INPUT  FOR M1 (# OF SPECIES), N1 (TOTAL # OF COM-
1004 REM PONENTS), AND K1 (# OF SPECIES FOR WHICH TOTAL CONCENTRATION
1005 REM IS KNOWN). THEN DATA ON EACH SPS FOLLOWS I.E. STOICHEMETRY
1006 REM KD AND SPECIES LAYEL. NEXT DATA STATMENTS CONTAINS COMPONENT
1007 REM CONCENTRATION FOLLOWED BY GUESS THEN SPECIES LIST. (SEE
OX.BAS
1008 REM FOR EXAMPLE. IF PH IS FIXED, PROGRAM ASKS FOR [H].
1009 REM
1010 REM
1011 REM
1012 REM ********************************************
3050 OPEN "scrn:" FOR OUTPUT AS #1
4000  GOSUB 5220: REM SETUP
4100 IF K1=N1 THEN 4200
4101 FOR I = 1 TO N1-K1  :A$=CMP$(N1+1-I)
4110 PRINT "INPUT LOG(";A$;"): "; : INPUT X(N1+1-I): T(N1+1-I)=0
4140 NEXT I
4200 GOSUB 5470: REM SOLVE
4400 GOSUB 6630: REM OUTPUT
4600 FOR I=1 TO 10 : BEEP: NEXT I:STOP
5210 REM ********************
5220 REM STORAGE
5221 REM ********************
5225 DEFDBL Z,R,E,C,V
5230 READ M1,N1,K1: EF=0
5240 DIM Z0(K1,K1), Y0(K1),D0(K1)
5260 DIM T(N1),E(N1),X(N1),Y(N1),Z(N1,N1)
5280 DIM A(M1,N1),K(M1),C(M1)
5281 DIM IDEN(K1,K1),MATRIX(K1,K1)
5282 DIM IR(K1), SP$(M1), CMP$(N1)
5299 REM ********************
5300 REM INPUT
5301 REM ********************
5340 FOR I=1 TO M1: FOR J=1 TO N1
5345 READ A(I,J): NEXT J: READ K(I), SP$(I) :NEXT I
5380 FOR J=1 TO N1: READ T(J): NEXT J
5400 FOR J=1 TO N1: READ X(J): NEXT J
5415 FOR J= 1 TO N1  : READ CMP$(J) : NEXT J
5420 REM PROGRAM CONTROL
5440 I9=20
5460 E9=.000005
5465 RETURN
5469 REM ********************
5470 REM BEGIN A CASE HERE
5471 REM ********************
5480 I8=0
5500 REM complexes ***********
5510 FOR I=1 TO M1:C(I)=0!: NEXT I
5520 FOR I=1 TO M1 '********************
5524 FOR J=1 TO N1                '*
5525 C(I)=C(I)+(A(I,J)*X(J))
5526 NEXT J:
5530 C(I)=C(I)+K(I)                    '* CALCULATE SPECIES CONC. FROM
5540 NEXT I                            '* COMPONENT CONCENTRATION
```

Table A.1. *Continued*

```
5560 FOR I=1 TO M1                        '*
5580 C(I)=10^C(I)                         '*
5600 NEXT I        '********************
5620 FOR J=1 TO N1 '********************
5640 LET E(J)=10^(X(J))                    '* CALCULATE LOG(COMPONENT)
CONC.
5660 NEXT J        '********************
5780 REM MASS BALANCE ********
5800 I7=0
5810 IF I8=0 THEN PRINT "v8/v9 ratio"
5820 FOR J=1 TO N1 '********************
5840 V8=-T(J)                              '*
5860 V9=ABS(T(J))                          '*
5880 FOR I=1 TO M1                         '*
5900 V8=V8+(A(I,J)*C(I))                   '*
5920 V9=V9+(ABS(A(I,J))*C(I))              '* CALCULATE ERROR FOR EACH
5940 NEXT I                                '* COMPONENT
5960 IF J>K1 THEN 6000                     '*
5980 IF ABS(V8)/V9>E9 THEN I7=1            '* SET FLAG IF ERROR EXCEEDS
LIMIT
5982 PRINT USING "##.##^^^^ "; ABS(V8)/V9;
6000 Y(J)=V8                               '*
6020 NEXT J:PRINT  '********************
6040 REM COMPUTE Z *******
6060 FOR J=1 TO N1 '********************
6080 FOR K=1 TO N1                         '*
6100 V9=0!                                 '*
6120 FOR I=1 TO M1                         '*
6140 V9=V9+A(I,J)*A(I,K)*C(I)/E(K)         '* COMPUTE JACOBIAN
6160 NEXT I                                '*
6180 Z(J,K)=V9                             '*
6200 NEXT K                                '*
6220 NEXT J         '********************
6260 IF I7=0 THEN 6620
6280 I8=I8+1
6300 IF I8>I9 THEN STOP
6320 REM SOLUTION *********
6340 FOR J=1 TO K1 '********************
6360 FOR K=1 TO K1                         '*
6380 Z0(J,K)=Z(J,K)                        '* READ JACOBIAN AND ERROR
VECTOR
6400 NEXT K                                '* INTO ANOTHER MATRIX AND
VECTOR
6420 Y0(J)=Y(J)                            '*
6440 NEXT J         '********************
6441 IF EF<>0 THEN GOSUB 7000
6460 GOSUB 8000: REM FIND THE INVERSE OF JACOBIAN AND READ BACK INTO
Z0
6470 REM FOR I=1 TO 3: PRINT D0(I): NEXT I
6480 FOR I=1 TO K1: FOR J=1 TO K1:D0(I)=D0(I)+(Z0(I,J)*Y0(J)) '*
COMPONENT
6490 NEXT J: NEXT I                        '*
CHANGE
6500 FOR J=1 TO K1      '********************     '*
6520 E(J)=E(J)-D0(J)                       '*
6540 IF E(J)=<0 THEN E(J)=(E(J)+D0(J))/10  '* NEW COMPONENTS
6560 X(J)=(LOG(E(J)))/(LOG(10))            '*
6580 NEXT J                                '*
6581 IF EF<>0 THEN GOSUB 7000
6600 GOTO 5500                             '*
6620 REM THIS IS A GOTO - DON'T DELETE     '*
6625 RETURN         '********************
```

Table A.1. *Continued*

```
6629 REM ****************
6630 REM OUTPUT
6631 REM ****************
6635 PRINT #1,
6640 PRINT #1, "SPECIES       ";:PRINT #1, "MATRIX A";
6641 FOR J = 1 TO N1-1: PRINT #1, "   ";:NEXT J:
6642 PRINT #1, "LOG(K) ";"    cONCENTRATION      ";"LOG(C):
6645 FOR I=1 TO M1: PRINT #1, I;: PRINT #1, USING "\       \"; SP$(I);
6650 FOR J=1 TO N1 : PRINT #1, A(I,J);: NEXT J
6660 PRINT #1, USING "####.##"; K(I);
6662 PRINT #1, USING "   ##.##^^^^"; C(I), LOG(C(I))/LOG(10)
6665 PRINT #1,
6670 NEXT I
6675 PRINT #1,: PRINT #1, "COMPONENTS":PRINT #1, " MATRIX T"," LOG
X"," Y":PRINT #1,
6680 FOR J=1 TO N1:PRINT #1, J;
6682 PRINT #1, USING "\      \"; CMP$(J);: PRINT #1,
T(J),X(J),Y(J):PRINT #1,:NEXT J
6685 PRINT #1,
6690 RETURN
7000 REM PRINT OUT INTERMEDIATE CALCULATIONS
7010 IF EF>1 GOTO 7200
7020 LPRINT "ITERATION #";I8
7030 LPRINT "   C(I)   ";"  E(I)   ";"   Y(I)   ";
7040 LPRINT "        ";"  Z(I,J) ";"          ";
7050 LPRINT " Z0(I,J)"
7060 FOR Q=1 TO M1
7070 IF Q<=N1 THEN 7085
7075 LPRINT USING "##.#^^^^ ";C(Q)
7080 GOTO 7170
7085 LPRINT USING "##.#^^^^ ";C(Q);
7090 LPRINT USING "##.#^^^^ ";E(Q);Y(Q);
7100 FOR Q1=1 TO N1
7110 LPRINT USING "##.#^^^^ ";Z(Q,Q1);
7120 NEXT Q1
7130 IF Q>K1 THEN 7170
7140 FOR Q1=1 TO K1
7150 LPRINT USING "##.#^^^^ ";Z0(Q,Q1);
7160 NEXT Q1
7165 LPRINT
7170 NEXT Q
7180 EF=2
7190 RETURN
7200 LPRINT:LPRINT:LPRINT
7210 LPRINT "         ";"Z0(I,I)-1";
7215 LPRINT "         ";"  D(I)   ";"  E(I)   ";"   X(I)  "
7220 FOR Q=1 TO K1
7230 FOR Q1=1 TO K1
7240 LPRINT USING"##.#^^^^ ";Z0(Q,Q1);
7250 NEXT Q1
7260 LPRINT USING "##.#^^^^ ";D0(Q); E(Q);X(Q)
7270 NEXT Q
7280 EF=1
7290 LPRINT:LPRINT:LPRINT:
7300 RETURN
8000 REM GENERATE IDENITY MATRIX
8020 NN=K1: FOR II=1 TO NN
8030 FOR JJ= 1 TO NN
8035 R(JJ)=0!:IR(JJ)=0!:D0(JJ)=0!
8040 IF II=JJ THEN IDEN(II,JJ)=1 ELSE IDEN(II,JJ)=0
8050 NEXT JJ
8060 NEXT II
```

Table A.1. *Continued*

```
8061 IF EF<>0 THEN PRINT "z0 going into inverse: iteration";I8: GOSUB
8501
8062 REM FOR I=1 TO M1: C(I)=0!: NEXT I
8070 REM CHECK FOR ZERO IN DIAGONAL POSITION
8080 FOR II=1 TO NN
8090 IF Z0(II,II) <> 0 GOTO 8230: REM if not zero diag.- proceed
8100 FOR III=II+1 TO NN
8110 IF Z0(III,II)=0 GOTO 8220: REM if diag.=0 - look next row
8120 FOR JJ=1 TO NN
8130 R(JJ)=Z0(III,JJ): REM read non-0 diag row into R()
8140 Z0(III,JJ)=Z0(II,JJ):REM place zero-diag row into former non-zero
row
8150 Z0(II,JJ)=R(JJ):REM put non-0 diag row in place
8160 PRINT II;III
8170 IR(JJ)=IDEN(III,JJ)
8180 IDEN(III,JJ)=IDEN(II,JJ)
8190 IDEN(II,JJ)=IR(JJ)
8200 NEXT JJ
8210 GOTO 8230
8220 NEXT III
8230 IF Z0(II,II)=1 THEN GOTO 8310
8240 REM SCALE DAIGONAL ROW TO 1
8241 IF Z0(II,II)=0 THEN 8242 ELSE 8250
8242 PRINT "z0 at point of faliure in inverse"
8243 GOSUB 8501
8248 STOP
8250 C=1/(Z0(II,II))
8260 FOR JJ=1 TO NN
8270 Z0(II,JJ)=C*Z0(II,JJ)
8280 IDEN(II,JJ)=C*IDEN(II,JJ)
8290 NEXT JJ
8300 REM ZERO COLUMNS ABOVE AND BELOW DIAGONAL
8310 FOR KK=1 TO NN
8320 IF KK=II GOTO 8400
8330 C=-Z0(KK,II)/Z0(II,II)
8340 FOR JJ=1 TO NN
8350 R(JJ)=C*Z0(II,JJ)
8360 Z0(KK,JJ)=Z0(KK,JJ)+R(JJ)
8370 IR(JJ)=C*IDEN(II,JJ)
8380 IDEN(KK,JJ)=IDEN(KK,JJ)+ IR(JJ)
8381 REM LPRINT "ZERO COLUMNS";"II=";II;"JJ=";JJ
8382 GOSUB 8601
8390 NEXT JJ
8400 NEXT KK
8410 NEXT II
8415 IF EF<>0 THEN PRINT "z0 transformed into idenity matrix":GOSUB
8501
8420 FOR II=1 TO NN: FOR JJ=1 TO NN : Z0(II,JJ)=(IDEN(II,JJ))
8430 NEXT JJ: NEXT II
8432 IF EF=0 THEN 8440
8435 PRINT "inverse of z0 - exit inverse subroutine"
8436 GOSUB 8501
8438 PRINT "the z matrix multiplied by the z0 matrix": GOSUB 8701
8440 RETURN
8501 FOR Q=1 TO NN
8502 FOR Q1=1 TO NN
8503 PRINT USING "##.####^^^^ ";Z0(Q,Q1),
8504 NEXT Q1
8508 PRINT
8509 NEXT Q
8510 RETURN
8601 RETURN:FOR Q=1 TO NN
8602 FOR Q1=1 TO NN
```

Table A.1. *Continued*

```
8603 LPRINT ZO(Q,Q1);
8604 NEXT Q1
8605 FOR Q1=1 TO NN
8606 LPRINT IDEN(Q,Q1);
8607 NEXT Q1
8608 LPRINT
8609 NEXT Q
8610 RETURN
8701 FOR IM=1 TO NN
8704 FOR JM=1 TO NN
8705 MAT=0!
8706 FOR KM=1 TO NN
8708 MAT=MAT + Z(IM,KM)*ZO(KM,JM)
8710 NEXT KM
8712 MATRIX(IM,JM)=MAT:PRINT USING "##.###^^^^ ";MATRIX(IM,JM);
8714 NEXT JM
8716 PRINT
8718 NEXT IM
8720 RETURN
9000 REM
9050 REM ****************************************
9052 REM DATA SET FOR USE WITH CHEMSP2. CONTAINS ALL SPS
9054 REM MUST BE EDITED ON 1ST LINE (#SPS ETC.) AND LAST
9056 REM LINES (COMP. CONC, -LOG GUESS, LIST)
9058 REM
9060 REM
9062 REM
9064 REM ***********************************************
9066 DATA   29,8,7 :REM SPS #,  COMP #,  FREE COMP #
9101 DATA   1,  0,0,0,0,0,1,0,   10.57,  CaEDTA
9102 DATA   1,  0,0,0,0,0,1,1,   14.08,  CaHEDTA
9103 DATA   1,  0,0,1,0,0,0,0,    8.69,  MgEDTA
9104 DATA   1,  0,0,1,0,0,0,1,   10.97,  MgHEDTA
9105 DATA   1,  0,0,0,0,0,0,1,   10.23,  HEDTA
9106 DATA   1,  0,0,0,0,0,0,2,   16.39,  H2EDTA
9107 DATA   1,  0,0,0,0,0,0,3,   19.06,  H3EDTA
9108 DATA   1,  0,0,0,0,0,0,4,   21.05,  H4EDTA
9115 DATA   0,  1,0,0,0,0,1,0,    3.00,  CaOX
9116 DATA   0,  1,0,1,0,0,0,0,    2.55,  MgOX
9117 DATA   0,  1,0,0,0,0,0,1,    3.80,  HOX
9118 DATA   0,  1,0,0,0,0,0,2,    4.94,  H2OX
9119 DATA   0,  0,1,0,0,0,1,0,    3.77,  CaATP
9120 DATA   0,  0,1,0,0,0,1,1,    8.43,  CaHATP
9121 DATA   0,  0,1,1,0,0,0,0,    4.06,  MgATP
9122 DATA   0,  0,1,1,0,0,0,1,    8.61,  MgHATP
9123 DATA   0,  0,1,0,0,0,0,1,    6.51,  HATP
9124 DATA   0,  0,1,0,0,0,0,2,   10.57,  H2ATP
9125 DATA   0,  0,1,0,1,0,0,0,    1.10,  NaATP
9126 DATA   0,  0,0,0,0,1,0,0,    1.00,  KATP
9127 DATA   0,  0,0,0,0,0,1,-1,-12.20,  CaOH
9128 DATA   1,  0,0,0,0,0,0,0,    0.00,  EDTA
9130 DATA   0,  1,0,0,0,0,0,0,    0.00,  OX
9131 DATA   0,  0,1,0,0,0,0,0,    0.00,  ATP
9132 DATA   0,  0,0,1,0,0,0,0,    0.00,  Mg
9133 DATA   0,  0,0,0,1,0,0,0,    0.00,  Na
9134 DATA   0,  0,0,0,0,1,0,0,    0.00,  K
9135 DATA   0,  0,0,0,0,0,1,0,    0.00,  Ca
9136 DATA   0,  0,0,0,0,0,0,1,    0.00,  H
9138 DATA   1e-3,1e-2,2e-3,3e-3,2e-3,5e-2,1e-5,1e-7:REM COMPONENT CONC.
9140 DATA   -8,-3,-3,-3,-3,-2,-6,-7 :REM -LOG GUESS
9142 DATA EDTA, OX, AP, Mg, Na, K, Ca, H
9999 END
```

Table A.2. Requirements for computing the Ca^{2+} concentration of solutions and a list of the components of a typical reaction mixture used in Ca^{2+} transport experiments.

1. The number of components and their chemical species in solution must be known. For example, in a solution containing EGTA and Ca^{2+} there are three components; EGTA, Ca^{2+}, and H^+. These components combine in different proportions, thus in a solution containing only EGTA, Ca^{2+}, and H^+ there are nine possible chemical species.
2. The stoichiometry in which the components of a solution combine must be known.
3. The concentration of each component of the solution must be known. It is important to distinguish between the concentration of a component and its activity. This is generally not difficult except for species such as H^+ that are usually measured on the basis of activity using a pH electrode. The actual concentration of H can be estimated from activity determinations on the basis of the ionic strength of the solution. For most physiological solutions where ionic strength is around 0.1, $-\log[H]$ is equal to pH − 0.15. The relation between pH and H concentration at other ionic strengths is given by Martell and Smith (16).

Components (moles of components/mole of species)								**Species**
EDTA	**Oxalate**	**ATP**	**Mg**	**Na**	**K**	**Ca**	**H**	
1	0	0	0	0	0	1	0	CaEDTA
1	0	0	0	0	0	1	1	CaHEDTA
1	0	0	1	0	0	0	0	MgEDTA
1	0	0	1	0	0	0	1	MgHEDTA
1	0	0	0	0	0	0	1	HEDTA
1	0	0	0	0	0	0	2	H₂EDTA
1	0	0	0	0	0	0	3	H₃EDTA
1	0	0	0	0	0	0	4	H₄EDTA
0	1	0	0	0	0	1	0	CaOX
0	1	0	1	0	0	0	0	MgOX
0	1	0	0	0	0	0	1	HOX
0	1	0	0	0	0	0	2	H₂OX
0	0	1	0	0	0	1	0	CaATP
0	0	1	0	0	0	1	0	CaHATP
0	0	1	1	0	0	0	0	MgATP
0	0	1	1	0	0	0	1	MgHATP
0	0	1	0	0	0	0	1	HATP
0	0	1	0	0	0	0	2	H₂ATP
0	0	1	0	1	0	0	0	NaATP
0	0	1	0	0	1	0	0	KATP
0	0	0	0	0	0	1	−1	CaOH
1	0	0	0	0	0	0	0	EDTA
0	1	0	0	0	0	0	0	OX
0	0	1	0	0	0	0	0	ATP
0	0	0	1	0	0	0	0	Mg
0	0	0	0	1	0	0	0	Na
0	0	0	0	0	1	0	0	K
0	0	0	0	0	0	1	0	Ca
0	0	0	0	0	0	0	1	H

These data are entered in lines 9101 to 9138 *Table A.1*. Also included on these lines are the stability constants for the formation of each species from the components.

Table A.3. Output of computer program given in *Table A.1*

The output is divided into two parts. The first lists the species (Species) as moles of each component that comprise each species (Matrix A), the stability constant (LOG(K)), the calculated concentration of each species (Concentration), and $\log_{10}$ of the calculated concentration (LOG(C)). The second part of the output lists the components (Components Matrix T), their concentration (X), the log of the concentration (LOG X), and the error in the calculation as the difference between the calculated total concentration of each component and the given total concentration (Y).

Output of CHEM-SP2

Species Log (C)	Matrix A								Log (K)	Concentration
1 CaEDTA $-5.01E+00$	1	0	0	0	0	0	1	0	10.57	9.68D-6
2 CaHEDTA $-8.90E+00$	1	0	0	0	0	0	1	1	14.08	1.25D-09
3 MgEDTA $-3.01E+00$	1	0	0	1	0	0	0	0	8.69	9.82D-04
4 MgHEDTA $-8.13E+00$	1	0	0	1	0	0	0	1	10.97	7.45D-09
5 HEDTA $-5.11E+00$	1	0	0	0	0	0	0	1	10.23	7.83D-06
6 H$_2$EDTA $-6.35E-00$	1	0	0	0	0	0	0	2	16.39	4.50D-07
7 H$_3$EDTA $-1.11E+01$	1	0	0	0	0	0	0	3	19.06	8.39D-12
8 H$_4$EDTA $-1.65E+01$	1	0	0	0	0	0	0	4	21.05	3.26D-17
9 CaOX $-6.67E+00$	0	1	0	0	0	0	1	0	3.00	2.12D-07
10 MgOX $-3.24E+00$	0	1	0	1	0	0	0	0	2.55	5.79D-04
11 HOX $-5.63E+00$	0	1	0	0	0	0	0	1	3.80	2.37D-06
12 H$_2$OX $-1.19E+01$	0	1	0	0	0	0	0	2	4.94	1.30D-12
13 CaATP $-7.07E+00$	0	0	1	0	0	0	1	0	3.77	8.43D-08
14 CaHATP $-9.81E+00$	0	0	1	0	0	0	1	1	8.43	1.53D-10
15 MgATP $-2.90E+00$	0	0	1	1	0	0	0	0	4.06	1.26D-03
16 MgHATP $-5.75E+00$	0	0	1	1	0	0	0	1	8.61	1.79D-06
17 HATP $-4.09E+00$	0	0	1	0	0	0	0	1	6.51	8.19D-05

Table A.3. *Continued*

Species Log (C)	Output of CHEM-SP2 Matrix A								Log (K)	Concentration
18 H2ATP −7.43E+00	0	0	1	0	0	0	0	2	10.57	3.74D-08
19 NaATP −4.80E+00	0	0	1	0	1	0	0	0	1.10	1.59D-05
20 KATP −1.34E+00	0	0	0	0	0	1	0	0	1.00	4.55D-02
21 CaOH −1.24E+01	0	0	0	0	0	0	1	−1	−12.20	3.57D-13
22 EDTA −7.94E+00	1	0	0	0	0	0	0	0	0.00	1.16D-08
23 OX −2.03E+00	0	1	0	0	0	0	0	0	0.00	9.42D-03
24 ATP −3.20E+00	0	0	1	0	0	0	0	0	0.00	6.36D-04
25 Mg −3.76E+00	0	0	0	1	0	0	0	0	0.00	1.73D-04
26 Na −2.70E+00	0	0	0	0	1	0	0	0	0.00	1.98D-03
27 K −2.34E+00	0	0	0	0	0	1	0	0	0.00	4.55D-03
28 Ca −7.65E+00	0	0	0	0	0	0	1	0	0.00	2.25D-08
29 H −7.40E+00	0	0	0	0	0	0	0	1	0.00	3.98D-08

Components Matrix T	X	Log X	Y
1 EDTA	.001	−7.936327	1.875627E-09
2 OX	.01	−2.026007	3.285656-09
3 ATP	.002	−3.196584	−2.126357E-10
4 Mg	.003	−3.761552	1.906038E-09
5 Na	.002	−2.702433	1.618901E-09
6 K	.005	−2.342423	4.656613E-10
7 Ca	.00001	−7.647802	4.766559E-12
8 H	0	−7.4	9.492976E-05

References

1. Punja, Z. K., Carter, J. D., Campbell, G. M., and Rossell, E. L. (1986). *Plant dis.*, **70**, 819.
2. Ferguson, I. B. (1984). *Plant. Cell Environ.*, **7**, 477.
3. Poovaiah, B. W., Glenn, G. M., and Reddy, A. S. N. (1988). *Hortic. Rev.*, **10**, 107.
4. Kotoujansky, A. (1987). *Annu. Rev. Phytopathol.*, **25**, 405.
5. Hepler, P. K. and Wayne, R. O. (1985). *Annu. Rev. Plant Physiol.*, **36**, 397.
6. Ferguson, I. B. and Drøbak, B. K. (1988). *Hort. Sci.*, **23**, 262.
7. Kauss, H. (1987). *Annu. Rev. Plant Physiol.*, **38**, 47.
8. Kurosaki, F., Tsurusawa, Y., and Nishi, A. (1987). *Phytochemistry*, **26**, 1919.
9. Stab, M. R. and Ebel, J. (1987). *Arch. Biochem. Biophys.*, **257**, 416.
10. Zook, M. N., Rush, J. S., and Kuc, J. A. (1987). *Plant Physiol.*, **84**, 520.
11. Gabriel, D. W., Loschke, D. C., and Rolfe, B. G. (1988). In *Molecular genetics of plant microbe interactions* (ed. R. Palacios and D. P. S. Verma), p. 3. APS Press, St Paul, Minnesota, USA.
12. Ferguson, I. B. and Watkins, C. B. (1989). *Hortic. Rev.*, **11**, 289.
13. Tsien, R. Y. and Rink, T. J. (1981). *J. Neurosci. Meth.*, **4**, 73.
14. Bers, D. M. (1982). *Am. J. Physiol.*, **242**, C404.
15. Westall, J., Zachary, J., and Morel, F. (1976). *Technical Note No. 18*, Ralph M. Parsons Laboratory, MIT, Cambridge, MA.
16. Martel, A. E. and Smith, R. M. (1974). *Critical stability constants.* Plenum Press, NY.
17. Johnson, C. M. and Ulrich, A. (1959). *Calif. Agric. Exp. Stn. Bull.*, **766**, 25.
18. Turner, N. A., Ferguson, I. B., and Sharples, R. O. (1977). *NZ J. Agric. Res.*, **20**, 525.
19. Perring, M. A. and Pearson, K. (1986). *J. Sci. Food Agric.*, **37**, 709.
20. Ferguson, I. B., Turner, N. A., and Bollard, E. G. (1980). *J. Sci. Food Agric.*, **31**, 7.
21. Punja, Z. K. (1985). *Annu. Rev. Phytopathol.*, **23**, 97.
22. Clark, C. J., Smith, G. S., and Walker, G. D. (1987). *New Phytol.*, **105**, 477.
23. Perring, M. A. (1984). *J. Sci. Food Agric.*, **35**, 182.
24. Drøbak, B. K. and Ferguson, I. B. (1985). *Biochem. Biophys. Res. Commun.*, **130**, 1241.
25. Bush, D. S., Biswas, A. K., and Jones, R. L. (1989). *Planta*, **178**, 411.
26. Jones, R. L. (1985). In *Modern methods of plant analysis, new series*, Vol. 1, *Cell components* (ed. H. F. Linskens and J. F. Jackson), p. 304. Springer-Verlag, Berlin.
27. Morré, D. J., Brightman, A. O., and Sandelius, A. S. (1987). In *Biological membranes: a practical approach* (ed. J. B. C. Findlay and W. H. Evans), p. 37. IRL Press, Oxford.
28. DuPont, F. M., Tanaka, C. K., and Hurkman, W. J. (1988). *Plant Physiol.*, **86**, 717.
29. Bush, D. R. and Sze, H. (1986). *Plant Physiol.*, **80**, 549.
30. Gross, J. and Marme, D. (1978). *Proc. Natl. Acad. Sci. (USA)*, **75**, 1232.
31. Klein, J. D. and Ferguson, I. B. (1987). *Plant Physiol.*, **84**, 153.
32. Simon, W., Ammann, D., Oehme, M., and Morf, W. E. (1978). *Ann. NY Acad. Sci.*, **307**, 52.

33. Dawson, A. P., Klingenberg, M., and Kramer, R. (1987). In *Mitochondria: a practical approach* (ed. V. M. Darley-Usmar, D. Rickwood, and M. T. Wilson), p. 35. IRL Press, Oxford.
34. Clapper, D. L. and Lee, H. C. (1985). *J. Biol. Chem.*, **260**, 13 947.
35. Allan, E., Dawson, A. P., Drøbak, B. K., and Roberts, K. (1989). *Cell. Signal.*, **1**, 23.
36. Grynkiewicz, G., Poenie, M. and Tsien, R. Y. (1985). *J. Biol. Chem.*, **260**, 3440.
37. Felle, H. (1988). *Planta*, **176**, 248.
38. Brownlee, C., Wood, J. W., and Briton, D. (1987). *Protoplasma*, **140**, 188.
39. Keith, C. H., Ratan, R., Maxfield, F. R., Bajer, A., and Shelanski, M. L. (1985). *Nature*, **316**, 848.
40. Tsien, R. Y., Rink, T. J., and Poenie, M. (1985). *Cell Calcium*, **6**, 145.
41. Gilroy, S., Hughes, W. A., and Trewavas, A. J. (1986). *FEBS Lett.*, **199**, 217.
42. Gilroy, S., Hughes, W. A., and Trewavas, A. J. (1989). *Plant Physiol.*, **90**, 482.
43. Clarkson, D. T., Brownlee, C., and Ayling, S. M. (1988). *J. Cell Science*, **91**, 71.
44. Hepler, P. K. and Callaham, D. A. (1989). *J. Cell Biol.*, **105**, 2137.
45. Bush, D. S. and Jones, R. L. (1988). *Cell Calcium*, **8**, 455.
46. Bush, D. S. and Jones, R. L. (1988). *Eur. J. Cell Biol.*, **46**, 466.
47. Lynch, J., Polito, V., and Lauchli, A. (1989). *Plant Physiol.*, **90**, 1274.
48. Lomax, T. L. and Melhorn, R. J. (1985). *Biochem. Biophys. Acta*, **72**, 1401.

11

Analysis of components of the plant phosphoinositide system

BJØRN K. DRØBAK and KEITH ROBERTS

1. Introduction: the phosphoinositide system in animals and plants

A key-role for Ca^{2+} as an intracellular messenger is now well established in a vast number of organisms ranging from slime-moulds to humans. Both mammalian and plant research in the Ca^{2+} field has, until recently, been somewhat hampered by the lack of understanding of *how* external signals initiate the responses mediated by the Ca^{2+} ion. This situation has rapidly changed with the discovery, in the early 1980s, by Berridge, Irvine, Michell, and their co-workers, that in many cell types a discrete pool of inositol-containing phospholipids constitute the missing link between extracellular signals and the modulation of intracellular Ca^{2+} fluxes (1–3). The pioneering work on this system (now known as the phosphoinositide signal transducing system—or the 'PI-system') carried out in mammalian cells and tissues has inspired plant scientists to investigate whether a similar system is involved in the transduction of signals in plant cells. A brief outline of how this system functions is given in Section 1.1.

1.1 The mammalian PI-system

When certain agonists (extracellular signals) arrive at the surface of the plasma membrane they bind to specific receptors. The agonist–receptor association activates phospholipase(s) C (phosphoinositidase C, PIC)—probably via interaction with one or more guanine nucleotide binding proteins (G-proteins) associated with the membrane. The primary target for PIC action is phosphatidylinositol(4,5)bisphosphate (PtdIns(4,5)P$_2$) a phosphorylated form of phosphatidylinositol (PtdIns) which resides in the inner leaflet of the plasma membrane. PtdIns(4,5)P$_2$ is cleaved by PIC giving the products inositol(1,4,5)trisphosphate (Ins(1,4,5)P$_3$) and 1,2-diacylglycerol (DG). Ins(1,4,5)P$_3$ is released into the cytosol where it mobilizes Ca^{2+} from intracellular stores (presumably a compartment associated with the endoplasmic reticulum (ER)). The Ca^{2+} release results from Ins(1,4,5)P$_3$ binding to specific receptors; an association which triggers the opening of Ca^{2+} channels. The transiently increased cytosolic Ca^{2+} levels cause modulation of a multitude of regulatory processes dependent either on Ca^{2+}- and/or $Ca–Ca^{2+}$-binding protein complexes. The other hydrolysis product, DG, remains in the membrane matrix where it modulates the activity of a Ca^{2+}/phospholipid dependent protein kinase, i.e. protein kinase C (PKC; for recent reviews see references 4–6). These reactions are summarized in *Figure 1*.

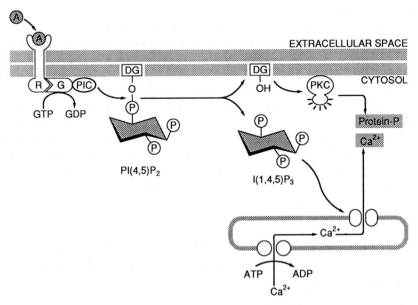

Figure 1. Outline of the mammalian PtdIns signalling system. A, Agonist; R, Membrane-associated receptor; G, Regulatory heterotrimeric GTP-binding protein (G_q); PIC, Phosphoinositidase C; PI(4,5)P$_2$, Phosphatidylinositol(4,5)bisphosphate; DG, 1,2-diacylglycerol; PKC, Protein kinase C; protein-P, Phosphorylated protein; I(1,4,5)P$_3$, D-myo-Inositol(1,4,5)trisphosphate. For further details please consult the text.

Many features of the mammalian PI-system have been found to exist in plant tissues, including:

- the ability of inositol(1,4,5)trisphosphate to release Ca^{2+} from intracellular stores (7–9)

- the presence of phosphatidylinositol (PtdIns), phosphatidylinositol(4)-monophosphate (PtdIns(4)P), and phosphatidylinositol(4,5)bisphosphate (PtdIns(4,5)P$_2$) in membranes (10–13)

- the existence of enzymes resembling mammalian phosphoinositidase C (14, 15) and protein kinase C (16–18)

1.2 The plant PI-system

Several reports have recently appeared which also point towards a functional role for the plant PI-system in signal transduction events, such as the response to light (19) and to plant hormones (20). It is important, however, to emphasize that research in this area is still in its infancy. The plant PI-system may have a similar function in general terms to that of the mammalian PI-system, but kinetic characteristics and cellular specificities may differ. To date, only very limited studies have been carried out to investigate a possible role for the PI-system in host–pathogen interactions, and results are conflicting. In a comparatively recent investigation of elicitor-induced phytoalexin synthesis in cultured parsley and soybean cells, no evidence was found for an involvement of the PI-system (21), but in a similar series of experiments on elicitation of phytoalexin production in cultured carrot cells the opposite conclusion was reached (22). In both of these studies techniques developed for mammalian research were used, and it is now clear that many problems exist in the direct adaption of 'mammalian' PI-techniques to the study of the plant PI-system. We shall highlight some of these difficulties and suggest possible solutions, and we hope that the methods outlined in this chapter will prove useful to researchers who wish to address the question of whether or not a PI-system is involved in plant–pathogen interactions.

2. Some problems associated with plant signal transduction research—with special reference to the PI-system

Problems associated with plant signal transduction research are numerous, and our intention is not to detail the routine problems but rather to point out a few possible pitfalls and aspects which need to be considered before experiments are planned.

The main problem is easily stated: we simply know too little about how external signals are transduced in plant cells.

The widespread involvement of the PI-system in transmembrane signalling in eukaryotic cells and the presence of many components of a PI-system in plants currently makes this system one of the most likely candidates for mediating the transmission of a number of signals across the plant plasma membrane.

The plant PI-system clearly has many features in common with, for example, its mammalian counterpart, but it is equally clear that sufficient differences exist between these systems to warrant caution in the design of experiments.

An important problem facing experimenters engaged in PI-research is the *timing* of experiments. Although signal transducing events generally are thought to be rapid, a considerable variation exists amongst cells. Even if the event is rapid, knowledge of the kinetics of the reaction is often of major importance. To give a practical example: in some responsive animal cells pre-labelled with, for example, [^{32}P]orthophosphate, the addition of the appropriate agonist results in a rapid (seconds) and drastic *decrease* of label in PtdIns(4,5)P$_2$ accompanying its hydrolysis. After a few minutes the lost label is replenished due to re-synthesis of PtdIns(4,5)P$_2$. However, the opposite situation could equally well exist. If PtdIns(4,5)P$_2$ is not labelled to radio-active equilibrium, an event involving rapid hydrolysis/*de novo* synthesis could result in a rapid *increase* in label in PtdIns(4,5)P$_2$. The worst scenario, from the point of view of the experimenter, is the situation where the PtdIns-(4,5)P$_2$-molecules hydrolysed by PIC have specific activities close to that of the PtdIns(4,5)P$_2$-molecules produced *de novo*. In this situation, assuming that there is no net change in the size of the PtdIns(4,5)P$_2$ pool, little or no change in labelling of PtdIns(4,5)P$_2$ would be apparent, despite the increased rate of hydrolysis.

Since ATP is the substrate for both the phosphatidylinositol/phosphatidylinositol(4)monophosphate (PtdIns(4)P) kinases and inositol phosphate (InsP) kinases, another problem is that transduction events leading to rapid changes in cellular ATP levels (specific activity) are likely to become manifest in the labelling intensity of polyphosphoinositides (PPtdInsPs) and InsPs.

Such considerations, and the findings that tracer incorporation into PtdIns-(4,5)P$_2$ in many plant tissues is very low, complicate the study of the plant PI-system.

Direct analysis of inositol phosphate production may, to some extent, overcome the labelling/kinetic problems but not the timing difficulties; the major obstacles to this approach are economy and specificity. As is discussed in Section 5.2, several simple methods exist for the separation of plant inositol phosphates, but currently only HPLC provides the level of resolution needed for routine separation of individual isomers. It is likely that strict compartmentation of various inositol phosphates (and their isomers) exists within plant cells, but current extraction procedures yield whole-cell extracts which contain the entire cellular pool of InsPs. In all of the plant cell types we have so far studied, significant levels of inositol phosphates are present in 'unstimulated' cells, so any phosphoinositide-derived messengers have to be detected against this background. It should perhaps be mentioned that 20 different isomers of D-myo-inositol-trisphosphate (InsP$_3$) are possible, and it is not unlikely that several of these are present in plant tissues.

3. Labelling of phosphoinositides and inositol phosphates

The polyphosphoinositides constitute only a minor fraction of membrane lipids. Current extraction techniques for phosphoinositides and inositol phosphates yield mixtures of low concentrations of both types of compounds together with many contaminants having somewhat similar chemical and/or physical properties. The use of radioactive labelling is, therefore, virtually indispensable in current PI-research.

The incorporation of isotopes into phosphoinositides and inositol phosphates has made the analysis and metabolic studies easier than relying on 'chemical' methods alone, but it has also made data interpretation more complex. A good fundamental knowledge of the metabolism of the various radioactive precursors and their incorporation pathways into the compound under study is essential for proper interpretation of experimental data. *Figure 2* shows the structure of phosphatidylinositol(4,5)bisphosphate (PtdIns(4,5) P$_2$), highlighting positions in the molecule where a radioactive label is normally introduced.

In the following section we have listed some of the radiolabelled precursors we have employed in the study of the plant PI-system with a short description of their advantages and drawbacks.

3.1 [^{32}P]orthophosphate

When [^{32}P]orthophosphate (^{32}Pi) is added to cultured plant cell suspensions (or indeed to most other plant tissues/cells) it is rapidly accumulated and incorporated into cellular phosphate-containing compounds. The total amount of label incorporated into a given compound at any time is largely determined by its chemical quantity, synthetic pathway, and rate of metab-

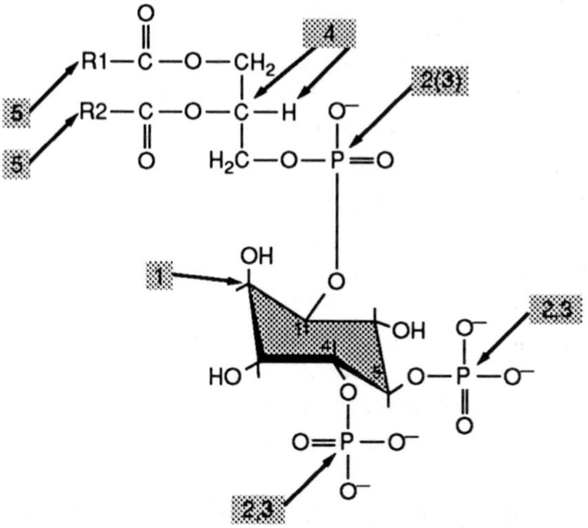

Figure 2. Structure of phosphatidylinositol(4,5)bisphosphate. Arrows indicate positions in the PtdIns(4,5)P$_2$ molecule where radiolabel is commonly introduced. Numbers beside the arrows indicate the precursors used for radiolabelling. 1, [2-^{3}H]Inositol; 2/3, ^{32}P-orthophosphate/[γ-^{32}P]ATP; 4, [^{14}C/^{3}H]Glycerol; 5, e.g. [^{14}C/^{3}H]acetate.

olism. For the labelling of polyphosphoinositides and inositol phosphates, the important step is the incorporation of ^{32}Pi into the gamma-phosphate of ATP. This process occurs rapidly in most plant cells (23). The terminal ^{32}P-phosphate of ATP is transferred to PtdIns and PtdInsP through the action of specific PtdIns- and PtdInsP-kinases. In all eukaryotic cells where the PI-system has been studied in detail, the phosphorylation of PtdIns and PtdInsP is accompanied by phosphomonoesterases simultaneously removing the 4 and 5 phosphates of PtdIns(4)P/PtdIns(4,5)P$_2$. This interconversion between PtdIns, PtdInsP, and PtdInsP$_2$ has been named a 'futile cycle'—although it is likely to be anything but futile! As the incorporation of ^{32}Pi into the diester-P of PtdIns occurs via a multi-step biosynthetic pathway of a major membrane phospholipid, it generally proceeds at a slower rate than the kinase-mediated incorporation of ^{32}P into the PPtdIns phosphomonoesters. It is, however, important to remember that the chemical quantities of the poly-phosphoinositides are small compared with those of the major membrane phospholipids. So, although tracer incorporation into PPtdIns may be fast, the total level of incorporated radioactivity normally only constitutes a modest proportion of total phospholipid label—particularly after longer incubation times when isotopic equilibrium is approached.

The high-energy beta-radiation of the ^{32}P isotope combined with its easy incorporation into most plant tissues (and membrane phospholipids) makes it perfectly suited for labelling phosphoinositides, especially if detection relies on autoradiography.

Little is known about the metabolism of inositol phosphates in plant tissues. However it is likely that ^{32}P-label in plant InsPs originates from either:

- kinase-mediated phosphorylation of inositol/inositol monophosphate (derived from glucose-6-phosphate) using ATP as P-donor

- hydrolysis of phosphoinositides

Many of the water-soluble contaminants co-extracted with InsPs contain considerably higher levels of tracer than the InsPs (for example ATP, GTP, P-esters, etc.). This makes identification of individual inositol phosphates very difficult.

3.2 [^{32}P]ATP

The use of [^{32}P]ATP circumvents the metabolic steps responsible for ^{32}P incorporation into the gamma-phosphate of ATP. However, the cost of [^{32}P]ATP and the likely presence of the major pool(s) of plant polyphosphoinositides in the *inner* leaflet of the plasma membrane double layer means that [^{32}P]ATP has been used exclusively for labelling poly-phosphoinositides in isolated membrane-fractions (for example microsomes/plasmalemma vesicles, etc.). [^{32}P]ATP is ideally suited for this purpose as the rapid turnover rate of PtdIns(4)P [and possibly PtdIns(4,5)P$_2$] in plant membranes means that this/(these) lipid(s) will have a relatively high specific activity compared to the lipids in the main phospholipid pool if short labelling times are employed. It should also be remembered that many enzymes/substrates involved in *de novo* lipid synthesis may be absent in membrane preparations, whereas membrane-bound kinases remain active. [^{32}P]ATP has not, so far, been employed for labelling inositol phosphates in plant systems.

3.3 myo[2-^{3}H]Inositol

myo[2-^{3}H]Inositol is one of the most useful compounds in the study of PPtdIns/InsP metabolism. It is easily taken up by cultured cells, leaf discs, and excised roots, and is selectively incorporated into compounds having inositol moieties. However, in plants several inositol-containing compounds exist, other than phosphoinositides and inositol phosphates. Enzyme systems are present in most plant cells which are able to convert [2-^{3}H]inositol into other inositol derivatives without the loss of the ^{3}H label in the 2'-position of the ring. Such derivatives are known precursors of, for example, cell wall components (24). While [2-^{3}H]inositol provides a valuable tool for the selective labelling of components of the PI-system, great care should be taken before conclusions are made about the structure/origin of a ^{3}H-inositol labelled compound extracted from plant tissues. The main drawback of [2-^{3}H]inositol is the very low energy of the β-emission of the ^{3}H-isotope which necessitates detection by direct liquid scintillation counting, or fluorography, or the use of expensive film sensitive to low-level β-radiation (e.g. Hyperfilm -^{3}H, Amersham International).

D-myo-inositol is also available in a ^{14}C-labelled form, although its cost and low specific activity make it of limited use for routine experiments.

3.4 [^{14}C]/[^{3}H]Glycerol and [^{14}C]/[^{3}H] fatty acids

These compounds can be employed only for labelling phosphoinositides as the label is lost by hydrolysis to inositol phosphates, but nevertheless they can be very useful in metabolic studies of plant phosphoinositides (see, for example, reference 12). The diacyl-moiety of plant phosphoinositides appears to be far less rich in arachidonic acid than is the case in many mammalian cells (Drøbak and Ferguson, unpublished data) so, for example, the use of [^{14}C]-arachidonate as a 'semi-specific' precursor for plant phosphoinositide labelling is likely to be of limited merit.

4. Analysis of plant phosphoinositides

Many techniques currently employed for the extraction and analysis of phosphoinositides are based on methods originally developed for general membrane phospholipid biochemistry and modified to account for particular features of polyphosphoinositides.

4.1 **Extraction of phosphoinositides from plant tissues**

The procedure most widely used for extraction of phosphoinositides from various tissue types is based on the Folch extraction (25). This method involves solubilization of membrane phospholipids in mixtures of chloroform ($CHCl_3$) and methanol (MeOH) followed by removal of the polar cell components by washing the non-polar phase with aqueous salt solutions. Despite their phospholipid nature polyphosphoinositides are comparatively polar due to the presence of the mono-ester phosphate groups. Special precautions must be taken to ensure that PPtdIns [in particular $PtdIns(4,5)P_2$] remain in the non-polar organic phase throughout the extraction and washing procedures. This is normally achieved by inclusion of hydrochloric acid (HCl) in both the extraction and wash media. The acidic conditions suppress protonation of both the monoester-P and diester-P groups.

The extraction procedure described in *Protocol 1* is modified from the original Folch method and has been extensively used in our laboratory to extract phosphoinositides (phospholipids) from a wide range of plant tissues.

Protocol 1. Extraction of phosphoinositides from plant tissues

CAUTION: All steps in this extraction procedure should be carried out in a fume hood as chloroform is a recognized carcinogen.

1. Prepare the tissue to be extracted by, for example, labelling with a suitable radioactive precursor.

2. Remove the incubation medium rapidly and quench the tissue by the immediate addition of ice-cold $CHCl_3$–methanol–conc HCl. (100:100:0.7, v/v/v). The volume of $CHCl_3$–methanol–HCl used depends on the amount of the tissue to be extracted. Use around 10 ml to extract approx. 0.5 g (fresh weight) of cultured cells or an equivalent amount of other tissue.

2. Transfer to a cooled homogenizer, for example a Kondes glass-grinder (40 ml capacity), and homogenize the tissue. A 1-phase system should be present after homogenization. If the tissue has a very high water-content or incubation medium is carried over forming a 2-phase system, it may be necessary to add a small amount of methanol at this stage.

4. Add 0.6 M HCl (2 ml per 10 ml of $CHCl_3$–methanol–conc HCl) to the tissue homogenate. This creates a two-phase system; the bottom phase is predominantly $CHCl_3$ (with small amounts of methanol and HCl) and contains non-polar compounds, while the top-phase is a polar H_2O–methanol–HCl phase containing water-soluble compounds.

5. Remove and discard the top-phase. **Note:** When ^{32}Pi, $[\gamma\text{-}^{32}P]ATP$, $[^3H]$inositol, or other water-soluble radioactive precursors are used, the top-phase contains the majority of unincorporated tracer, and it is thus important to dispose of this phase in a suitable manner.

6. Wash the bottom-phase 3 times with 4 ml aliquots of $CHCl_3$–methanol–0.6 M HCl (3:48:47, v/v/v), each time removing and discarding the top-phase. The volume of wash-solution stated here is based on using 10 ml of initial quench-solution and so should be adjusted accordingly.

7. Transfer the washed bottom-phase and any inter-phase material to glass centrifuge tubes and centrifuge at 2000 g for 5 min. If the bottom-phase is not clear repeat the centrifugation step.

8. Recover the clear bottom-phase containing the extracted lipids with a Pasteur pipette and transfer this to small glass storage containers.

9. Evaporate the extract to dryness under a stream of O_2-free N_2.

10. Add a small volume of $CHCl_3$–methanol (1:1, v/v) to the dry extract.

11. Flush the storage container briefly with N_2 and seal it.

12. Store the extract in a sealed container at $-20°C$. Some oxidation of fatty acids and degradation of phospholipids to their lyso- and glycero-phosphoryl-form does occur over time. For accurate quantitative analysis the best results are obtained if the extract is analysed as soon after extraction as possible. If appreciable amounts of highly unsaturated lipids are present in the extract, 0.05% butylated hydroxytoluene should be added as an anti-oxidant.

4.2 Analysis of radiolabelled phosphoinositides

Many techniques have been developed to separate and analyse phospho-inositides, including TLC (1- and 2-dimensional), silic acid column chroma-tography, affinity chromatography on neomycin columns, HPLC, and chemical conversion of phosphoinositides followed by analysis of their derivatives. We will describe only two methods

- the most commonly used i.e. TLC (Section 4.3)
- the method that we routinely use for more detailed analysis of plant phospho-inositides i.e. O → N transacylation followed by chromatography of glycero-phosphorylinositides (GroPtdInsPs) (Section 4.4). In the following discussion it is assumed that the extracted phosphoinositides are radioactively labelled.

4.3 Thin layer chromatography (TLC)

Research into plant phosphoinositides has, until recently, relied almost exclusively on TLC which has many virtues: it is cheap, comparatively rapid, easy to use, and gives reproducible and accurate results. Several excellent reviews containing detailed information about TLC techniques have been published (see, for example, references 26–28) so we will concentrate on those aspects of TLC which are of particular relevance to the analysis of plant phosphoinositides.

Divalent cations (in particular Ca^{2+}) affect the TLC separation of polyphos-phoinositides, thus influencing the choice of plate type. If plates can be manufactured in the laboratory, the mixed-layer plates (Silica Gel H/cellulose) described by Drøbak *et al.* (12) give excellent separation. If commercial plates are used, it is important to choose Silica Gel H plates as the G-type plates contain large amounts of Ca^{2+}. The inclusion of Ca^{2+}-binding agents into the layer of commercial plates is often necessary for optimal results, this is most easily achieved by spraying the plates with 1% potassium oxalate to saturation before activation.

A multitude of solvent systems have been employed for the separation of inositol lipids. We describe two solvent systems that have proved to be particularly useful (see *Table 1*). *Protocol 2* describes the procedure for the separation of plant phosphoinositides by one-dimensional TLC. *Figure 3* shows the approximate positions of plant phospholipids separated by TLC using the solvent systems described in *Table 1*.

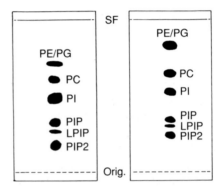

Figure 3. TLC separation of phospholipids. The approximate positions of phospholipids separated on silica gel H plates using CMEWA (left lane) and CAMAW (right lane) as running solvents are shown. PIP$_2$, phosphatidylinositol(4,5)bisphosphate; LPIP, lysophos-phatidylinositol(4)phosphate; PIP, phosphatidylinositol(4)phosphate; PI, phosphatidyl-inositol; PC, phosphatidylcholine; PE, phosphatidylethanolamine; PG, phosphatidyl-glycerol. The exact position of the individual phospholipid may vary depending on experimental conditions (e.g. humidity, temperature, etc.).

Protocol 2. TLC analysis of plant phosphoinositides

1. Remove excess CHCl$_3$–methanol from the stored samples by evaporation, and reconstitute the dried lipid extract in 50–250 μl CHCl$_3$–methanol (1:1, v/v). **CAUTION**: note warning in *Protocol 1*.)

2. Activate the potassium oxalate sprayed plate(s) at 110°C for 45 min. Remove the plate(s) from the oven and leave to cool in a desiccator containing water-absorbing material.

3. Make up the solvent of choice (e.g. *Table 1*). It is very important that solvents for TLC are freshly made just before use.

4. Transfer the solvent to a paper-lined TLC tank to a level approx. 1 cm above the bottom of the tank. Leave the solvent in the tank to equilibrate for at least 15–30 min before the plates are run.

5. Apply aliquots of the lipid extract to the plate in spots (5–50 μl) or streaks approx. 2–2.5 cm from the bottom of the plate.[a]

6. Gently dry the applied lipid sample (for example with a hair-dryer). Place the plate in the TLC tank and allow it to develop until the solvent has reached approx. 2–3 cm from the top edge of the plate.

7. Remove the plate and let the solvent evaporate. Rapid drying can be achieved by use of a hair-dryer.

8. Visualize the separated lipids. Assay the chemical amounts, radiotracer incorporation, etc. in individual lipids.

---

[a] The exact amount of lipid to be used depends both on the chemical amount and type of lipids in the extract, and their specific activity. The amounts to be used are best determined by trial and error.

4.3.1 Visualization of phosphoinositides separated by TLC

For a detailed and comprehensive description of methods for the visualization of phospholipids (phosphoinositides) on TLC plates see references 29, 30. In most cases the agents used for visualization are destructive (i.e. the phospholipids cannot be used for further analysis). In the plant tissues we have so far studied, the chemical levels of PtdIns(4)P have been low (a few per cent of total phospholipids) and the levels of PtdIns(4,5)P$_2$ *very* low (less than 0.1% of total phospholipid). The former can only be visualized using current chemical techniques if high levels of phospholipids are loaded on to the TLC plates. Where internal PtdInsP/PtdInsP$_2$ standards are used it is necessary to employ a non-destructive method of visualization if the underlying (radio-labelled) polyphosphoinositides are to be quantified. In our experience this is

---

Table 1. Two solvent systems for the separation of plant phosphoinositides by TLC

CAMAW

Chloroform–acetone–methanol–acetic acid (glacial)–water (40:15:13:12:8, v/v/v/v/v)

CMEWA

Chloroform–methanol–water–ammonia (saturated) (45:35:8:2, v/v/v/v)

---

best achieved by the exposure of separated phospholipids to iodine vapour. Leave the TLC plates for 15 min to 1 h after exposure before setting up for autoradiography since excessive iodine vapour can cause blurring of the auto-radiogram. Some TLC systems show 'double-running' of polyphosphoinositides (PPtdInsP) from different sources. This is probably due to differences in fatty acid composition. At present, all commercially available PPtdInsP standards are prepared from non-plant sources, so this problem should be borne in mind. When using the CAMAW system (see *Table 1*) we have seldom had problems with PPtdIns 'double-running'.

Although TLC is often quite adequate for the analysis of PtdInsPs from animal sources, it is not always adequate for plants. There are two major reasons for this. First, only three major inositol containing lipids [i.e. PtdIns, PtdIns(4)P and PtdIns(4,5)P$_2$)] are present in most mammalian tissues, whereas a considerable number of ill-characterized phosphate- and inositol-containing lipids exist in plant tissues (12, 31). Some of these lipids have chromatographic properties resembling those of PtdIns(4,5)P$_2$ in several of the most commonly used TLC-systems. Second, the low tracer incorporation into PtdIns(4,5)P$_2$ in plant tissues means that unless very high levels of precursors or internal standards, are employed, the search for the PtdInsP$_2$ spot on a TLC plate is much like looking for the proverbial 'needle in the haystack'.

In summary, if due care is taken in choosing both the tracer type and incubation times, the metabolism of both PtdIns and PtdInsP in plant cells can be investigated by current TLC-techniques. If, however, an overall pic-ture of the relative distribution between PtdIns/PtdInsP and PtdInsP$_2$ is needed, other techniques must be employed. We have found that deacylation (O → N transacylation) of phosphoinositides followed by analysis of the re-sulting glycerophosphorylinositides (GroPtdInsPs) provides a useful alterna-tive to TLC for this purpose.

4.4 O → N transacylation of phospholipids

4.4.1 Procedure

Several methods have been developed for the deacylation of phospholipids. Most of these methods, based on alkaline methanolysis (alkali metal hydroxide alcolysis), are only semi-quantitative and lead to side formation of com-pounds other than glycerophosphoryl derivatives. The O → N transacylation method developed by Clarke and Dawson (32) using monomethylamine as both alkali and acyl acceptor is, however, quantitative and little side-reaction occurs. the nature of the reaction is illustrated in *Figure 4*.

This method does require some preparation but once set up it is both quick and easy to use. *Protocol 3* describes the preparation of the monomethylamine reagent, while the method for O → N transacylation is outlined in *Protocol 4*.

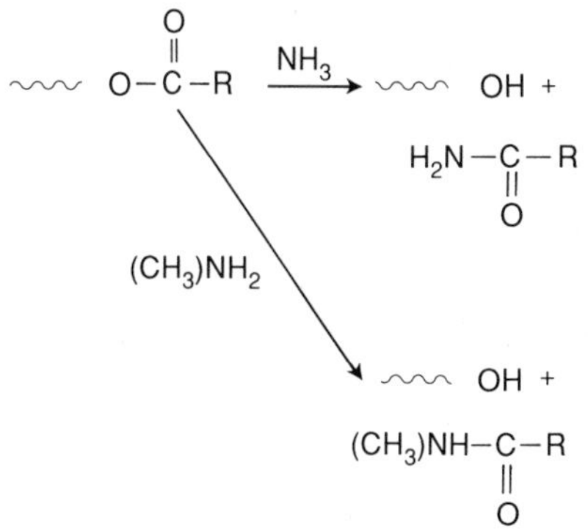

Figure 4. O → N transacylation using ammonia or monomethylamine. The formation of fatty acid amides and monomethyl derivatives is illustrated. (R = fatty acid.)

Protocol 3. Preparation of the monomethylamine reagent

1. Obtain monomethylamine as liquefied gas in small cannisters (e.g. 250 ml) from a commercial source (e.g. BDH or Fluka).

2. Prepare a mixture of 144 ml methanol and 108 ml H_2O in a 1 litre plastic bottle and mark the level of liquid in the bottle.

3. Place the plastic bottle containing methanol–H_2O on dry ice.

4. Place the monomethylamine cannister in water and connect it to a trap.

5. Connect the trap outlet to a glass tube which is placed in the methanol–H_2O mixture.

6. Gently bubble monomethylamine gas through the methanol–H_2O mixture.

7. When the volume of the mixture has about doubled stop the bubbling of monomethylamine gas and add 36 ml of *n*-butanol.

8. Store the monomethylamine reagent in a tightly-closed plastic container at −15 to −20°C. The reagent stores well under these conditions and remains usable for many months.

Protocol 4. O → N transacylation of plant lipids

1. Transfer the lipid sample in a small volume of $CHCl_3$–methanol (1:1, v/v) to a long-necked pear-shaped glass flask with a tapering bottom. (**CAUTION**: note warning in *Protocol 1.*)

2. Evaporate the $CHCl_3$–methanol under a stream of O_2-free N_2.

Protocol 4. *Continued*

3. Quickly add 3 ml of cold monomethylamine reagent and stopper the flask immediately. Secure the stopper with strong adhesive tape.

4. Swirl the flask and warm gently to facilitate solubilization of the lipids.

5. Incubate the stoppered flask in a water-bath at 53°C for 45 min.

6. Transfer the flask to ice and leave for 5–10 min to cool.

7. Remove the stopper (*with care!*) and evaporate *in vacuo* using a splash head. Keep the bottom of the flask cool by submerging in a water–ice mixture.

8. After a few minutes (when any sign of frothing has ceased) increase the temperature by substituting the ice–water mixture with tap-water and continue the evaporation.

9. Repeat step 8, but substitute lukewarm water for tap-water.

10. Repeat step 9, but increase the temperature of the water to approx. 50°C and continue evaporation to dryness.

11. Add 1 ml of water and 1.2 ml of a mixture containing *n*-butanol–light petroleum (b.p. 40–60°C)–ethyl formate (20:4:1, v/v/v) to the dry residue.

12. Shake the flask several times and leave for 3–5 min for the phases to settle.

13. Remove and discard the upper phase which contains the fatty acid derivatives.

14. Wash the bottom-phase, which contains the glycerophospholipids, with 750 μl of the solvent mixture described in step 11 above. Discard the top-phase.

15. Repeat step 14, but leave approximately one-third of the top-phase.

16. Transfer the bottom-phase and the remaining top-phase to microcentrifuge tube(s).

17. Centrifuge the tubes at 12 000 g for 2 min in a microcentrifuge.

18. Recover the bottom-phase and transfer this to fresh tubes and store glycerophospholipids at −20°C. GroPtdInsPs are comparatively stable if stored at −20°C in neutral solutions.

4.4.2 Analysis of GroPtdInsPs by simple anion exchange chromatography and HPLC

We currently employ two different methods for the separation of glycerophosphorylinositides (GroPtdInsPs). For routine experiments we use small trimethylaminopropyl anion exchange columns, while for more detailed analysis HPLC is used.

i. Anion exchange (Ampreps)

We use Amprep columns (Amprep, Trimethylaminopropyl SAX, 100 mg, Amersham International Ltd) for the separation of GroPtdInsPs (*Protocol 5*).

Protocol 5. Analysis of GroPtdInsPs by anion exchange chromatography on Amprep[a] columns

1. Convert Amprep columns to the formate form by washing with five 1 ml aliquots of ammonium formate–formic acid (1.0 M: 0.1 M).[b] Some positive pressure is needed to maintain a suitable flow rate. Amprep manifolds are commercially available, but we routinely use an aquarium pump with variable speed; a flow rate of 1 ml/15 sec is appropriate.

2. Wash each column with 10 ml of water.

3. Load the deacylated lipid sample on to the column.

4. Elute GroPtdInsPs with the following amounts of AMFA:
   - GPI: twelve 1 ml aliquots of 0.04 M AMFA[b]
   - GPIP: twelve 1 ml aliquots of 0.20 M AMFA[b]
   - GPIP$_2$: twelve 1 ml aliquots of 0.45 M AMFA[b]

5. Elute the column with five 1 ml aliquots of 1.0 M AMFA to ensure that all labelled compounds have been eluted.

6. Assay radioactivity in the collected fractions by liquid scintillation spectrometry.

[a] Other types of anion exchange columns (for example DOWEX) can be employed for the separation of GroPtdInsPs. Recently we have noted some variation between batches of Amprep columns, so it is strongly recommended that authentic standards are always used to check the specific elution characteristics of each new batch.

[b] AMFA; where only one molarity is shown it refers to that of the ammonium formate, formic acid is diluted proportionately.

---

ii. HPLC

For the separation of GroPtdInsPs, by HPLC, we use a Partisil 10-SAX column (0.46 × 25 cm, Whatman). The gradient system is described in *Table 2* and details of running conditions are given in the legend to *Figure 5*.

5. Analysis of inositol phosphates

Two methods for the extraction of plant inositol phosphates together with a number of methods currently used for their separation are described in this section.

Table 2. Gradient system for separation of GroPtdInsPs by HPLC using a Partisil 10 SAX column.

Time (min)	Ammonium formate (M) (pH 3.7 with H_3PO_4)
0	0 m
14	0 m
60	0.92 m
75	2.50 m

5.1 Extraction of InsPs

None of the methods employed in the extraction of inositol phosphates (InsPs) from tissues/cells have the ability to extract InsPs selectively. In most cases the extracted InsPs are present only in small amounts in the crude extract which contains high levels of a variety of water-soluble cellular components with both sugar and/or phosphate-moieties (for example sugar-alcohols, nucleotides, P-esters, etc.). This problem can best be overcome by the use of appropriate radioactive precursor(s) followed by optimal downstream separation techniques.

As stated in the introduction to this section, two main methods for InsP extraction are currently employed. Both methods are based on the use of a

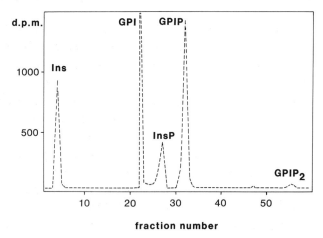

Figure 5. Separation of glycerophosphorylinositides by HPLC. The gradient system used is decribed in *Table 2*. Flow rate : 1.25 ml/min. Samples were collected at 1 min intervals. Sampling was started at $t = 6$ min (sample 1). Ins, inositol; GPI, glycerophosphorylinositol; InsP, inositolphosphate; GPIP, glycerophosphorylinositol(4)phosphate; GPIP$_2$, glycerophosphorylinositol(4,5)bisphosphate.

quenching agent which precipitates protein. One method employs trichloro-acetic acid (TCA) as the quenching agent followed by removal of the TCA from the extract (neutralization) by water-saturated diethyl ether. In the other method, TCA is replaced by perchloric acid (PCA), and the PCA is removed from the extract by the Sharps and McCarl-phase partitioning against freon and tri-*n*-octylamine method (33). In our experience both methods give a quantitative and reproducible recovery of InsPs from a variety of plant tissues/cells. While the TCA-method is very easy to use, it is slightly more time-consuming than the PCA-method. The PCA-method does require a little 'hands-on' experience, but it is time well-spent. Whenever many small-volume samples have to be processed in a short space of time, we employ the PCA-method. Both methods are described in *Protocol 6*.

Protocol 6. Extraction of inositol phosphates from plant tissues

A. *PCA method*

1. Prepare the plant tissue for extraction (for example by suitable radio-labelling).

2. Quench the reaction by rapidly adjusting the incubation medium to 5.0% (w/v) PCA by adding the appropriate amount of ice-cold PCA.[a] Depending on tissue type, homogenization may be needed at this stage.

3. Transfer 1.0 ml aliquots of the PCA/tissue homogenate to 1.5 ml Eppendorf/Sarstedt microcentrifuge tubes and leave on ice for 10 min.

4. Centrifuge at 12 000 g for 2 min in a microcentrifuge to pellet protein and cell debris.

5. Use a replacement tip pipette to transfer 400 µl aliquots of the supernatant to 1.5 ml screw-top Eppendorf/Sarstedt microcentrifuge tubes.

6. Prepare a fresh solution of freon and tri-*n*-octylamine (1:1, v/v) under cold conditions. Do *not* use old solutions of this reagent.

7. Add 650 µl of the freon and tri-*n*-octylamine solution to each of the microcentrifuge tubes containing the PCA-homogenate samples.

8. Vortex each tube very vigorously for at least 20–30 sec.

9. Centrifuge for 2 min at 12 000 g in a microcentrifuge. A 3-phase system is obtained. The bottom-phase consists of freon and unreacted tri-*n*-octylamine, the middle phase contains tri-*n*-octylamine perchlorate, and the top phase is the neutralized sample containing the inositol phosphates.

10. Recover the top-phase and store at −20°C until required for analysis.[b]

B. *TCA-method*

1. Prepare the tissue for extraction (as part A, step 1).

2. Quench the reaction by rapidly adjusting the incubation medium to 10%

Protocol 6. *Continued*

  TCA[a] (w/v). Again, depending on the tissue type, homogenization may be needed at this stage.

3. Transfer 1 ml aliquots to Eppendorf microcentrifuge tubes. Leave the tubes on ice for 15 min.

4. Centrifuge the tubes at $12\,000\,g$ for 2 min to pellet protein and cell debris.

5. Recover and pool all the supernatants in a stoppered glass test tube.

6. Add 1–1.5 vol. water-saturated diethyl ether to each extract, stopper the flask and vortex. A two-phase system is obtained. If the phase-boundary is not clear it may be necessary to centrifuge the sample briefly.
  NOTE! Diethyl ether is *very* flammable/explosive so caution should be taken to ensure that no sources of ignition are present.

7. Remove the top-phase and repeat the diethyl ether wash (step 6 above) 4–6 times to remove the majority of TCA.

8. After 4 washes, start to monitor the pH of the extract after each wash (e.g. by the use of pH indicator strips). Stop the washes when the extract is only slightly acidic.

9. Neutralize the extract by the addition of a small amount of dilute ammonia.

10. Store the extract at $-20\,°C$ until use.[b]

---

[a] It is important that the addition of PCA/TCA is accompanied by a rapid drop in temperature. The volume of ice-cold PCA/TCA to be added should thus not be too small unless other means of cooling are employed. Normally addition of an equal volume of 10% (w/v) PCA/20% TCA (w/v) to the incubation mixture is appropriate.

[b] Inositol phosphates are stable if stored frozen in neutral solutions.

5.2 Separation of InsPs

Simple column chromatographic techniques for InsP separation rely on differences in net negative charges created by the number of P-monoester groups attached to the inositol ring. The chromatographic separation of inositol phosphates into fractions containing $InsP_1$, $InsP_2$, $InsP_3$, etc. by the use of simple anion exchange columns (for example Dowex) has led to much new information about phosphoinositide and inositol phosphate metabolism. However, the high degree of complexity of inositol phosphate metabolism in cells and the large differences in physiological specificity between individual InsP isomers means that separation simply into 'phosphorylation-classes' is inadequate for most metabolic studies. Chromatographic techniques nevertheless remain extremely useful both for bulk preparation and desalting of InsP-fractions prior to further analysis.

5.2.1 Simple column chromatography

Separation of InsPs into phosphorylation classes by simple anion exchange column chromatography is described in *Protocol 7*.

Protocol 7. Dowex anion exchange chromatography of inositol phosphates

1. Convert a Dowex anion exchange column (1.5 × 0.6 cm, Dowex AG 1-X8) to the formate form by passing 15 bed-volumes of 1 M ammonium formate through the column.
2. Wash the column with 15 bed-volumes of distilled H_2O.
3. Dilute the inositol phosphate extract (from *Protocol 3*) 4-fold with H_2O.
4. Adjust the pH of the InsP extract to approx. pH 8.0 with dilute ammonia.
5. Apply the extract to the Dowex column and collect 1 or 2 ml fractions.
6. Elute the column sequentially with 12 ml of the following solutions:
   (a) Distilled water: check for radioactivity in the last fraction before changing to the next wash reagent in (b) below; the radioactivity should be near background levels.
   (b) 5 mM disodium tetraborate, 60 mM sodium formate
   (c) 0.1 M formic acid, 0.2 M ammonium formate
   (d) 0.1 M formic acid, 0.4 M ammonium formate
   (e) 0.1 M formic acid, 1.0 M ammonium formate

Using *Protocol 7* inositol phosphates will be eluted in the following order:

(a) (water) inositol
(b) glycerophosphoinositol
(c) inositol phosphate
(d) inositol bisphosphate
(e) inositol tris/tetrakis phosphates

5.2.2 Separation of inositol phosphates by HPLC

The potential presence of 63 isomers of phosphorylated D-myo inositol, several stereoisomers of inositol, and the presence of the phytic acid ($InsP_6$) pathway makes it obvious that very specific techniques need to be employed whenever plant inositol phosphates are analysed in the context of phospho-inositide-mediated signal transduction.

The most efficient technique for plant-derived inositol phosphate separation is by HPLC. With this technique advantage can be taken of the *position* of the phosphomonoesters around the inositol ring allowing individual isomers, in many cases, to be separated. The degree of separation which can be achieved by HPLC is determined by a combination of factors such as: type of column, solvent system, and distribution/number of P-monoesters around the inositol ring. Even with the degree of separation obtained by HPLC, in some cases further analysis may be necessary before unequivocal statements can be made about the exact structure of a specific inositol phosphate isomer.

The HPLC-procedure described in *Protocol 8* was developed with the aim of providing an overall picture of *all* inositol phosphates ($InsP–InsP_6$) in plant extracts in *one* HPLC-run. Clearly when a spectrum of compounds with large differences in polarity are separated in one step, the degree of resolution of individual compounds with very similar polarities (adsorption characteristics) is reduced.

Protocol 8. Separation of inositol phosphates by HPLC

A. *Sample preparation*

1. Freeze-dry the extracted inositol phosphates if the volumes are too large. Samples are normally applied to HPLC columns in 25–1000 µl aliquots.

2. Reconstitute each sample in an appropriate volume of water. Ensure that the pH is near neutral.

3. Filter each sample through an HPLC certified filter to avoid column blockage.

4. Apply the sample to the HPLC column.

B. *Separation*

1. Use a Partisil 10-SAX (0.46 × 25 cm) column (Whatman).

2. Prepare the gradient system:
   - A = water
   - B = 3.0 M ammonium formate, pH 3.7 with H_3PO_4

3. Adjust the flow rate to 1.25 ml/min and collect 1.2 ml fractions, according to the following regime.

Time (min)	% B
0	0
14	0
60	23
89	100
100	100
110	0

4. Monitor absorbance, if possible, at 254 nm to check that the run is progressing as expected. AMP, ADP, and ATP can be used as internal markers.

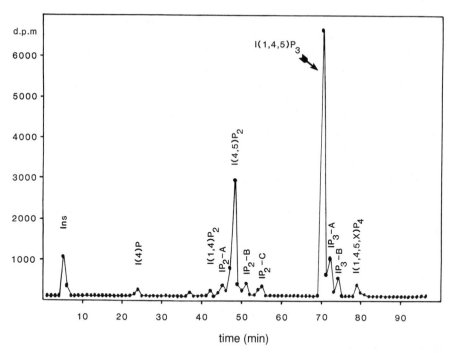

Figure 6. HPLC profile of water-soluble ³H-labelled compounds obtained after 1 minute incubation of [2-³H]Ins(1,4,5)P₃ with soluble enzymes from pea roots. Sampling was started immediately after injection of the sample ($t = 0$) and 1.25 ml samples were collected. The isomeric configuration is indicated for selected compounds. (Figure reproduced with permission from Drøbak *et al.*, *Plant Physiology*, In press.)

Figure 6 shows the separation of ³H-labelled inositol phosphates obtained from incubations of soluble enzymes from pea roots with [2-³H]Ins(1,4,5)P₃. The position of elution of authentic InsP isomers is indicated.

Problems can occur in liquid scintillation counting (LSC) analysis of samples from the latter part of the run due to the high salt content. To overcome this problem we prefer to use a special salt tolerant scintillation fluid (for example Hionic Fluor, Flo-Scint, Packard).

Recently, an elegant system for the improved separation of individual InsP isomers, which combines separations by Amprep/Dowex columns and isocratic HPLC, has been developed by Wreggett and Irvine (34). For further information on InsP-isomer separation see references 35, 36.

5.2.3 Other methods for inositol phosphate separation

Many other methods for the separation of InsPs have been developed, such as paper chromatography (37) and ionophoresis (38, 39). Several types of anion-exchange columns other than Dowex have been employed, for example

Ampreps (Amersham) and SEP-PAKS (Water Associates). Recently a number of assays for specific, physiologically important, InsP$_x$-isomers have been developed by utilizing cellular receptor proteins (40–42). An Ins(1,4,5)P$_3$ 'assay-kit' based on this principle is now commercially available from Amersham. The progress in development of biological assay systems with high specificity for selected InsP-isomers could shortly make the assay of Ins(1,4,5)P$_3$ and Ins(1,3,4,5)P$_4$, for example, a routine matter. However, it should be borne in mind that these assays have been developed for use in mammalian systems and the degree of cross-reactivity with potential unknown plant or pathogen InsPs has yet to be determined. For the structural (isomeric) analysis of InsP$_2$/InsP$_3$/(InsP$_4$), the 'periodate–borohydride–dephosphorylation-method' still remains the most direct approach. This technique utilizes the conversion into alditols of InsPs with vicinal C-atoms without P-monoesters. The D/L-configuration and the type of alditol obtained directly reflects the positions of P-monoesters in the parent InsP$_x$. For a detailed description of this method see references 13, 43, 44.

6. Diacylglycerol and protein kinase C

As outlined in Section 1, an integral feature of the birfurcated pathway of the phosphoinositide system in mammalian cells is the activation of a specialized protein kinase, called protein kinase C. This protein kinase, discovered in 1977 by Nishizuka and co-workers (6), is a Ca^{2+}- and phospholipid-dependent enzyme. A key role in transmembrane signalling for this kinase became apparent when it was discovered that 1,2-diacylglycerol (DG) produced by agonist-induced hydrolysis of PtdIns(4,5)P$_2$ strongly increased the Ca^{2+} affinity of PKC (2), thereby rendering it fully active even at low Ca^{2+} activities. Of the diacylglycerols containing the 1,2-*sn* configuration, molecular species containing an unsaturated fatty acid are the most effective for PKC activation. The two stereoisomers 1,3-diacylglycerol and 2,3-diacylglycerol are nonfunctional, which suggests that the lipid–protein interaction necessary for activation is highly specific. By analogy with the rapid metabolism of Ins(1,4,5)P$_3$, 1,2-DG only has a short residence time in the membrane and is either broken down yielding arachidonic acid, or converted back to phosphoinositides after phosphorylation to phosphatidic acid. The recent interest in PKC in mammalian tissues has been stimulated by the finding that this enzyme is activated by tumour-promoting phorbol esters (2, 6). Phorbol esters such as 12-*O*-tetradecanoyl-phorbol-13-acetate (TPA), appear to be structural analogues of diacylglycerol, and have been demonstrated to activate PKC both *in vitro* and *in vivo*. In contrast to DG, phorbol esters are very slowly degraded, and may thus lead to a permanent activation of PKC. More than 40 likely substrates for PKC have been proposed and it is now clear that at least seven PKC-isozymes, having slightly differing kinetic characteristics, exist in mammalian tissues. Some evidence exists for the presence in plant tissues of enzyme activity which in several respects resembles mammalian PKC. Schäfer *et al.* (18) demonstrated the presence of a Ca^{2+}- and phospholipid-dependent protein kinase in a soluble fraction from zucchini hypocotyls, and evidence for a similar kinase in a number of other plant tissues has been reported by Elliott and Skinner (17). The kinase investigated by Elliott and Skinner showed a similar behaviour on DE 52 cellulose columns to that of animal PKC, and exhibited substrate specificity and Ca^{2+}-dependency comparable to that of the animal enzyme. These authors, however, did not find any specific binding of phorbol esters to the enzyme, and concluded that complete identity between the plant enzyme and animal PKC is doubtful. The presence of a Ca^{2+}-dependent protein kinase in wheat cell homogenates, specifically stimulated by phosphatidylserine and phorbol esters (TPA), has been investigated by Olah and Kiss (16).

In summary, present evidence suggests that an enzyme (or group of isozymes) resembling mammalian PKC exists in plant cells. Whether this enzyme (or enzymes) has a physiological role similar to PKC in animal cells remains to be elucidated. Several approaches to the study of a putative DG/PKC signalling pathway involved in plant–pathogen interaction can be envisaged. One approach is to monitor the possible signal-induced DG-production, another is to assay the activity and/or activation of PKC-like enzymes (directly of indirectly). Third, compounds interfering with the normal function of PKC ´(for example phorbol esters) could potentially prove useful for studying the effects related to signal-induced PKC-activation. In *Protocols 9* and *10* methods are described for the assay of DG and PKC-activity respectively.

6.1 Diacylglycerol

The assay for diacylglycerol outlined in *Protocol 9* has kindly been provided by Drs Richard C. Crain and Lynne M. Quarmby of the University of Connecticut and is based on the method of Preiss *et al.* (45). Recently, this procedure was successfully applied to the study of cellular diacylglycerol levels during light-stimulated phosphoinositide-turnover in *Samanea saman* pulvini (46). The method uses the quantitation of radioactivity in [^{32}P]phosphatidic acid formed from diacylglycerol as a measure of produced DG, giving linear incorporation of ^{32}P[Pi] between 100 pmol and 2 nmol of diacylglycerol. The amount of diacylglycerol can be calculated when the specific activity of [^{32}P]ATP is known. It should be pointed out that, until recently, the phosphoinositides have been considered to be the main substrates for lipase-mediated DG-production in signalling events, but new evidence suggests that other lipids, notably PC, may play a similar role. In detailed metabolic studies of signal-induced DG-production, it may thus be necessary to identify the exact nature of the parent lipid(s).

Protocol 9. Assay of *sn*-1,2-diacylglycerol

1. Transfer the lipid extract from *Protocol 1* in chloroform–methanol (1:1, v/v) to fresh test tubes (see **CAUTION** warning in *Protocol 1*).

2. Evaporate the $CHCl_3$–methanol under a stream of nitrogen.

3. Prepare a micellar substrate suspension by sonicating the lipids in a bath-type sonicator for 15 sec after the addition of 20 μl of a solution containing 7.5% octyl-β-D-glucoside, 5 mM cardiolipin, and 1 mM diethylene-triaminepentaacetic acid (DETAPAC).[a]

4. Add 50 μl of 100 mM imidazole–HCl, 100 mM NaCl, 25 mM $MgCl_2$, 2 mM EGTA (pH 6.6), and then add 10 μl 20 mM freshly prepared DTT.

5. Add 10 μl diacylglycerol kinase rich membrane (for example Lipid EX) diluted 10-fold with 10 mM imidazole, 1 mM diethylenetriaminepenta-acetic acid (pH 6.6).

6. Start the phosphorylation reaction by the addition of 10 μl 10 mM [γ-^{32}P]ATP (specific activity: 50 000–500 000 c.p.m./nmol). Incubate at 25 °C for 30–60 min.

7. Stop the reaction by the addition of 3.75 ml $CHCl_3$–methanol (2:1, v/v) and then add 0.84 ml of 0.15 M HCl containing 0.58% NaCl.

8. Vortex the mixture and then separate the phases by centrifugation at 500 g for 10 min. Discard the upper aqueous layer.

9. Wash the mixture 3–4 times with 1.0 ml aliquots of $CHCl_3$–methanol–0.7 HCl (3:47:48, v/v/v) followed each time by vortexing and centrifugation to separate the phases. Remove and discard the top-phase after each wash.

10. Transfer the bottom-phase to a scintillation vial and evaporate the $CHCl_3$–methanol. Determine radioactivity by liquid scintillation counting.

---

[a] This solution is prepared by first making a 7.5% solution of octyl glucoside in 1 mM DETAPAC which is then added to a test tube containing cardiolipin dried to a thin film.

6.2 Assay of protein kinase C-like enzyme activity in plant extracts

The method for the assay of protein kinase C-like activity outlined in *Protocol 10* is slightly modified from that of Schäfer *et al.* (18). Several protein kinases other than PKC are able to phosphorylate histones. Before conclusions are made with regard to PKC-like enzymes, it is thus important to ensure that the enzyme(s) under investigation is indeed Ca^{2+}- and phospholipid dependent and *can* be activated by DG (DG-analogues).

Protocol 10. Assay of PKC-like activity

1. Make up PKC assay-medium:
   - 20 mM Tris–HCl, pH 7.5
   - 5 mM $MgCl_2$
   - 670 µg/ml histone H1

2. Adjust the ATP concentration of the assay medium to 10 µM and add [γ^{32}P]ATP to approx. 1–5 × 10^7 d.p.m./ml. Transfer 150 µl aliquots of this mixture to Eppendorf to microcentrifuge tubes.

3. Adjust the Ca^{2+} and/or phosphatidylserine concentration of the reaction mixture as necessary.[a]

4. Add approx. 50 µl enzyme-solution. The amount of protein to be added is best determined by 'trial and error' experiments.

5. Incubate the tubes at 30°C for 5 min.

6. Stop the reaction by the addition of 1.2 ml ice-cold trichloroacetic acid (TCA) and leave the tubes on ice for 10 min.

7. Collect the protein precipitate on nitrocellulose filters (pore size 0.45 µm) by vacuum filtration and wash the filters twice with 5 ml aliquots of ice-cold TCA.

8. Dry the filters and assay radioactivity by liquid scintillation spectrometry.

---

[a] For preparation of Ca^{2+}/EGTA buffers see Chapter 10, *Table 1*. Phosphatidylserine (PS) suspensions can be prepared as follows:

- Evaporate PS in $CHCl_3$–methanol to dryness under a stream of oxygen-free N_2.
- Reconstitute in ethanol and 20 mM Tris, pH 7.5.
- Vortex for 5 min.

---

7. Concluding remarks

In this chapter, we have described some particularly useful methods for the analysis of components of the plant PI-system. However, very little is known about the involvement of the PI-system in plant–pathogen interactions and an increased understanding of processes will go 'hand in hand' with refinement of the methodology employed to study them. Therefore, the methods in this section should not be considered definitive—but rather constitute a 'pioneering kit'.

Without knowing the extent or timing of change in the metabolism of plant PI-system components it is often difficult to decide the most meaningful way in which to present/express data. So far, we have not found a final solution to this problem and the answer may vary from experiment to experiment. The best recommendation is probably to consider carefully all the basic metabolic steps involved in the process under study and then apply a good proportion of common sense. The approach we find most attractive, and often also most convenient, is to express data on the basis of a component(s) which remains constant during the experiment and is carried through the entire procedure of preparation of tissue extraction and analysis, that is, to use internal standards.

References

1. Berridge, M. J. and Irvine, R. F. (1984). *Nature*, **312**, 315.
2. Nishizuka, Y. (1984). *Nature*, **308**, 693.
3. Majerus, P. W., Neufeld, E. J., and Wilson, D. B. (1984). *Cell*, **37**, 701.
4. Shears, S. B. (1989). *Biochem. J.*, **260**, 313.
5. Berridge, M. J. and Irvine, R. F. (1989). *Nature*, **341**, 197.
6. Nishizuka, Y. (1986). *Nature*, **334**, 661.
7. Drøbak, B. K. and Ferguson, I. B. (1985). *Biochem. Biophys. Res. Commun.*, **130**, 1241.
8. Schumaker, K. S. and Sze, H. (1987). *J. Biol. Chem.*, **262**, 3944.
9. Ranjeva, R., Carrasco, A., and Boudet, A. M. (1988). *FEBS Lett.*, **230**, 137.
10. Boss, W. F. and Massel, M. (1985). *Biochem. Biophys. Res. Commun.*, **132**, 1018.
11. Heim, S. and Wagner, K. G. (1986). *Biochem. Biophys. Res. Commun.*, **134**, 1175.
12. Drøbak, B. K., Ferguson, I. B., Dawson, A. P., and Irvine, R. F. (1988). *Plant Physiol.*, **87**, 217.
13. Irvine, R. F., Letcher, A. J., Lander, D. J., Drøbak, B. K., Dawson, A. P., and Musgrave, A. (1989). *Plant Physiol.*, **89**, 888.
14. McMurray, W. C. and Irvine, R. F. (1988). *Biochem. J.*, **249**, 877.
15. Melin, P. M., Sommarin, M., Sandelius, A. S., and Jergil, B. (1987). *FEBS Lett.*, **223**, 87.
16. Olah, Z. and Kiss, Z. (1986). *FEBS Lett.*, **195**, 33.
17. Elliott, D. C. and Skinner, J. D. (1986). *Phytochemistry*, **25**, 39.
18. Schäfer, A., Bygrave, F., Matzenauer, S., and Marmé, D. (1985). *FEBS Lett.*, **187**, 25.
19. Morse, M. J., Crain, R. C., and Satter, R. L. (1987). *Proc. Natl. Acad. Sci. (USA)*, **84**, 7075.
20. Ettlinger, C. and Lehle, L. (1988). *Nature*, **331**, 176.
21. Strasser, H., Hoffmann, C., Grisebach, H., and Matern, U. (1986). *Z. Naturforsch.*, **41c**, 717.
22. Kurosaki, F., Tsurusawa, Y., and Nishi, A. (1987). *Plant Physiology*, **85**, 601.
23. Bieleski, R. L. and Ferguson, I. B. (1983). In *Encyclopedia of plant physiology*, Vol. 15 (ed. A. Lauchli and R. L. Bieleski), p. 422. Springer-Verlag, Berlin.
24. Harran, S. and Dickinson, D. B. (1978). *Planta*, **141**, 77.
25. Folch, J., Lees, M., and Sloane Stanley, G. H. (1957). *J. Biol. Chem.*, **226**, 497.
26. Kates, M. (1986). *Techniques of lipidology. Isolation, analysis and identification of lipids.* North-Holland Publishing Company, Amsterdam, Oxford.
27. Christie, W. W. (1982). *Lipid analysis* (2nd edn). Pergamon Press, Oxford.
28. Higgins, J. A. (1987). Separation and analysis of membrane lipid components. In

Biological membranes: a practical approach (ed. J. B. C. Findlay and W. H. Evans), p. 103. IRL Press, Oxford.

29. Krebs, K. G., Heusser, D., and Wimmer, H. (1969). Spray reagents. In *Thin-layer chromatography, a laboratory handbook* (ed. E. Stahl), p. 854. Springer-Verlag, Berlin.

30. Marinetti, G. V. (1962). *J. Lipid Res.*, **3**, 1.

31. Kaul, K. and Lester, R. L. (1975). *Plant Physiol.*, **55**, 120.

32. Clarke, N. G. and Dawson, R. M. C. (1981). *Biochem. J.*, **195**, 301.

33. Sharps, E. S. and McCarl, R. L. (1982). *Anal. Biochem.*, **124**, 421.

34. Wreggett, K. A. and Irvine, R. F. (1989). *Biochem. J.*, **262**, 997.

35. Dean, N. M. and Moyer, J. D. (1988). *Biochem. J.*, **250**, 493.

36. Irvine, R. F., Anggard, E. E., Letcher, A. J., and Downes, C. P. (1985). *Biochem. J.*, **229**, 505.

37. Brockerhoff, H. and Ballou, C. E. (1961). *J. Biol. Chem.*, **236**, 1907.

38. Seiffert, U. B. and Agranoff, B. W. (1965). *Biochim. Biophys. Acta*, **98**, 574.

39. Tate, M. E. (1968). *Anal. Biochem.*, **23**, 141.

40. Ross, C. A., Meldolesi, J., Milner, T. A., Satoh, T., Supattapone, S., and Snyder, S. H. (1989). *Nature*, **339**, 468.

41. Willcocks, A. L., Cooke, A. M., Potter, B. V. L., and Nahorski, S. R. (1987). *Biochem. Biophys. Res. Commun.*, **146**, 1071.

42. Bradford, P. G. and Irvine, R. F. (1987). *Biochem. Biophys. Res. Commun.*, **149**, 680.

43. Irvine, R. F., Letcher, A. J., Lander, D. J., and Downes, C. P. (1984). *Biochem. J.*, **223**, 237.

44. Stephens, L., Hawkins, P. T., Carter, N., Chahwala, S. B., Morris, A. J., Whetton, A. D., and Downes, C. P. (1988). *Biochem. J.*, **249**, 271.

45. Preiss, J., Loomis, C. R., Bishop, W. R., Stein, R., Niedel, J. E., and Bell, R. M. (1986). *J. Biol. Chem.*, **261**, 8597.

46. Morse, M. J., Crain, R. C., Cote, G. G., and Satter, R. L. (1989). *Plant Physiol.*, **89**, 724.

12

Metabolic changes following infection of leaves by fungi and viruses

RICHARD C. LEEGOOD and JULIE D. SCHOLES

1. Introduction

Pathogenic organisms can broadly be divided into two categories, based on their mode of nutrition. Nectrotrophic pathogens, such as the damping-off fungi and soft-rot bacteria, rapidly kill and degrade the host tissue by the production of large quantities of toxins and cell-wall degrading and proteolytic enzymes. These pathogens derive the majority of their nutrients from the dead host tissue and subtle changes in the regulation of metabolism will be transient. In contrast, the biotrophs (for example viruses, rust fungi, downy and powdery mildews) are highly specialized pathogens. They require a continual supply of nutrients from the host in order to develop and reproduce. They therefore co-exist with the host for a certain period of time. This co-existence must be accompanied by regulated changes in the physiology and metabolism of the host.

Nevertheless, infection of leaves by biotrophs has a dramatic effect on crop yields and is often characterized by visible symptoms, such as chlorosis and stunting. In the very early stages of infection the nutrient fluxes characteristic of an uninfected leaf may be sufficient to sustain the growth of the pathogen. However, as the amount of the fungus or virus increases, a redirection of host assimilates to the pathogen occurs. Infection often results in a reduction in soluble carbohydrate, an increase in starch accumulation (1–4), and an increase in CO_2 fixation into amino acids, organic acids, and protein (1, 2). In leaves infected with biotrophic fungi, alterations in the translocation and metabolism of carbohydrates are complex and depend upon the pathogen involved. The powdery mildews are ectoparasites forming dense white mycelia on the leaf surface, with haustoria confined to epidermal cells. In contrast, the rusts and downy mildews form intercellular mycelia with haustoria in epidermal, mesophyll, bundle sheath (and, in some cases, phloem) cells. Other factors such as the density of infection, time after infection, and the response of uninvaded cells to the presence of the fungus will add to the heterogeneity of carbohydrate and nutrient availability.

Besides effects on carbon partitioning, infection may be accompanied by a number of other changes. In some rust, mildew, or viral infections, photosynthesis (per unit chlorophyll) may be initially enhanced (1, 2, 5, 6), however, in most systems photosynthesis per unit area decreases (although, on a chlorophyll basis, photosynthesis may remain high) (1, 2) and chlorophyll is lost from the tissue as the disease progresses (5, 6). The rate of respiration of the host often increases in response to infection (1, 2, 7, 8). Nitrogen nutrition can also influence the response to infection (1, 2).

Although these general effects are well documented, the precise sequence of events and the mechanisms underlying these changes remain poorly understood. Such observations do not reveal whether the pathogens actively intervene in metabolism at a molecular level, as has been suggested for sugar-beet leaves infected with powdery mildew (9) and tobacco and spinach leaves infected with Tobacco mosaic virus (TMV) (10), or whether they act indirectly by creating an additional sink for carbon, nitrogen, and phosphorus (3, 4). Difficulties arise because experiments are often done late in the infection cycle when symptoms are visible and it is impossible at this stage to distinguish primary from secondary events.

There are two advantages in studying viral, as opposed to fungal, infections. Firstly, it is possible to quantify the relationship between changes in partitioning and the amount of a virus produced (determined using antibody and nucleic acid probes) and, secondly, we are dealing with, what is for many practical purposes, one organism. There is no comparable procedure for analysing quantitatively, the biomass of the fungus on or within the host. The main methods which give an indication of biomass are based upon the degradation of the fungal cell wall polymers, chitin and mannan, and the subsequent measurement of glucosamine (from the chitin) by colorimetric assay and mannose (from the mannan) by gas–liquid chromatography (GLC) (11, 12). However, these measurements are subject to inaccuracy as the proportion of chitin and mannan change with age, growth rate, environmental conditions, or the developmental stage of the fungus (11, 13, 14). Further complications which must be addressed when considering analysis of the concentration of metabolites and the activity of enzymes from fungus-infected tissue include (a) tissue heterogeneity and (b) the problem of the large number of metabolites and enzymes common to both fungus and host. These problems can be minimized in a number of ways. For example, inoculation of leaves with both rusts and mildews can be manipulated to give discrete areas of uniform infection and thus reduce the heterogeneity of the tissue. The contamination of the extracts by metabolites from the fungus is negligible if experiments are carried out during the early stages of infection, before symptoms are visible (this can be checked by simultaneously analysing the samples for the presence of fungal polyols, for example mannitol, arabitol, erythritol, etc.). Although not yet developed for such complex systems, non-aqueous fractionation (15) is likely to be the only technique for learning more about the precise intercellular compartmentation of metabolites between the fungus and the host.

2. Assessing the impact of infection on metabolism

Complex phenomena, like the metabolic response to infection by a pathogen, cannot be interpreted simply in terms of isolated measurements of gas-exchange, rates of chloroplast electron transport, or single metabolites such as ATP. The entire system requires thorough characterization by:

- physiological measurements of photosynthesis and respiration
- biochemical measurements, to identify sites of regulation
- measuring changes in carbon partitioning between starch, soluble carbohydrate, and organic and amino acids

In Section 2.1, we have strongly emphasized the effects on carbon metabolism, but the techniques of extraction, etc. are just as applicable to other studies.

2.1 Identification of metabolic limitations in photosynthesis through physiological measurements

Previous studies *in vivo* (for example, references 16–18) have shown that it is possible to select conditions of gas phase, temperature, etc. so as to impose limitations on photosynthesis by electron transport (low light, high CO_2), by carboxylation (low CO_2, high light), or by the capacity for sucrose synthesis (low temperature, high light, and high CO_2) (16, 17). We can then study the susceptibility of photosynthesis following infection by pathogens under conditions which lead to these limitations. If, for example, infection leads to a change in the initial slope of the assimilation rate vs. the intercellular CO_2 concentration curve in high light, then it will have led to an alteration in the activity of ribulose-bis-phosphate (RuBP) carboxylase. On the other hand, if infection leads to a selective restriction of sucrose synthesis, this will be evident as an enhanced sensitivity of photosynthesis to conditions of low temperature, high light and CO_2, and to stimulation by feeding inorganic phosphate (Pi) (13). In this regard it should be noted that leaves infected with many biotrophic fungi accumulate phosphorus (as polyphosphate bodies and Pi) (3, 4). It has been suggested that fungal sequestration of Pi from the host cytoplasm leads to a reduction in the export of triose-phosphate (triose-P) from the chloroplast, favouring the partitioning of photosynthate into starch and decreasing the rate of photosynthesis (3).

The use of non-intrusive probes, such as analysis of chlorophyll fluorescence by the use of light doubling and the measurement of $P700^+$, together with measurements of the rate of photosynthesis, can also yield information about the regulation of metabolism *in vivo*, since both photosystems I and II are subject to regulation which enables them to adapt to variations in the metabolic demands of the leaf (19, 20).

2.2 Identification of sites of regulation

A number of approaches may be used in the identification of regulatory reactions (21, 22).

- A comparison of the mass-action ratio, from *in vivo* contents of metabolites, with the equilibrium constant. This method can be used to identify reactions which are strongly removed from equilibrium *in vivo* and which are, therefore, likely to be regulatory (although this does not mean that enzymes which catalyse reactions which are close to equilibrium are not regulatory).

- Identification of those enzymes which have maximum catalytic activities of the same order of magnitude as the fluxes which they catalyse.

- Measurement of changes in metabolite pools and fluxes. The observation that the substrate changes in a direction opposite to that of the flux is sufficient for identification of a regulatory reaction, although this does not necessarily mean that the enzyme is regulatory if other substrates or co-factors are involved.

- Measurement of changes in metabolite ratios can serve two purposes. First, they can be used to indicate the compartmentation of metabolites such as hexose phosphate (15). Second, metabolite ratios can be employed to estimate other intracellular metabolite concentrations or ratios if it is assumed that a particular enzyme catalyses a reaction which is close to equilibrium. Thus electron transport provides ATP and NADPH for the reduction of glycerate-3-P (PGA) to triose-P, and the ratio between the amounts of triose-P and PGA can be used as an indication of the combined phosphorylation and redox potential (23).

- A knowledge of total amounts of phosphorylated intermediates (i.e. all glycolytic intermediates, plus intermediates of the Calvin cycle and adenylates) can be employed to estimate changes which occur in the amount of Pi in the extravacuolar compartments.

2.2.1 Preparation of leaf extracts for metabolite assays

When measuring any metabolite, it is important both to stop leaf metabolism immediately and to prevent the subsequent action of endogenous enzymes (especially phosphatases) on metabolites. The most effective method of stopping metabolism is to freeze the leaf in liquid N_2. This is best done by freeze-clamping (22, 24), which causes both rapid cooling and effective disruption of the tissue. Leaf samples can temporarily be stored in aluminium foil envelopes in liquid N_2. When samples are frozen in liquid N_2, enzymes are *not* inactivated. This must be done by grinding the frozen leaf sample into a small quantity of $HClO_4$ (*Protocol 1*) or $CHCl_3/CH_3OH$ (*Protocol 2*), both of which inactivate enzymes in the tissue and precipitate the protein.

Protocol 1. Extraction of leaf material in perchloric acid

1. Prepare frozen pellets of 1 M $HClO_4$ by pipetting the required amount of $HClO_4$ (1.0 ml for a 10 cm^2 leaf disc) into Parafilm 'boats'. Immerse in liquid N_2, and store the pellets in liquid N_2.

2. Pre-cool the mortar and pestle with liquid N_2. Add the leaf material to the mortar and, when all the liquid N_2 has evaporated, grind with a pellet of $HClO_4$ to a pale green powder. The mixture should take about 20 min to thaw. Grind it at regular intervals to ensure thorough mixing.

3. When the mixture has thawed, transfer it to a plastic centrifuge tube (using an automatic pipette with the narrow end of the tip cut off at an angle), and wash out the pestle and mortar with 1.0 ml cold 0.1 M $HClO_4$. Centrifuge the extract at 2000 g for 2 min to remove the protein, starch, phaeophytin, etc. Decant the supernatant into a plastic tube. The pellet can be reserved for the determination of phaeophytin (see Section 2.2.4).

4. Neutralize the ice-cold supernatant carefully, stirring constantly, with 5 M K_2CO_3 (this will be approximately one-tenth of the volume of $HClO_4$ added). The final pH should be between pH 6 and pH 7 (using pH paper, but note that most plant extracts turn a bright yellow colour under alkaline conditions and, therefore, already contain a pH indicator). Remove the $KClO_4$ precipitate by centrifugation at 2000 g for 2 min.

5. Add sufficient charcoal, following neutralization, to render the extract completely colourless (except where adenylates are to be measured, since they are absorbed by the charcoal). First add about 10 mg charcoal, from a 100 mg/ml suspension in water, but considerably more may be required to decolorize the extract. Centrifuge the extract at 2000 g for 2 min.

6. Measure the volume of the extract at this point. Measured aliquots of the supernatant can be stored in liquid N_2 or at −80°C, or they can be lyophilized.

The abundance of secondary products, such as phenolics and tannins, in leaves, particularly in infected tissues, means that it is extremely important to establish their biochemical tractability. Measurement of metabolites in any plant tissue should be accompanied by a rigorous check that they have not been degraded during the inactivation and extraction procedure. The best method is to prepare and extract duplicate samples of tissue, adding a measured amount of the metabolite to one pestle and mortar. The amount of the compound added should be comparable to that present in the tissue. Both samples are extracted and measured and the recovery of the metabolite estimated. This should normally be within 10% of the expected value.

An alternative to extraction in $HClO_4$ is the use of a chloroform–methanol mixture (25). This is necessary for the measurement of acid-labile compounds, such as fructose-2,6-bisphosphate, and is useful for the extraction of sucrose. However, in principle, extraction may be performed at any pH value. Care should be taken to check the final pH of the extract if these labile compounds are to be measured, as the mixture often becomes considerably more acid during extraction. Extraction in ether–trichloroacetic acid has been used for the estimation of inorganic pyrophosphate (PPi) (26).

Protocol 2. Extraction of leaf material in chloroform–methanol (25)

1. Mix 0.2 g frozen leaf material in a pestle and mortar (pre-cooled with liquid N_2) with 4.2 ml extraction medium:
   - 1.2 ml $CHCl_3$
   - 2.4 ml CH_3OH
   - 0.6 ml buffer containing 50 mM NaF. 10 mM EGTA, 50 mM Hepes (pH 8.5) (NaF prevents the possible action of phosphatases)

   Hold the paste at 4°C until it thaws.

2. Transfer to a glass tube and add 4 ml H_2O, shake the mixture vigorously, and centrifuge at 2000 g for 2 min. Two phases form a lower (chloroform) phase containing the chlorophyll, etc. and an upper (methanol–water) phase containing metabolites, sugars, etc. (This distribution should be checked through the recovery procedures, as metabolites may not always partition in the expected fashion in different plant material. If necessary, the lower phase can be re-extracted with water.) Starch, cell wall material, protein, etc. lies in between the two phases.

3. Remove the supernatant and dry at 35°C. Redissolve in 1 ml water and store in liquid N_2 or at −80°C or lyophilize.

4. Charcoal may be added as described in *Protocol 1*.

2.2.2 Metabolite assays

The above procedures are suitable for the extraction of starch, sugars, phosphorylated metabolites, and both amino and organic acids. There are, of course, numerous methods for the analysis of these compounds, including HPLC, GLC, and amino acid analysis. Many of these compounds can be assayed enzymatically in pyridine nucleotide-linked assays using a dual-

Table 1. References to methods for the estimation of metabolites, etc. in plant extracts. All assays are simple pyridine nucleotide-linked enzymic assays unless otherwise shown. Assays for these metabolites and many others may also be found in reference 41.

Compound	Reference (and alternative method)
Glucose, fructose	27
Glucosamine	12
Mannose	12
Polyols	38
Sucrose	28
Starch	29 (precipitation as starch–I_2 complex)
Starch + glycogen	41
Hexose-P (G-6-P, G-1-P, Fru-6-P)	27
UDPGlc, Fru1,6bisP$_2$, Triose-P, PEP, Glycerol-1P, Gluconate-6P	27
3-PGA	18
Ribulose1,5bisP$_2$	30
	24 (incorporation of $^{14}CO_2$)
Sed1,7bisP$_2$	31
Fru2,6bisP$_2$	32, 40 (stimulation of PPi-Fru-6-P kinase)
PPi	26, 33
Pi	27
	34, 35 (colorimetric, (35) gives greatest sensitivity)
Organic acids (oxaloacetate, malate, citrate, *iso*-citrate, fumarate, lactate pyruvate, 2-oxoglutarate)	27
Amino acids (glutamate, aspartate, alanine)	27
Adenine nucleotides (ATP, ADP, AMP)	27
	36 (luciferin/luciferase)
Uridine nucleotides (UTP, UDP, UMP)	37
Pyridine nucleotides (NAD(H), NADP(H))	27, 39 (cycling assay)

wavelength spectrophotometer (340–400 nm). A summary of references to these and some alternative assays is given in *Table 1*. In infected tissues which are particularly rich in phenolics, tannins, etc. it may be found that charcoal treatment is insufficient for removing all interfering substances from enzyme-linked assays. Treatment either of the extract or the addition of bovine serum albumin (BSA) (5–10 mg/ml) or dithioreitol (10 mM) to the cuvette can often overcome problems due to inactivation of coupling enzymes or drift in the assays. Note that such interference can be particularly severe in fluorimetric or luminometric (3) assays.

The most widely used methods for determining the concentration of α-glucans in plant tissue are based on acid or enzymic hydrolysis to, and assay of, glucose (41). To our knowledge there is no technique that permits the independent hydrolysis of starch and glycogen. To distinguish the two α-glucans when they occur together, as in leaves infected with biotrophic fungi, an extraction procedure based on their different water solubilities can be used (42).

2.2.3 Assay of enzymes

Apart from influencing the direction of metabolism, pathogens are likely to alter both the amounts of enzymes and their activation states. It would, however, be invidious to indicate that changes in any particular enzyme are likely to be more important. Methods for the assay of a wide range of enzymes in plants may be found in compendia (for example see Lea, reference 43).

For many enzymes already studied it is entirely satisfactory to freeze-clamp the leaves, in order to arrest changes in the activation state of enzymes, and to store samples in liquid N_2 prior to measurement (see Section 2.2.1). The measurement of changes in the activation state, for example of sucrose-P synthase (44) or of light-activated enzymes of the Calvin cycle (45), requires rapid extraction procedures in appropriate media. However, difficulties with phenolics, tannins, etc. will be encountered as with metabolite assays. Recoveries should be checked when enzymes are measured in plant tissues, either by adding known quantities of commercially prepared enzymes, or by mixing the plant tissue under study with another tissue, or with tissue at different stages of infection, in which enzyme lability is believed to be low. Enzyme inactivation by phenolics, etc. may then be countered by the addition of compounds such as polyvinylpyrrolidone or polyethylene glycol (which complex phenols), thiols, or mercapto compounds (for example dithiothreitol, 2-mercaptoethanol), BSA, and the use of reducing compounds (for example *iso*-ascorbate) or anaerobic media.

2.2.4 Chlorophyll determination

For chlorophyll estimation in leaves, the simplest method is to freeze the leaf in liquid N_2, grind it to a powder, extract the chlorophyll with 80% acetone in the dark, and then centrifuge. The absorbance is measured at 652 nm and the chlorophyll concentration (in mg/l) is given by the expression: $(A_{652})1000/34.5$ (46). For the accurate determination of chlorophyll *a* and *b*, the extinction coefficients of Graan and Ort (47) can be used, based on absorbance readings made in 80% acetone at 664 nm and 647 nm.

chlorophyll *a* (μM) $= 13.19\ A_{664} - 2.57\ A_{647}$

chlorophyll *b* (μM) $= 22.10\ A_{647} - 5.26\ A_{664}$

total chlorophyll (μM) $= 7.93\ \ A_{664} + 19.53\ A_{647}$

Although extraction in acidic conditions results in chlorophyll destruction, the product, phaeophytin, can be measured (48). The pellet resulting from extraction in $HClO_4$ should be homogenized in water, and acetone added to a final concentration of 80% (v/v). Samples may require overnight extraction. Absorbance readings are taken at 536 nm, 655 nm, and 666 nm. In 80% acetone, the content of phaeophytin (P) (in mg/ml) is given by one of two expressions. Either:

$$P = (77.58\ A_{536} - 0.33\ A_{666})$$

or

$$P = (6.75\ A_{666} + 26.03\ A_{655}).$$

References

1. Goodman, R. N., Kiraly, Z., and Zaitlin, M. (1964). *The biochemistry and physiology of infectious plant disease.* Van Nostrand, Princeton.
2. Goodman, R. N., Kiraly, Z., and Wood, K. R. (1986). *The biochemistry and physiology of plant disease.* Univ. Missouri Press, Columbia.
3. Whipps, J. M. and Lewis, D. H. (1981). In *Effects of disease on the physiology of the growing plant* (ed. P. G. Ayres), p. 47. Cambridge University Press.
4. Farrar, J. F. and Lewis, D. H. (1987). In *Fungal infection of plants* (ed. G. F. Pegg and P. G. Ayres), p. 92. Cambridge University Press.
5. Scholes, J. D. and Farrar, J. F. (1986). *New Phytol.*, **104**, 601.
6. Scholes, J. D., Lee, P., Horton, P., and Lewis, D. H. (1989). In *Proceedings of the VIII International Congress on Phytosynthesis.* Martinus Nijhoff/Dr W. Junk, The Hague.
7. Kosuge, T. and Kimpel, J. A. (1981). In *Effects of disease on the physiology of the growing plant* (ed. P. G. Ayres), p. 29. Cambridge University Press.
8. Daly, J. M. (1976). In *Physiological plant pathology* (ed. R. Heitfuss and P. H. Williams), p. 450. Springer-Verlag, Berlin.
9. Magyarosy, A. C. and Malkin, R. (1978). *Physiol. Plant Path.*, **13**, 183.
10. Hodgson, R., Beachy, R., and Pakrasi, H. (1989). *Physiol. Plant.*, **76**, 66.
11. Whipps, J. M., Haselwanter, H., McGee, E. E. M., and Lewis, D. H. (1982). *Trans. Brit. Mycol. Soc.*, **79**, 385.
12. Ride, J. P. and Drysdale (1972). *Physiol. Plant Path.*, **2**, 7.
13. Ride, J. P. and Drysdale (1971). *Physiol. Plant Path.*, **1**, 409.
14. Sharma, P. D., Fisher, P. J., and Webster, J. (1977). *Trans. Brit. Mycol. Soc.*, **69**, 479.
15. Gerhardt, R., Stitt, M., and Heldt, H. W. (1987). *Plant Physiol.*, **83**, 399.
16. Stitt, M. and Grosse (1986). *J. Plant Physiol.*, **133**, 392.
17. Leegood, R. C. and Furbank, R. T. (1986). *Planta*, **168**, 84.
18. Leegood, R. C., Labate, C. A., Huber, S. C., Neuhaus, H. E., and Stitt, M. (1988). *Planta*, **176**, 117.
19. Horton, P. (1986). In *Progress in photosynthesis research* (ed. J. Biggins), Vol. 2, p. 681. Martinus Nijhoff, Dordrecht.
20. Horton, P., Oxborough, K., Rees, D., and Scholes, J. D. (1988). *Plant Physiol. Biochem.*, **26**, 453.
21. Newsholme, E. A. and Start, C. (1973). *Regulation in metabolism.* John Wiley, London.

22. Leegod, R. C., Adcock, M. D., and Doncaster, H. D. (1989). *Phil. Trans. R. Soc. Lond. B.*, **323**, 339.
23. Heber, U., Neimanis, S., Dietz, K.-J., and Viil, J. (1986). *Biochim. Biophys. Acta*, **852**, 144.
24. Badger, M. R., Sharkey, T. D., and von Caemmerer, S. (1984). *Planta*, **160**, 305.
25. Stitt, M., Gerhardt, R., Kürzel, B., and Heldt, H. W. (1983). *Plant Physiol.*, **72**, 1139.
26. Weiner, H., Stitt, M., and Heldt, H. W. (1987). *Biochim. Biophys. Acta*, **893**, 13.
27. Lowry, O. H. and Passonnneau, J. V. (1972). *A flexible system of enzymatic analysis*. Academic Press, New York.
28. Jones, M. G. K., Outlaw, W. J., Jr., and Lowry, O. H. (1977). *Plant Physiol.*, **60**, 379.
29. Lustinec, J., Hadacova, V., Kaminek, M., and Prochazaka, Z. (1983). *Anal. Biochem.*, **132**, 265.
30. Doncaster, H. D., Adcock, M. D., and Leegood, R. C. (1989). *Biochim. Biophys. Acta*, **973**, 176.
31. Leegood, R. C. and Walker, D. A. (1980). *Arch. Biochem. Biophys.*, **200**, 575.
32. van Schaftigen, E. (1985). In *Methods in enzymatic analysis* (ed. H. U. Bergmeyer), Vol. 6, p. 335. Verlag Chemie, Weinheim.
33. Dancer, J. E. and ap Rees, T. (1989). *Planta*, **177**, 261.
34. Taussky, H. H., Shorr, E, E., and Kurzmann, G. (1953). *J. Biol. Chem.*, **202**, 675.
35. Itaya, K. and Ui, M. (1966). *Clin. Chim. Acta*, **14**, 361.
36. Carver, K. A. and Walker, D. A. (1983). *Science Tools*, **30**, 1.
37. Quick, P., Neuhaus, E., Feil, R., and Stitt, M. (1989). *Biochim. Biophys. Acta*, **973**, 263.
38. Holligan, P. M. and Drew, E. A. (1971). *New Phytol.*, **70**, 271.
39. Slater, T. F. and Sawyer, B. (1962). *Nature*, **193**, 454.
40. Stitt, M. (1990). In *Methods in plant biochemistry* (ed. P. J. Lea), Vol. 3, p. 87. Academic Press, London.
41. Bergmeyer, H. U. (ed.) (1985). *Methods in enzymatic analysis*, Vols. 6–8. Verlag Chemie, Weinheim.
42. Holligan, P. M., McGee, E. E. M., and Lewis, D. H. (1974). *New Phytol.*, **73**, 873.
43. Lea, P. J. (ed.) (1990). *Methods in plant biochemistry*, Vol. 3. Academic Press, London.
44. Stitt, M., Wilke, I., Feil, R., and Heldt, H. W. (1988). *Planta*, **174**, 217.
45. Leegood, R. C. (1990). In *Methods in plant biochemistry* (ed. P. J. Lea), Vol. 3, p. 15. Academic Press, London.
46. Arnon, D. I. (1949). *Plant Physiol.*, **24**, 1.
47. Graan, T. and Ort, D. R. (1984). *J. Biol. Chem.*, **259**, 14003.
48. Vernon, L. P. (1960). *Anal. Chem.*, **32**, 1144.

13

Strategies for cloning plant disease resistance genes

RICHARD W. MICHELMORE, RICHARD V. KESSELI,
DAVID M. FRANCIS, MARC G. FORTIN, ILAN PARAN,
and CHANG-HSIEN YANG

1. Introduction

1.1 Present status

Many single genes determining resistance to diverse pathogens have been identified by geneticists and plant breeders. Despite the wide use of these genes in agriculture, they have so far proved recalcitrant to molecular isolation, with one exception (see Section 5.3.1). The molecular bases and biochemical modes of action of their products remain one of the major unknowns in plant biology. Neither the pattern of expression nor the nature of resistance gene products are known. It is suspected that they may be constitutively expressed at a low level; therefore, many of the strategies used to clone more highly-expressed or induced plant genes are inappropriate. Successful cloning strategies will probably utilize the best characterized feature of resistance genes which is their Mendelian nature.

Many genes induced during the resistance response have already been cloned and their function in resistance is being determined. A distinction must be made between resistance genes identified at the genetic level and genes induced during the resistance response. For the purposes of this chapter, resistance genes are those that determine whether an interaction between a plant and potential pathogen is compatible (susceptible plant and virulent pathogen) or incompatible (resistant plant and avirulent pathogen). Response genes are those that are capable of being expressed in both resistant and susceptible plants and which may be involved in limiting invasion by the pathogen. A similar range of response genes seems to be induced regardless of the challenging pathogen (1, 2). Therefore, such genes are unlikely to be the determinants of intra-specific incompatibility between host and pathogen.

Cloning and characterization of disease resistance genes will have profound fundamental and applied consequences. Therefore, the efforts of several laboratories are increasingly focused on isolating resistance genes. Groups attempting to clone genetically well-defined resistance genes in agriculturally important crops have spent much of the past five years developing the required technology and genotypes. These efforts are now at a point, in several species, where success is likely within the next few years. In contrast, the necessary tools for several approaches have become available more rapidly in *Arabidopsis thaliana*. However, disease resistance genes in *A. thaliana* are only just becoming sufficiently defined genetically. It is likely that resistance genes will be cloned from several species within the next few years using a variety of approaches.

1.2 Implications of classical genetics to cloning strategies

Genetic studies of disease specificity have demonstrated that resistance is often (although not always) conditioned by single dominant genes. Such genes should be amenable to analysis and manipulation at the molecular level. Genetic analyses of the pathogens have shown that intra-specific (and some inter-specific) differences in the ability to cause disease is often determined by dominant avirulence genes complementary to the individual resistance genes. Such gene-for-gene interactions have now been demonstrated in diverse plant–pathogen interactions (reviewed in reference 3). This has led to the suggestion that some resistance genes encode recognition molecules. In contrast, the primary function of avirulence genes in the pathogen may not be to condition avirulence; they may be a heterogeneous group of genes whose common denominator is that they have been genetically characterized as avirulence genes due to the host's ability to detect their direct or indirect products. It is unlikely that all resistance genes are of one type. Resistance to some diseases is recessive, but resistance to other diseases may involve insensitivity to a toxin rather than recognition. This chapter focuses on the cloning of single dominant genes for resistance as these should be the most tractable.

Classical genetics has shown that resistance genes in several plants tend to be clustered either as apparent allelic series or as groups of distinct genes (4–6). Even genes for resistance to taxonomically diverse pathogens may be in the same linkage group. This clustering of functionally related genes has led to the suggestion that resistance genes are related at the sequence level. This hypothesis is supported by the identification of sequences duplicated between different genomic regions containing clusters of resistance genes in lettuce (7). Also, instability at the *Rp1* locus in corn was frequently associated with exchange of flanking markers, indicating that the apparent instability was due to unequal cross-over in this complex region (8).

Our underlying model is that many resistance genes are members of multigene families that have evolved to have different specificities. As such they may have domains that are highly variable (receptor domains?) and other domains that are more conserved (structural and effector domains?). While there is no molecular evidence for this, there are precedents in animal systems (9) and the *S* locus in *Brassica* spp. and *Nicotiana* spp. (10, 11). Whether this is the true nature of plant disease resistance genes awaits their cloning and characterization. Nevertheless, this model has important consequences for the design of cloning strategies. For example, if related sequences are present, genomic subtraction approaches are unlikely to be successful. Also, chromosome walking through such regions may be complex. Similarly, if new resistance genes evolve by duplication followed by sequence divergence, some resistance genes may be duplicated sequences that still have the same specificity; such genes will be nearly impossble to inactivate by transposon mutagenesis. In the absence of any contrary evidence, strategies should be designed to minimize the number of assumptions that have to be made and control experiments should be conducted to check that individual components of the strategy are feasible.

1.3 Prerequisites for cloning

The targeted resistance gene should have an unambiguous phenotype. All the strategies require the precise determination of genotype, usually at several stages. Although this is not possible for many resistance genes as they may show incomplete penetrance. Reliable and routine screening procedures are required, and it is preferable that resistance should be scored in young plants allowing the screening of large numbers of seedlings in defined environmental conditions.

A theme running through this chapter is the importance of classical genetics. It is critical that the targeted resistance is encoded by a single gene. This precludes analysis of many disease resistances whose inheritance is more complex and is not always easy to ascertain. A monogenic Mendelian segregation indicates that resistance to a particular pathogen isolate is determined by a single gene **or** a cluster of linked genes. As pathogen avirulence genes usually do not show a similar clustering, segregation analysis of avirulence in the pathogen can be used to determine how many different genes may be in a particular region (12). This still will not indicate whether there are duplicated genes of the same specificity.

The targeted gene should be determined by a dominant allele with complete penetrance since, regardless of the method of isolation, the cloning of a resistance gene will ultimately have to be confirmed by complementation of the susceptible phenotype. Not all disease resistance is determined by single genes; molecular analysis of more genetically complex diseases will have to await further developments in technology and our ability to dissect the genome for quantitative traits.

Classical genetic studies provide defined genetic stocks of host and pathogen for molecular analysis. The identification of several human genes has involved recombinants or mutations (deletions or point mutations) in the targeted region (for example references 13, 14). An advantage of plant genetics over human genetics is the opportunity to generate large segregating or mutated populations of specific genotypes. As recombinants are the most informative individuals, their identification at the phenotypic level considerably reduces the numbers of individuals that must be analysed at the molecular level. Mutants are also likely to be very useful in the identification and characterization of resistance genes. Deletion mutants will greatly aid in the localization of targeted genes. Base-substitution mutants will be useful in the subsequent characterization of functional domains in cloned genes. Generation of such mutant populations is a lengthy process and should be initiated well before they are needed.

1.4 Aims of this chapter

This chapter reviews approaches currently being used by various laboratories in attempts to clone disease resistance genes. These efforts focus on single dominant genes. For each approach, we will try to review the experimental options and designs and their inherent assumptions.

To some extent, in places, the contents of this chapter are speculative as most of the strategies described have yet to result in cloning of a resistance gene. Furthermore, if resistance genes are members of multigene families, cloning of one resistance gene should lead to cloning of others by sequence similarity, and much of the chapter may be superseded. However, not all resistance genes are likely to be of one type and many of the strategies described will be relevant to any plant gene whose biochemical nature and pattern of expression is unknown.

The chapter emphasizes strategies for targeting genes from specific regions, rather than a 'brute force' approach of large scale genome characterization and sequencing of relevance to few plants and relatively few laboratories. We have provided detailed protocols for techniques with which we are most familiar. However, for several strategies, detailed descriptions already exist and appropriate references are given; discussion of strategies is limited to current status and to aspects of the technique that may be relevant to cloning resistance genes. For cloning genes for downy mildew resistance from lettuce, we currently favour a map-based cloning approach; although we have also been pursuing a transposon mutagenesis approach. Consequently, map-based cloning receives the greatest attention here. Other approaches are also reviewed since resistance genes will probably be cloned by several alternative strategies within the next few years.

2. Product-orientated approaches

2.1 Via avirulence genes

Initially, it was hoped that the cloning of avirulence genes from bacterial pathogens would rapidly lead to the cloning of the corresponding plant

resistance genes by identifying what host component the avirulence gene product was interacting with. The approach has yet to live up to early expecta-
tions. In large part, this failure may be due to the lack of an obvious role of
the primary products of avirulence genes in determining specificity. Most
seem to be cytosolic rather than secreted or membrane associated, while the
molecules interacting with the host seem to be secondary consequences of
avirulence gene expression. This is understood in greatest detail for *AvrD*
from *Pseudomonas syringae* pv *glycinea*. Small fatty acid derivatives that are
capable of eliciting the hypersensitive response on resistant soybean plant are
produced when *AvrD* is expressed in Gram-negative bacteria such as *P. s.* pv
glycinea or *Escherichia coli* (15 and N. Keen, personal communication).
Recently, the *A9* gene (that encodes a 28 amino acid polypeptide) has been
cloned from the fungus, *Fulvia fulvum* (125). This polypeptide is induced
during infection and is found in the intercellular spaces of the leaf; it is
capable of eliciting the hypersensitive response only on *Cf9*-expressing tomato
plants. Similarly, the coat protein of TMV has been shown to interact specifically with N' genotypes of tobacco (17). There is good evidence from
extensive studies that the biochemical determinant of avirulence has been
identified in each case, and studies are now underway to isolate the plant
component(s) that interact with them.

As more such biochemical determinants are characterized, studies on their
interactions with host factors will become increasingly productive. Time will
tell whether the host components identified in this way are the products of
resistance gene or whether the host–pathogen interaction is mediated indirectly by host components not encoded by resistance genes. Either way
these studies are a critical component to the characterization of resistance
gene action.

2.2 Via host-specific toxins

Specificity in some plant–pathogen interactions is mediated by host-specific
toxins (18). Characterization of such a toxin presents the opportunity to identify
the host components that interact with the toxin. One of the best characterized
interactions[a] involves sensitivity in oats to the toxin, victorin, produced by the
fungus, *Cochliobolus victoriae*. Sensitivity is determined by the dominant
Vb allele. This seems to be allelic to resistance to *Puccinia coronata*, conditioned
by the *Pc-2* gene. Oats are usually either sensitive to victorin and resistant to
certain races of crown rust or vice versa. Victorin may be mimicking the
elicitor from *P. coronata* that results in the resistance response conditioned by
Pc-2 (19). Therefore, the apparent allelism of *Vb/Pc-2* gene may provide a
special opportunity for cloning race-specific resistance genes.

---

[a] Note added in proof: recently the biochemical basis of resistance in corn conferred by *Hm1* to
Cochliobolus carbonum has been determined (129). This complements the cloning of the *Hm1*
locus by transposon tagging (Section 5.3.1).

Victorin has stringent structural requirements for toxic activity, suggesting that a specific recognition site exists in the cells of sensitive genotypes (20). It can be labelled with ^{125}I to high specific activity while retaining its *in vivo* binding specificity. This characteristic has been used to identify a 100 kDa recognition protein from oats (21). Antibodies were generated to the protein (22) and used to select a cDNA from an expression library in lambda gt11. The identity of the cDNA was confirmed by immunopeptide mapping between the recombinant protein expressed in *E. coli* and the native protein from plants (T. Wolpert, personal communication).

The relationship between a cDNA isolated using a biochemical approach, such as the example above, and a resistance/sensitivity gene can be determined either functionally or genetically. Experiments to investigate whether the cDNA isolated by Wolpert and Macko is related to the *Vb/Pc-2* gene are underway. A transformation system for oats is not currently available and, therefore, a functional test using a genomic clone to complement a *Vb/Pc-2* genotype is not feasible. It may be possible to test the ability of a gene to confer toxin sensitivity using a heterologous system much as the chimeric *T-urf13* gene from T-cytoplasm corn was shown to confer T-toxin and methomyl sensitivity in *E. coli* (23) and yeast (24). However, this approach would not confirm the relationship between the isolated cDNA and the gene identified genetically. Segregation analysis is a more powerful test. If the cDNA detects a restriction fragment length polymorphism (RFLP) that genetically absolutely co-segregates with the resistance/sensitivity gene, then it is highly likely that the cDNA is derived from that gene unless there are several cross hybridizing sequences in the genome. If the cDNA does detect a multigene family, the selected cDNA may not determine the targeted specificity but may be derived instead from a related gene. Even so, studies of such a family may provide important information as to resistance gene structure and function. In view of these possible genetic complexities, genetic analyses should precede functional tests to ensure the sequence encoding the targeted specificity has been isolated.

2.3 Conserved domains in resistance genes

As the products of resistance genes are currently unidentified, most cloning approaches dependent on knowledge of the gene product are inapplicable. However, resistance genes may encode products similar to molecules involved in the initial steps of signal transduction in mammalian systems, such as protein kinases. If so, it may be possible to employ degenerate oligonucleotide primers encoding conserved protein domains to isolate plant analogues. This strategy has been used to isolate sequences similar to animal serine/threonine protein kinases from bean (25) and corn (26). However, it is highly speculative as it assumes the nature of the resistance gene product. It may be successful but it is more likely to identify either nothing at all or members of complex multigene families. The probability of detecting the sequence encoding the targeted specificity is small and would require extensive genetic and complementation analysis to confirm that the required sequence had been identified. Defined mutations in resistance genes would be highly informative for such studies. Even if these approaches do not yield a resistance gene, however, they may identify genes and molecules involved in signal transduction in plants, some of which may be intermediaries between resistance genes and the induced response genes.

3. Subtractive hybridization approaches

3.1 Basic technique

Subtractive approaches rely on the utilization of DNA annealing kinetics to enrich for sequences present in one sample of DNA but absent in another. The starting material can be either cDNAs, to select for genes that are differentially expressed between the source tissues, or genomic DNA, to select sequences present in one genome but not another.

In outline, subtractive approaches involve annealing excess driver DNA with the target DNA sample. Sequences present in both samples, or just in the driver, will anneal to form duplexes. Sequences present only in the target sample will form duplexes slowly due to their low concentration, and will remain predominantly single-stranded under the conditions of the experiment. A variety of techniques have been used to remove the unwanted sequences. Early experiments involved running the reannealed samples down hydroxyapetite columns to remove double-stranded DNA. More recently, the driver DNA has been labelled with biotinylated nucleotides. Biotinylated sequences will bind to streptavidin-coated beads (27). This allows the removal of all driver DNA sequences regardless of whether it has formed duplexes, but it does not remove any sequences unique to the target sample whether or not they have formed duplexes with each other. Multiple rounds of denaturation, reannealment with excess driver DNA, and removal of common/driver specific sequences results in the increasing enrichment of sequences specific to the target DNA sample. After the final round of enrichment, the sequences are cloned and tested to confirm their origin. The easiest test is to assay for genetic co-segregation of the cloned sequence with the targeted phenotype.

The advantage of subtractive approaches is that they require relatively little labour and time compared to other methods. Unfortunately, at present it is probably at or beyond the limit of this technology to clone resistance genes from genomes the size of most plant species. However, technological advances, particularly the use of PCR to amplify rare sequences, are making this approach increasingly applicable to complex genomes. A more critical constraint is that subtractive hybridization will fail if there are many related sequences in both target and driver samples, as may be the case for certain types of resistance genes (see Section 1.2).

3.2 Differentially expressed genes

While it is now possible to select differentially expressed messages from small amounts of tissue (28), it is less clear whether subtractive hybridization of cDNA populations will help clone resistance genes. There are two alternatives, subtractive hybridization between cDNA populations derived from near-isogenic genotypes differing in the expression of a resistance gene, or subtractive hybridization between cDNA populations derived from challenged and unchallenged tissue of the same genotype.

The inter-genotype subtraction will select all sequences that are expressed differentially between the two genotypes. For near-isogenic lines produced by backcrossing, this may be many genes. For near-isogenic lines produced by mutation, those lines derived from treatments that tend to generate deletions (see Section 3.3) would probably prove the most useful (although deletions may span many loci). The selection of source tissue requires assumptions about the pattern of expression of the resistance gene. If uninoculated tissue is used, then the resistance genes must be constitutively expressed for a subtractive approach to be successful. Many resistance genes are assumed to be constitutive due to the speed of the resistance response, but this may not be the case for some resistance genes. The alternative of extracting RNA from challenged tissue adds the potential complication of differentially expressed genes from the challenging pathogen. Even if the resistance gene is not cloned, subtractive selection of cDNAs may supply low copy sequences tightly linked to the target gene which would be useful in other approaches, particularly chromosome walking.

Subtraction between challenged and unchallenged tissue assumes that the resistance gene is expressed at minimal levels in the unchallenged state and is significantly induced when the plant is challenged. However, the patterns of resistance gene expression will not be known until these genes are cloned and characterized. Subtraction approaches are more likely to enrich for genes that are induced as part of the signal transduction pathway and the resistance response. This will be useful in the characterization of resistance gene action.

3.3 Genomic subtraction

Recently, it has become feasible to identify sequences that are present in one genome but not another. The method has been used to isolate a sequence that was part of a 5 kb deletion in yeast, corresponding to 1/4000th of the genome (27). This approach relied on three rounds of subtraction with removal of biotinylated driver DNA by avidin-coated polystyrene beads. After the final round of subtraction, the remaining DNA was ligated to oligonucleotide adaptors, amplified by PCR and then labelled. The labelled DNA was then used as a probe to a library of the wildtype strain. The only clones detected were those that also hybridized to sequences that had been deleted in the mutant. In addition, a gene affecting gibberellin biosynthesis, *ga-1*, has recently been cloned by genomic subtraction from *Arabidopsis thaliana* (126).

Several aspects should be borne in mind when considering the use of genomic subtraction to isolate resistance genes.

- So far, as described above, this technique has been successful with genomes of medium complexity such as yeast and *A. thaliana*. The reannealing times required for the application of the published protocol to more complex genomes would be prohibitive as it would take months of hybridizations to reach the required C_0t. Whilst hybridization times may be reduced by additives such as phenol that enhance reannealment (29), technical advances are required before genomic subtraction is applicable to cloning of resistance genes from most plant species.

- Genomic subtraction requires small, well-defined deletions. Certain types of mutagens [for example fast neutrons (30), gamma rays (31), and some chemicals such as diepoxybutane (32)] have been shown predominantly to generate deletions in model organisms. However, their effects have not been defined extensively in plants. Most large deletions are lethal when homozygous (30), and so selfing will tend to select for small deletions in non-essential regions.

- Prior to embarking on a subtraction approach, the mutant should be characterized genetically using flanking markers (see Section 6.2). This will confirm that the mutation is in the targeted gene and may indicate the genetic size of the deletion. If at least one flanking marker is missing, then the mutation is probably due to a deletion, but it is also likely that more than one gene has been deleted. If a long-range map is available (see Section 6.3.2), the deletions could be characterized using the flanking marker(s) to determine the physical size of the deletion.

- The smallest deletion lacking the resistance gene should be used for genomic subtraction to maximize the chances that the resistance gene will be isolated. Even if this approach did not isolate the resistance gene, it might provide tightly-linked sequences that would be useful in other approaches.

4. Shotgun complementation

Shotgun cloning has been highly successful in prokaryotes and yeast for isolating genes by complementation for which there was little a priori information. Avirulence genes have been isolated in this manner from several phytopathogenic bacteria (33, 34). The approaches were successful as the genome complexity was low, the rate of transformation/tansconjugation was high, and the phenotype could be quickly and readily assayed.

In plants, cloning by complementation has to overcome several major obstacles. The greatest barrier is genome size. Many thousands (probably tens of thousands) of transformants would have to be generated and tested to screen a genomic library. The exact number required is a function of genome size, the number of genes targeted, the average size of the cloned DNA, and the degree of redundancy required at each step. Current plant transformation procedures are not conducive to shotgun cloning. Even if libraries are made using cosmid-based *Agrobacterium* vectors with average inserts of 30–40 kb, the routine transfer and integration of such inserts to the plant is far from guaranteed (see Section 7.3). The rate of stable transformation and regeneration following microprojectile bombardment is very low. The resistance phenotype is not conducive to shotgun transformation as it usually requires the regeneration of whole plants. Even if generating an adequate number of plants was feasible, this approach still requires several additional assumptions; any one of which, if invalid, would obstruct progress. The approach assumes that expression of the introduced gene would be detectable, that the library is representative, that the sequences are stable in the library, and that the targeted gene is smaller than the average size of the cloned DNA. Several of these assumptions are inherent in other techniques but they are not made simultaneously and, therefore, are more likely to be recognized as problems if progress is blocked. Unless there are major improvements in the ability to introduce large DNA fragments into plants (for example, plant artificial chromosome vectors) and to select for resistance at the cell level, it seems unlikely that resistance genes will be cloned by complementation.

5. Insertional mutagenesis

5.1 Overview

Cloning by insertional mutagenesis relies on the introduction of a characterized piece of DNA into the target gene to generate a change in phenotype as well as to serve as a molecular tag for isolation of the flanking sequences. The flanking sequences are used in turn to isolate the wild-type gene. This has been very effective in prokaryotes, for example with Tn5 insertion (35), and has been used to generate numerous loss-of-function mutants that could be readily identified. Two types of insertion mutagenesis strategy have been used to clone genes in plants, T-DNA mutagenesis in *Arabidopsis thaliana* and transposon mutagenesis in corn and snapdragon. The biology of plant transposons has been extensively reviewed (36, 37). Furthermore, strategies have been reviewed for transposon tagging in general (38–40) and for tagging disease resistance genes in particular (41). Therefore, we restrict our comments to recent developments relating to tagging disease resistance genes.

5.2 T-DNA mutagenesis

When the T-DNA inserts into the genome during *Agrobacterium*-mediated transformation, it may interrupt expression of the genes at the point of integration. This potentially allows the T-DNA to be used for insertional mutagenesis. There are, however, several problems with this approach for all plants except *A. thaliana*. Once integrated, the T-DNA does not move so that each mutation event involves a separate transformation. Although *Agrobacterium*-mediated transformation is a remarkably efficient process, the generation of sufficient transformants to have even a moderate probability of detecting one of a broad class of mutants in an average plant genome is an impossible task if regeneration of plants is required. If the phenotype can be selected for in culture (which most disease resistances cannot) the approach may be more feasible. However, an equally important constraint is the background rate of mutation. Passing plants through tissue culture results in high rates of mutation due to a variety of genetic changes, so mutations caused by T-DNA insertions need to be distinguished from spontaneous background mutations. As the background rate may be significantly higher than the rate of insertional mutagenesis, the isolation of tagged genes is unlikely. If a mutation could be identified in culture, it would still require some type of segregation analysis to demonstrate co-segregation of T-DNA or flanking sequences and the mutation. Therefore, T-DNA mutagenesis for targeted isolation of genes is unlikely to be useful in most plants.

T-DNA mutagenesis has proven useful in *A. thaliana* for two reasons. *A. thaliana* can be readily transformed by incubating seeds in a bacterial suspension of *A. tumefaciens* without any step involving tissue culture. Therefore transformation does not result in high rates of background mutation. Also, the genome of *A. thaliana* is small and so the probability of integrating into a gene is greater than for other plants. Furthermore, there may be less gene duplication than in other plants resulting in potential exposure of more types of mutants. The procedure involves generating large numbers of transformants and screening the individual segregating T_3 progeny (third selfed generation after transformation) for mutants (42, 43). So far over 12 500 T_3 families have been made with an average of 1.4 T-DNA insertions per transformant (K. Feldmann, personal communication). This has resulted in the identification of a wide range of morphological and developmental mutants. The underlying mutations are currently being characterized at the molecular level (44, 45).

Initially, these studies were based on the recognition of serendipitous mutants that had been generated (and were being characterized) by other means. Subsequently screens have been made for multiple, easily recognizable targets within broad classes of mutants rather than for single genes. Theoretically it would require generating and screening of approximately 75 000 T_3 families to have a 95% chance of tagging a single gene, assuming random insertion of the T-DNA. This is significantly more families than have been generated to date. However, the success rate for tagging genes in broad classes of phenotypes has proved much higher than expected and, therefore, this may be a pessimistic estimate (K. Feldmann, personal communication). Also, it is not currently realistic to tag genes in genetic backgrounds other than the Wassilewskija genotype for which T_3 progeny already exist, because of the large amount of effort required to generate the T_3 families. Therefore, the cloning of disease resistance genes from *A. thaliana* by T-DNA tagging will depend on whether suitable target genes are identified in the Wassilewskija genotype, whether efficient screens can be developed to allow the screening of hundreds of thousands of seedlings for disease susceptibility (at least 12 from each T_3 family), and what probability of success one is prepared to accept (which will be determined in large part by the number of T_3 families available).

5.3 Transposon tagging

5.3.1 Background

Transposon tagging relies on a characterized transposable element inserting into the target gene causing a recognizable change of phenotype. It is an effective method for cloning genes that have been well characterized genetically but where one lacks any information as to their product. Transposon tagging is a routine cloning procedure in bacteria (35) and *Drosophila melanogaster* (46, 47), and has been used successfully in corn and snapdragon to isolate a range of genes (reviewed in references 39, 48). In plants, genes were initially isolated from existing mutants that had been shown, by extensive classical genetic analysis, to be unstable due to the presence of a characterized element. More recently, genes have been isolated in deliberate, targeted tagging efforts. These experiments involved more sophisticated strategies and screening of large populations; often at least 10^5–10^6 plants had to be screened to identify *de novo* mutations due to transposons (48).

Transposon tagging would, therefore, seem to be a promising method for isolating resistance genes. Despite early optimism, however, most resistance genes have remained recalcitrant to isolation by transposon tagging. The lack of success has probably been due to several reasons.

- The greatest confounding factor has been the high rate of instability at the targeted loci. The most detailed studies have been at the *Rp1* locus in corn. The spontaneous mutation rate of the different alleles varies from 8×10^{-3} to 1.5×10^{-4} (49, 50) which is higher than the mutation rate observed in experiments to tag various genes in corn, 10^{-4} to 10^{-6} (41, 48). This has resulted in a high background masking any mutations caused by transposon insertion.

- The mutant is susceptible to the disease and some potential mutants may be lost to the disease.

- Screening for the loss of resistance is not as straightforward as screening for morphological mutants. Resistance genes can exhibit partial dominance or penetrance and be subject to environmental influences.

- Screening may require large plants which will restrict the number that can be assessed effectively.

- Some individuals may escape infection even when susceptible.

- Mutation may not result in a completely susceptible phenotype so unambiguous identification of insertion events may be difficult.

Tagging strategies in corn have become increasingly sophisticated, and recently the gene for resistance to *Cochliobolus carbonum*, *Hml*, has been successfully isolated (S. Briggs *et al.*, personal communication). Key components were a stable target gene, the use of a highly active element (*Mul*), an effective field screen, and the use of several mutants generated with a variety of mutagens to confirm that the targeted locus had been tagged. Also, the sequences flanking the transposons were single copy in the maize genome, making molecular and genetic analysis straightforward.

The cloning of transposons from maize and other species provided the opportunity to develop tagging systems in species lacking characterized elements by using heterologous transposons, several of which have been introduced into plants. So far *Ac*, and *Spm*, both from corn and *Tam3* from snapdragon have been shown to mobilize in heterologous species (51–53). Most tagging strategies in heterologous species have utilized *Ac* (40). *Tam3* frequently causes rearrangements during transposition in both snapdragon and heterologous species (53, 54 and Michelmore *et al.*, unpublished data) and is therefore sub-optimal for transposon tagging. The study of *Spm* in heterologous species is increasing and this element may also prove useful for tagging.

The use of heterologous transposons has several advantages.

- The new host plant does not have sequences that cross-hybridize to the heterologous transposon, making analysis and cloning easier. There will not be related elements in the genome, some of which may be active but show little cross-hybridization.

- As the element becomes increasingly understood it becomes possible to separate and manipulate its various functions; transposition can be controlled and components can be added that make tagging or subsequent molecular isolation easier.

Despite extensive research on the activity of heterologous elements in several plant species, these efforts have been aimed mainly at fundamental studies on transposon function and at developing the technology required for tagging. No gene has yet been tagged using a heterologous transposon, although the technology is reaching a sufficient level of maturity that several genes are likely to be tagged soon. As some of the targets are resistance genes, tagging should prove a useful approach, providing the prerequisites described in Section 5.3.2 are met. While Sections 5.3.2 and 5.3.3 relate principally to tagging using heterologous elements, several components are also relevant to the use of endogenous transposons.

5.3.2 Requirements for transposon tagging with heterologous elements

In its simplest form, transposon tagging involves the introduction of an autonomous heterologous element into a species that is readily transformed (see *Figure 1*). The genotype transformed is homozygous for a well-defined dominant resistance gene, R. The transformant is then test-crossed to a homozygous susceptible (r) line. All of the F_1 progeny should be resistant (R/r) unless the dominant allele has been inactivated. Large numbers of test-cross progeny are, therefore, screened and susceptible individuals selected for further study. The mutants are characterized (see below) to determine which mutations are due to interruption by the known transposon. Sequences flanking the transposon in such mutants are isolated, checked for co-segregation with the target gene, and used to select the wildtype allele from the progenitor unmutated line.

Requirements for the successful application of this strategy are as follows:

(a) The targeted resistance gene must be genetically well-defined. As far as possible it must be demonstrated that the targeted resistance is determined by a single, mutable gene. This may not be straightforward to establish. Some resistance genes are clustered in the genome and, therefore, 3:1 F_2 segregation merely reflects a single cluster of genes. As the matching avirulence genes in pathogens are often unlinked, if pathogen genetics is feasible, it can be used to indicate how many genes are conferring resistance to a particular isolate in a particular host line (12). This analysis will still not indicate if there are duplicated, linked genes of the same specificity (see Section 1.3). Therefore, ideally, chemical mutagenesis with agents such as ethyl methane sulphonate (EMS) that tend to cause point mutations should be conducted to check that the gene is mutable. This will also provide mutants that will be useful later in the characterization of resistance gene function. Mutagens tending to cause deletions should not be used for this control experiment as they could delete duplicated genes.

(b) The screen for susceptibility must be efficient and unambiguous, since tens of thousands of plants will probably have to be screened. Ideally, it should be possible to screen young seedlings in controlled growth room or greenhouse conditions. The larger the plants, the fewer that can be screened and the greater chance of environmental influences. In some disease interactions, such as TMV resistance in tobacco (R. Hehl and B. Baker, personal communication) and *Pseudomonas syringae* pv *tomato* on *Arabidopsis thaliana* (A. Bent and B. Staskawicz, personal communication) elegant schemes have been developed that allow the survival of only those plants with mutated resistance genes. Checking potential mutants is time consuming and, if there are many false positives, true mutants may be missed. Therefore the difference between resistance and susceptibility must be unambiguous.

(c) The target gene must be sufficiently stable that spontaneous mutations do not obscure those caused by insertion of the known transposon. Instability at the *Rp1* locus in corn is much higher than the rate of mutation expected by transposon mutagenesis (see Section 5.3.1). It is unclear exactly which genotypes should be used in control experiments. Even if the genes are apparently stable in normal plants, the passage through tissue culture required for most transformation protocols may result in increased levels of instability over several subsequent generations. Further data does not support this statement.

(d) Transposon tagging requires an element that exhibits an adequate level of transposition. In Solanaceous species, *Ac* transposes readily. In some other species, for example *A. thaliana* (C. Dean, personal communication) and lettuce (C.-H. Yang and R. Michelmore, unpublished data), the rate of transposition is much lower, thus necessitating more sophisticated constructs. The observation of transposition of *Ac* in callus cultures, particularly if there was selection for excision from a drug resistance gene, is not a good indicator of how useful *Ac* will be in regenerated plants.

5.3.3 Experimental options and potential refinement

There is no single optimal strategy for transposon tagging, but there are a variety of experimental designs of differing complexity. One of the problems of transposon tagging, in contrast to chromosome walking, is that many of the decisions have to be made at the beginning of the experiment. As the

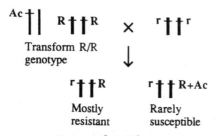

Figure 1. Transposon mutagenesis of a single resistance gene using an autonomous element and a test-cross population. Note that if the resistant parent is used as female, occasional selfing will not result in false-positives. *Ac*, autonomous heterologous transposon from corn; R, homozygous resistance allele; r, homozygous susceptibility.

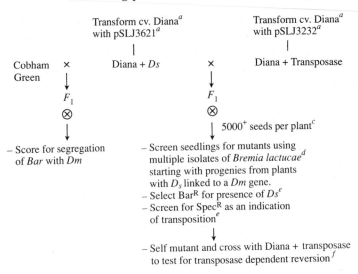

Figure 2. Potential transposon tagging of four genes for resistance to downy mildew in lettuce using a heterologous two element system. [a] cv. Diana expresses four dominant genes for resistance to downy mildew *Dm1, Dm3, Dm7* and *Dm5/8*, any one of which confers complete resistance when challenged by an isolate of *Bremia lactucae* expressing the matching avirulence gene (12). Cobham Green has no genes for resistance to downy mildew. [b] pSLJ3621 and pSLJ3232 were constructed and generously provided by B. Carroll, S. Scofield, and J. Jones (The Sainsbury Lab., UK). pSLJ3621 carries a *Ds* element containing *Nos :: Bar* (to confer resistance to the herbicide, Basta) and bacterial chloramphenicol resistance (to help recloning in *E. coli*). The *Ds* interrupts a chimeric *Nos ::* spectinomycin resistance gene to allow selection for excision. pSLJ3232 contains the 35S :: genomic coding region of *Ac* transposase and will be used to transactivate the *Ds* element. [c] A selfing scheme is preferred in lettuce as generating large test-cross progenies is difficult. [d] Multiple isolates will be used. Each isolate is avirulent on Diana due to the interaction of a different pair of resistance and avirulence genes. [e] Screening for disease resistance is easy at the seedling level. Herbicide and spectinomycin resistance will be assessed on explants once the mutant seedling has been rescued in case the chemical resistance genes have been inactivated. PCR analysis for the presence of the *Ds* element and the excision fragment would provide slower backup methods. [f] Assumes mutant has no transposase genes (1/4 of plants if transposase unlinked to target). If transposase gene is present, mutant will be just selfed and assayed for revertants.

technology is continually evolving, it is difficult to know when to commit to an experimental design and embark on an experiment that will take several years to complete and will be outmoded by the time it is finished. *Figure 2* shows a potential tagging scheme incorporating many of the refinements discussed below.

There are a variety of genetic schemes. The simplest involves a test cross between a homozygous resistant line carrying the transposon and a homozygous susceptible tester. Outcrossing will only expose mutations in resistance genes for which the tester is recessive. Alternatively, the resistant line carrying the transposon can be selfed and the progeny screened. Selfing will expose lesions in any gene determining resistance, not just those identified previously; this may complicate the analysis but lead to the identification of multiple genes involved in resistance. Selfing also avoids the problem of having to distinguish between the mutated allele and the recessive allele from the tester parent. In addition, for some plant species, generating selfed seed is much easier than making many test crosses. The disadvantage of a selfing strategy is that identification of mutants in selfed progeny is less efficient as two gametes carrying the mutated gene must fuse; therefore more individuals must be screened. The likelihood of fusion between two gametes carrying a tagged allele is influenced by the reproductive biology of the plant. If selfing is predominantly within a hermaphroditic flower (for example in lettuce), then up to a quarter of the seed should exhibit mutations that occurred prior to flower development. However, if male and female occur on widely separated parts of the plant (for example maize), only events that occur early in plant development are likely to become homozygous.

Tagging can be made more efficient by screening for mutations in more than one gene. This increases the chance that at least one gene will be tagged and spreads the risk of problems from insertion site specificity and target gene duplication. If test-cross progeny are screened, the test-cross parent must be homozygous resistant for all genes targeted. Targeting multiple disease resistance genes requires isolates of the pathogen, one for each resistance gene. The pathogen must be reisolated from infected plants and typed to determine which resistance was ineffective. Practically, it is usually difficult to synchronize more than three or four isolates and to include a susceptible check for each one.

Tagging can use one- or two-component systems. The *trans* and *cis* factors required for transposition can be separated. The simplest schemes use a one-component autonomous element, such as the complete *Ac* element, that encodes both *trans* and *cis* functions. Alternatively, the transposase functions can be encoded by a separate component and supplied in *trans* to a non-autonomous, *Ds*-like element that is capable of transposition only in the presence of the transposase-encoding component. This allows the two components to be manipulated and introduced separately. Therefore, the position of the non-autonomous element can be characterized genetically prior to mobilization by crossing in a transposase component. Mutants can be stabilized by segregating the transposase component away.

One of the major decisions to be made is whether to screen for mutants in lines in which the position of the transposon is unmapped, or whether to identify lines with the transposon linked to the target gene. Sometimes, but not always, *Ac* has a tendency to jump to nearby positions in maize (55, 56), and tobacco (57), often within 5 cM. This tendency can be used to provide many mutants from a nearby element and, therefore, reduces the number of individuals that have to be screened. The degree of bias for an element to jump to a linked position may depend on its position relative to a replicon organization (57). While transposon mutagenesis using a linked element may be more efficient as fewer progeny have to be screened for mutations, the linked element strategy requires the generation of many independent transformants and mapping of their integration sites to identify lines carrying transposons linked to the resistance gene. Mapping can be done by cosegregation analysis of a marker within the T-DNA and the target locus; this involves analysis of a separate progeny for each transformant. Alternatively, the sequences flanking the T-DNA can be isolated by inverse polymerase chain reaction (IPCR) and mapped on an existing genetic map (J. Jones, personal communication); only one segregating population need be analysed. However, not all insertions may be mapped due to the failure of (IPCR) to provide flanking sequences or the inability to conduct restriction fragment length polymorphism (RFLP) analysis with such sequences due to monomorphism or the presence of repeated sequences.

Marker genes can be placed within the element. The inclusion of screenable (for example β-glucuronidase, GUS) or selectable (for example herbicide resistance) markers that function in the whole plant allows the presence of the element to be monitored (58). The inclusion of markers that function in bacteria increase the efficiency of subsequently isolating genomic clones containing the transposon.

If the frequency of transposition is low, it may be advantageous to be able to screen only those plants in which excision has occurred. Excision assays rely on the restoration of a readily-scored phenotype. A visual assay has been used effectively in corn by monitoring excision of *Ac* from the *P* locus (J. Chen, S. Dellaporta, and S. Briggs, personal communication). In transgenic plants, it is possible to monitor excision from a herbicide or drug resistance gene (59). Once the initial transposition has occurred, however, the excision assay is uninformative as an indicator of further transposition.

Ideally, it would be useful to be able to control the frequency and timing of transposition. One scenario is to induce a burst of transposition just before or during pollen formation, so that each fertilization event contains a different transposition event. There would be no transposition somatically and, therefore, the mutants would be somatically stable during analysis of the resultant progeny. It should be possible to regulate transposition using heterologous promoters fused to either genomic or cDNA sequences of the transposase. However, the ideal promoter has yet to be identified. Environmentally-induced promoters are either too leaky or it is impossible to regulate their activity cleanly. Most pollen-specific promoters are active too late in development after the separation of the vegetative and generative nuclei into different cells. Promoters from some of the regulatory loci which are active earlier in flower development might be useful in those species that set few seed per flower (such as lettuce). Such promoters will be less useful in species that set many seed per flower, such as tomato, as a burst of transposition early in flower development might result in many progeny having the same mutation event; this would necessitate screening more progeny individuals.

One of the major challenges is the identification of which mutants are due to insertion of the known transposon. There are several options:

- The simplest is to determine if there is an RFLP in the sequences that flank the transposon that co-segregates with the mutated gene.

- Alternatively, if the transposon carries a marker gene (and there are few copies of the element present in the genome), co-segregation between the marker and the mutated locus can be analysed. When potential linkage has been identified, large populations must be analysed to confirm co-segregation, because if enough mutants are investigated tight linkage between the target gene and the transposon will occur fortuitously. Tight linkage may occur frequently if the transposon was linked at the beginning of the experiment.

- Reversions to resistance are a more reliable indication that the mutation is transposon-induced; this is particularly powerful if reversions only occur in the presence of the transposase component. However, not all transposon-induced mutants will revert.

The best evidence that the target gene has been inactivated is the analysis of multiple mutants. If multiple mutants have the overlapping flanking sequences there is a high probability that they belong to the target locus (S. Briggs, personal communication). Mutants generated by other means are also useful at this stage as sequences flanking the transposon may detect changes in the mutant genotype relative to its progenitor.

The optimal method for isolation of sequences flanking the transposon depends on the system. Analysis is complicated if there are many elements in the genome. Several strategies have been used in maize, including identification of genomic clones using hybridization with sequences only present in complete active elements, hypomethylation of regions with recent transposition, cross-hybridization of two sets of clones containing sequences mutated with different transposons, and cloning of specific RFLP bands that co-segregate with the mutation (38). Strategies utilizing heterologous elements could exploit selectable markers within the element such that only sequences containing the element are cloned.

5.3.4 Potential problems

Transposon tagging requires several years of work and multiple plant generations before a gene is successfully isolated. It is an all-or-nothing process and it is difficult to monitor progress. There is no way of determining if progress is blocked, for example by target site incompatibility (see below). Therefore, multiple targets should be pursued and strategies should be designed to minimize the chances that the effort will be wasted. It would also be advantageous to use different transposons, for example both *Ac* and *Spm*.

There is the potential for complicated crossing and marker selection schemes to manipulate and monitor transposition. However, not all components may function; in particular, some of the marker genes may be inactivated and therefore not provide the required phenotype. Strategies should not be totally dependent on all of the components working.

The background mutation rate may be too high. The gene must be stable but mutable, as discussed above. Not all mutations identified may be caused by the known transposon. Other elements in the genome may cause mutations. When a pathogen is used to test for resistance, the mutation rate in the pathogen must also be considered. The pathogen should be reisolated from susceptible mutants and checked to ensure that it still detects the targeted resistance.

Chemical mutagenesis of some resistance genes has resulted in only partial susceptibility (60). Also, insertion of a transposon can result in novel patterns of gene expression (61, 62). Therefore, the phenotype of a chemically-induced mutant of a resistance gene may not be a good predictor of a transposon interrupted resistance gene. Also, in tagging experiments, potential mutants that exhibit only partial susceptibility should be considered for analysis.

The frequency of transposition may be too low or too high. If the frequency of transposition is very low, many individuals will have to be screened to identify a mutant and tagged mutants may be obscured by the background mutation rate. In lettuce, the frequency of transposition can be increased by expression of the transposase from a heterologous promoter (Yang and Michelmore, unpublished data). Also, the use of a linked transposon may be more important in these species. If the rate of transposition is too high, the copy number of the element may increase and make isolation of the targeted sequences difficult. Also, frequent transposition increases the probability of an element inserting into a gene and then excising to leave a footprint that results in a frame-shift mutation; such a mutation would not be tagged by the element and would therefore be unidentifiable using probes for the element. This may be a problem with the high rates of transposition observed in tomato (63). In this case it may be necessary to modify the system to decrease the rate of transposition.

The element may become inactivated. In their native genomes, transposons become methylated; such DNA modification may be involved in the regulation of their activity and copy number. There is insufficient data to know how prevalent methylation will be in heterologous species. *Tam3* seems to become methylated in tobacco in contrast to *Ac* (53). *Ac* continues to transpose over several generations in tomato (63) and is rarely methylated in tobacco (64). However, the demonstration of excision in tissue culture may be a long way from a functional transposon mutagenesis system for whole plants (for example in lettuce; Yang and Michelmore, unpublished data).

Transposons are not sequence-independent in their integration sites (38, 65). There is no way of knowing a priori whether the transposon is amenable to insertion into the target gene. The risk can be minimized by using either multiple elements, which is more feasible in corn than in other species, or multiple targets (see above).

6. Map-based cloning

6.1 Overview

Map-based cloning of genes is receiving increasing attention in plants as techniques become refined for cloning and manipulating large fragments of genomic DNA. Much of the driving force for development of this technology is the human genome project where well-funded efforts explore different approaches and develop the basic technology. While this approach is no less expensive in plants than animals, adequate funding for 'brute-force' approaches in plants is rare. Therefore, plant researchers must develop more subtle, focused approaches. One important advantage of plant research over human studies is the ability to generate and screen large experimental populations (segregating or mutant). This allows the identification and use of highly informative individuals, thereby minimizing the molecular analysis required. Therefore, in plants it is probably expedient to emphasize genetic components of a map-based cloning strategy more than has been the case in human studies which have often emphasized physical characterization of the target region.

Map-based cloning involves:

- identification of markers genetically tightly linked to the target gene
- determination of the relationship between genetic and physical distance in the target region
- chromosome walking by identifying overlapping genomic clones starting from the closest marker
- introduction of clones potentially carrying the target gene to complement the recessive phenotype

Advantages of map-based cloning approaches over alternatives are that each step yields useful data, progress can be monitored, and potential problems identified. Also, map-based cloning requires no assumptions as to the gene product or the pattern of gene expression.

While conceptually straightforward, map-based cloning is still a labour-intensive long-term undertaking. Also, if resistance genes are members of multigene families, they may reside in complex regions of the genome that may contain many related sequences. Cross-hybridization between such sequences will complicate chromosome walking. Therefore, techniques may be needed that allow walking through complex regions.

6.2 Saturating target regions with closely-linked markers

6.2.1 Introduction

The starting points for any map-based cloning strategy are closely-linked molecular markers. These can be obtained by genetic or physical means. Genetic approaches include mapping of random markers, pre-selecting markers using near-isogenic lines or bulked segregant analysis (see Section 6.2.3). Physical approaches include preparative pulsed-field gel electrophoresis, subtractive hybridization (see Section 3.3), microdissection of chromosomes (66), and fluorescent sorting of chromosomes (67). Although theoretically possible, physical approaches have yet to be used in plants; this is partly because genetic approaches are technically simpler and have been very successful. However, several techniques are in the developmental stage (for example reference 68) that may facilitate the selection of individual large fragments from complex genomes. This would revolutionize map-based cloning strategies.

6.2.2 Types of markers

Two useful types of molecular markers can be identified in genetic studies as linked to the target gene, restriction fragment length polymorphisms (RFLPs) and random amplified polymorphic DNAs (RAPDs).

- RFLPs rely on hybridization of cloned low copy DNA fragments to restricted genomic DNA. RFLPs are due to either alterations in restriction enzyme recognition sequences or chromosomal alterations resulting in changes in the length of DNA between restriction sites. RFLP markers are often co-dominant, but probes are restricted to sequences homologous to low or single copy genomic DNA.

- Recently, RAPD markers have been developed (69, AP-PCR markers, 16; see also Volume I, Chapter 11). This technique relies on the differential enzymic amplification of small genomic DNA fragments using PCR with arbitrary oligonucleotide primers (usually 10-mers). Polymorphisms result from either base changes that alter primer binding or from insertions/deletions that alter the length of the amplified region. RAPD markers are readily identified. However, the initial cycles of amplification probably involve extensive mismatch, and rigorous standardization of reaction conditions are required for reliable, repeatable results. We use a method similar to that developed by Williams *et al.* (69) (*Protocol 1*). There reaction conditions and cycling parameters have worked well for a wide variety of organisms, including lettuce, horse, honey bee, and the fungus, *Bremia lactucae*.

Protocol 1. Identification of RAPD markers

1. Prepare the following solutions:

TE buffer:

- 10 mM Tris
- 1 mM EDTA, pH 8.0

10 × PCR buffer:

- 100 mM Tris–HCl, pH 8.3
- 500 mM KCl
- 15 mM $MgCl_2$
- 0.01% gelatin

dNTP mix:

- 1.25 mM dATP
- 1.25 mM dCTP
- 1.25 mM dGTP
- 1.25 mM dTTP

2. Prepare genomic DNA by using a modified CTAB method (70).

3. Store concentrated stocks of DNA frozen in TE buffer (see step 1) at approximately 0.5 µg/µl. Prior to use, dilute DNA stocks 1:100 in modified TE buffer containing only 0.1 mM EDTA. These dilutions can be stored at 4°C for several months.

4. Set up the DNA amplifications by mixing the following:

- water   13.2 µl
- 10 × PCR buffer (step 1)   2.5 µl
- dNTP mix (step 1)   2.0 µl
- 10 mM $MgCl_2$ 1.1 µl
- 7.5 µM primer   1.0 µl
- 5 units/µl *Taq* polymerase   0.2 µl
- 5 ng/µl genomic DNA   5.0 µl
  Total volume   25.0 µl

5. Overlay the reaction mix with mineral oil. Amplify the DNA for 30–40 cycles using the following cycling parameters (Perkin Elmer Cetus thermocycler):

Protocol 1. *Continued*

94°C	1.5 min	1 cycle
35°C	1 min	
72°C	2 min	40 cycles
94°C	1 min	
35°C	1 min	
72°C	5 min	1 cycle
10°C		Hold

6. Separate the amplification products on a 2% agarose gel. Visualize by staining with ethidium bromide (0.5 µg/ml) and observe by UV fluorescence.

Key parameters are the concentrations of DNA and primer. The precise ratio of primer to genomic DNA may have to be adjusted for each particular organism. The lack of clear amplified bands or a high level of background smear may be indicative of the suboptimal primer or DNA concentrations. For all the species we have tested, consistent results were usually obtained with 1–25 ng DNA per 25 µl reaction with 200–400 nM primer concentrations. Increasing the template DNA concentration generally decreases both the number and quality of amplified bands, although some bands may become more intense. Lowering the primer concentrations to 100 nM or less will, for some primer/DNA combinations, have little effect on the banding pattern; however, for other combinations it may significantly reduce the total number of scorable bands. Increasing the primer concentration to 400–800 nM has little effect on most combinations, but sometimes increases the number and quality of bands for some combinations that work poorly at lower concentrations. This effect seems to be specific to the template DNA; for example, primers that gave poor amplification with lettuce DNA at the standard concentrations worked well with horse DNA. Doubling the primer concentration to 15 µM gave good amplification with lettuce DNA. Reducing the annealing temperature may achieve the same result.

Changes in other components in the RAPD reaction, such as $MgCl_2$ concentrations, will affect banding patterns. However, there are no obvious interactions with the variable components (template DNA and primer). The gelatin is apparently not important and it is omitted from new versions of the buffer. If gelatin is used, the buffer solutions must be mixed well after frozen stock solutions are thawed.

The major advantage of RAPDs over RFLPs is the large number of loci that can be screened with minimal effort. Hundreds of loci can be screened in days as multiple loci (often 5–10) are detected by each primer. Five hundred arbitrary 10-mer oligonucleotide primers are currently available commercially (Operon Technologies); this number should increase to 1000 within a year.

The primers are 60% or 70% G + C and their sequences have been checked to avoid primer-dimers due to 3' complementarity. Also, as no Southern hybridization is required for RAPD analysis, polymorphisms can be detected in fragments containing highly repeated sequences, thereby, providing markers in regions previously inaccessible to genetic analysis. This may be an important attribute for obtaining markers in regions containing resistance genes if such regions contain complex multigene families.

The drawback to using RAPD markers is the usual occurrence of only two genotypes: presence or absence of an amplified product. Genetic resolution of such dominant loci is poor in segregation analysis of F_2 populations because heterozygotes cannot be distinguished from one of the homozygotes. Therefore, recombinant inbred or backcross populations are preferred for precise genetic localization of RAPD markers. Alternatively, the amplified fragment can be cloned and mapping of the locus attempted by RFLP analysis. However, this will not be possible if the amplified region contains repeated DNA or if no polymorphism is detected between parents of available mapping populations. Another drawback of RAPD markers is that they cannot be used directly for selecting genomic fragments from libraries, as a primer may amplify 5–10 bands each from different regions of the genome. The RAPD must either be cloned and used as a hybridization probe or sequenced and used to develop a sequence characterized amplified region (SCAR).

We are developing SCARs as an additional type of marker. These markers are most easily generated from RAPD markers. SCARs are similar to sequence tagged sites (STSs) proposed for the human genome (71) except that they do not have to contain single copy sequences. The amplified RAPD fragment is cloned and the termini sequenced using primers flanking the cloning site. Longer oligonucleotide primers (currently 24-mers) are made by extending the sequence inwards from the RAPD primer sequence. These primers are then used to amplify genomic DNA in a high stringency PCR reaction. This results in only one or a few amplified products. Sometimes the same sized fragment is amplified from both parents and the RAPD polymorphism is lost, presumably because the original polymorphism was due to mismatch at the primer binding site. We are currently investigating methods for detecting polymorphism within the amplified fragment which would turn these into co-dominant markers. PCR-based markers will be extremely useful in screening libraries during chromosome walking as, although the amplified fragment may contain repeated DNA, SCARs usually identify unique genomic sequences (see above).

6.2.3 Genetic analysis

Genetic analysis of randomly selected markers is useful for developing genetic maps of the whole genome but it represents an inefficient method for obtaining markers in a particular region. Closely-linked markers may be obtained by chance, but most markers will map to unlinked regions. For initiating a chromosome walking approach, markers as tightly linked as possible are required, preferably within 1 centiMorgan (cM) so that there is a reasonable probability of detecting physical linkage between markers and the target gene (see Section 6.3). Therefore, the total number of markers mapped throughout a genome is less important than the number of markers in the target region.

If near-isogenic lines (NILs) are available that differ for a resistance gene, they can be used to obtain markers in genomic regions that distinguish them. Markers that are polymorphic between the NILs will likely be linked to the resistance gene. This approach has been used successfully to identify markers linked to several resistance genes: *Tm2A* and *Pto* in tomato (72, 73), *mlo* in barley (74), and to *Dm* genes in lettuce (127). Between 5–10 RFLP probes can be screened simultaneously in the same hybridization reaction. Single oligonucleotide primers for RAPD markers detect several loci simultaneously. The efficiency of screening is increased if NILs are available with the same resistance gene backcrossed into different genetic backgrounds. This maximizes the chances that polymorphisms linked to the target gene will be identified, a marker may be monomorphic between one pair of NILs but polymorphic in another if the recurrent parents represent very different backgrounds. Also, if a polymorphism is detected within both pairs of NILs, it is highly likely that the polymorphism will map to the targeted region. The efficiency of obtaining polymorphic markers linked to the target region is dependent on the degree of isogenicity which is a function of the number of backcrosses used to generate the lines (75, 76). The efficiency of detecting markers in the target region is dependent on the degree of sequence divergence in the introgressed target region between the donor and recurrent parents. Not all probes in the region will detect polymorphism. Linkage of polymorphisms to the target gene are confirmed by segregation analysis; however, not all polymorphisms detected between the near-isogenic lines may segregate in mapping populations available to the researcher. The biggest disadvantage to near-isogenic lines is the time taken to generate them; a minimum of five backcrosses is usually required. This takes many years for most plant species.

Recently, we developed bulked segregant analysis as a method for rapidly identifying markers linked to any gene or genomic region (77, *Protocol 2*). Bulked segregant analysis involves screening for polymorphism between two bulked DNA samples derived from a segregating population (*Figures 3* and *4*). Each bulk comprises individuals selected to have identical genotypes or phenotypes for a particular region ('target region or locus') and are random for other loci. Therefore, loci tightly linked to the target region, if segregating in the population, will be polymorphic between the bulks, while unlinked loci will appear heterozygous and monomorphic. For loci linked to the target gene, there will be increasing similarity between the bulks with decreasing linkage between the target locus and the locus assayed. Empirical studies have shown that markers can be detected in a symmetrical 'genetic window' approximately 20 cM wide either side of the target locus (77 and Paran *et al.*, unpublished data). Bulked segregant analysis has several advantages over the

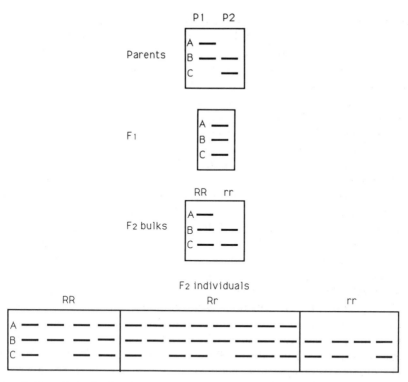

Figure 3. Genetic basis of bulked segregant analysis. This figure shows the genotypes of three RAPD loci (A through C) detected by a single primer in two parents (P1 and P2), their F_1 and F_2 progeny, and bulks derived from F_2 individuals homozygous for resistance (RR) or susceptibility (rr). The dominant allele at locus A is linked to the gene for resistance and, therefore, distinguishes the bulked DNA samples. Locus C is unlinked to the resistance gene and, therefore, the dominant allele is present in both bulks. This schematic is the genetic basis of the top three loci identified by primer OPK15 in *Figure 4*.

analysis of near-isogenic lines (77). The greatest advantage is that bulks can be made instantaneously from any segregating population and for any region of the genome for which there is a marker. Screening for RAPD markers is the most efficient way of identifying new loci linked to resistance genes. This approach has proved very successful in lettuce; within a few months, we utilized over 300 primers to screen over 2700 loci and identify eight markers linked to resistance genes (77 and Paran *et al.*, unpublished data).

Protocol 2. Bulked segregant analysis

1. Construct two bulks comprising 2.5 µg DNA from each of at least 10 individuals. Most of our bulks comprise 15–20 individuals. Fewer individuals can be used but this increases the chances that loci unlinked to the

Protocol 2. *Continued*

target gene will be detected. The individuals within a bulk are selected to have the same genotype at the resistance locus. If F_3 family data are available, it is most efficient to bulk individuals that are homozygous for each allele. Heterozygotes can then be excluded from the analysis. This allows RAPD markers to be identified that are in both *cis* and *trans* to the dominant allele. If only F_2 data are available and heterozygotes cannot be distinguished from dominant homozygotes, then only RAPD markers with a band in *cis* with the dominant allele will be identified.

2. Screen the two bulks for polymorphisms at RAPD or RFLP loci. Previously mapped RFLP or RAPD markers can be used to locate an unmapped resistance gene on the genetic map. The RFLP probes can be bulked to increase the speed of the analysis. To identify new markers around a resistance gene, screen the two bulks for differences in RAPD loci using arbitrary 10-mer oligonucleotide primers and the conditions in *Protocol 1*. Thousands of RAPD loci can be screened within a few months utilizing the RAPD primers that are currently available commercially.

3. When markers that distinguish the bulks have been identified, determine their genetic position relative to the target locus by segregation analysis of the source population used to generate the bulks. This can be done by RAPD analysis of the segregating individuals; it is not necessary to clone the amplified fragment and to try to identify and map an RFLP. However, if an RFLP or co-dominant SCAR can be generated from the RAPD marker, it greatly increases the genetic resolution of the analysis.

Markers selected using near-isogenic lines or bulked segregant analysis must be mapped using segregating populations to define their precise genetic position relative to the targeted resistance gene. When linkage is tight, large populations of several hundred individuals have to be analysed to detect a sufficient number of recombinants. Particularly large populations are necessary if dominant loci such as resistance genes and RAPD markers are being analysed. Alternatively, as only recombinants in the region are informative, many informative individuals can be identified by selecting for recombination between the resistance gene and flanking markers that are scorable at the whole plant level. Only recombinant individuals then need be analysed at the molecular level, thereby greatly reducing the number of DNA extractions required. In addition, the distribution of RAPDs in recombinants identified in several segregating populations can be monitored. F_3 family analysis of recombinant individuals further increases the genetic resolution with dominant markers. Such recombinant individuals are also useful in correlating the genetic and physical maps (see Section 6.3) and in chromosome walking (see Section 6.4).

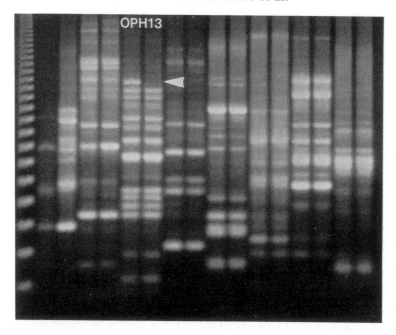

Figure 4. Identification of loci linked to a resistance gene by bulked segregant analysis. DNA samples from F_2 individuals from a cross between two lettuce cultivars were pooled on the basis of their genotype at *Dm5/8* and amplified using seven different random 10-mer oligonucleotide primers. The left-hand lane in each pair contained homozygous *Dm5/8* individuals; the right-hand lane contained homozygous *Dm5/8* individuals. The primer OPH13 identified a polymorphic locus (arrowed). The leftmost lane contains a 123 bp ladder.

6.3 Estimating genetic versus physical distances

6.3.1 Introduction

Some knowledge of the relationship between genetic and physical distance in the target region is a prerequisite for progressing from linkage analysis to chromosome walking. The little data available on variation in recombination over the genome indicate considerable non-linearity. Average values obtained by dividing the physical genome size by the number of centi-Morgans in the genome are irrelevant to chromosome walking strategies. For example, intragenic recombination at the *Bronze* locus in *Zea mays* was equivalent to 1 cM per 14 kb, but was greatly reduced in flanking regions (78, and H. Dooner personal communication); the average value for maize over the whole genome is 1 cM per 2000 kb. Five RFLP markers have been identified within 1.2 ± 0.6 cM close to the *Tm2A* locus in tomato; however, they are distributed over at least 4 Mb (79); the average value over the tomato genome is approximately 1 cM per 500 kb. Studies in mouse (80), yeast (81), and *Drosophila* (82) have also shown that rates of recombination can vary over short physical distances along the chromosome. The non-linearity of recombination with physical distance may be particularly acute when genes, such as resistance genes, have been introgressed from a wild species. Extreme sequence divergence or hemizygosity would likely result in greatly reduced rates of recombination in the introgressed region. Even when a marker is genetically tightly linked to the target gene, it is not necessarily physically close. Genetic data alone is, therefore, not sufficient to identify suitable starting points for chromosome walking.

Advances in the preparation and digestion of high molecular weight DNA in agarose and in the electrophoretic separation of megabase sized DNA fragments using pulsed-field gel electrophoresis (PFGE) (124) has allowed the correlation of genetic data with the physical distance along the chromosomes. These techniques have been used in humans to study chromosomal rearrangements and to make physical maps of complex loci using long-range restriction mapping (83–86).

The minimum required for chromosome walking is the localization of two markers that genetically flank the targeted resistance gene to the same genomic fragment in multiple digests. Beyond determining the maximum physical distance separating the flanking markers, little further is necessary if the distance is perceived as walkable using the libraries available to the investigator. In the absence of flanking markers that clearly hybridize to a single fragment, it is necessary to continue selecting for markers in the target region and to develop maps around single markers. This should continue until a marker potentially within a 'walkable distance' is identified. The relationship between genetic and physical distance may vary within the target region and an understanding of this relationship will be continually refined as each step in the walk is monitored and the walk progresses.

6.3.2 Long-range restriction mapping

Long-range restriction mapping involves the generation of large fragments of genomic DNA by digestion with rare-cutting restriction enzymes, fractionation of the fragments by pulse-field gel electrophoresis, and Southern analysis with DNA probes. Southern blot analysis of single, double, and partial digestion products separated using PFGE provides the information to construct a restriction map that covers hundreds of kilobases on either side of a RFLP probe (83). In mammalian systems, long-range maps of complex regions have been established using characterized clones (85). Alternatively, analysis with PFGE can be used to establish long-range maps based on anomymous markers that have been shown to be genetically linked (79). Ideally, the targeted region is saturated with markers, allowing the construction of several overlapping maps. However, for plant genomes, markers are rarely clustered with sufficient density to allow the immediate construction of detailed maps based on several markers.

The use of classical genetics in combination with PFGE to determine the relationship between genetic and physical distance around a single marker was first suggested by Meagher *et al.* (87). Their assay relied on detecting novel fragments generated by recombination between large fragments with rare-cutter restriction site polymorphisms at both ends. This approach however, is limited by the requirement for restriction site polymorphism at both ends of the fragment analysed. Therefore, this approach can only be used when there is a high level of polymorphism. We have refined this approach to make it more broadly applicable.

Long-range restriction maps can be made using a single probe to detect polymorphisms with enzymes that cut both frequently and infrequently in the genome. An enzyme that cuts frequently in the genome will define a locus close to the sequences homologous to the probe (often within 10 kb); an enzyme that cuts infrequently can detect loci up to several megabases from the sequences homologous to the probe. Therefore, a single probe can detect loci separated by hundreds of kilobases. To correlate the genetic and physical maps requires polymorphisms between parents with both types of enzymes. The co-inheritance of these polymorphisms is analysed in segregating populations.

One of the limitations of long-range restriction mapping is that it presently requires hybridization of low copy number probes. Therefore, some of the sequences identified as RAPD markers that contain high copy DNA are not readily located on the map. One of the current challenges is to develop strategies for integrating PCR-based markers on to the long-range map.

Preparation of high molecular weight DNA from plants requires tissue with a minimal amount of structural carbohydrate. Protoplasts from suspension cultures and young leaves (79, 88) have proved to be useful source tissues for high molecular weight DNA that is sensitive to a variety of restriction endonucleases. The protoplasting procedure and source tissue will have to be optimized for each species. However, the integration of genetic and physical maps requires the analysis of many different individuals and it is not feasible to derive suspension cultures for each one. We developed a mini-preparation protocol to produce high molecular weight DNA from many individuals (*Protocol 3*) that had been selected as recombinants in earlier genetic studies. This method has provided DNA of sufficient quality and quantity to provide long-range maps of a number of cultivars and recombinant families (89). The procedures for single and double restriction digests of the high molecular weight DNA in agarose are given in *Protocol 4*.

Protocol 3. Preparation of high molecular weight plant DNA for long-range restriction site mapping

A. *Protoplast isolation*[a]

1. Prepare the following solutions:

    10 × Protoplasting salts (filter-sterilized):

    • KH_2PO_4 272 mg/l

Protocol 3. *Continued*

- KNO$_3$ 1 g/l
- CaCl$_2$·2H$_2$O   1.5 g/ml
- MgSO$_4$·7H$_2$O   2.5 g/l
- Fe SO$_4$·7H$_2$O   2.5 mg/l
- KI   1.6 mg/l

Protoplast solution:

- 13% mannitol, 1× protoplasting salts. pH 5.8 (autoclaved) (see below)
- 0.25% Macerase (CalBiochem), added just prior to use
- 0.25% cellulase (Worthington), added just prior to use

2. Grow the plants in sterile soil in a growth chamber and fertilize once per week to provide vigorous, clean, and uniform plants.

3. Keep the plants in the dark for 24–72 h prior to harvesting to minimize starch in the leaves.

4. Harvest the leaves from 4–6-week-old plants.

5. Cut the tissue from 2–3 young leaves (1–2 g) into 2 mm strips then immerse in 30 ml of protoplasting solution (0.2–0.5 g/20 ml) in 50 ml screw-top plastic tubes.

6. Wrap the tubes in foil and gently agitate them (50 r.p.m.) on their sides for 6–12 h at approximately 25°C, or 12–24 h at 15°C.

7. Filter the protoplasts through 70 µm nylon mesh into 15 ml graduated conical tubes and centrifuge them at 800 g to form a loose pellet.

8. Estimate the volume of protoplasts from the graduations on the 15 ml tube.

B. *Embedding protoplasts and preparing high molecular weight DNA in agarose blocks*[b, c]

1. Prepare the following solutions:

   ESP (make fresh):

   - 0.5 m EDTA, pH 8.0
   - 1% *N*-lauroylsarcosine
   - 1 mg/ml proteinase K (Boehringer–Manheim)
   - Adjust the pH of the 0.5 M EDTA stock (pH 9.0) to pH 8.0 with 1/100 vol. of 2 M Tris, pH 7.5

   LMP agarose stock:

   - 1% InCert (FMC) agarose dissolved in 0.125 M EDTA

2. Add an equal volume of 1% LMP agarose stock to the protoplasts.

3. Mix and transfer the suspension to sample moulds (Histoprep Base Moulds, Fisher Scientific), and place at 4°C until the agarose has solidified.

4. Remove the agarose plugs from the moulds and cut them into 100 μl blocks using a sterile glass coverslip.

5. Add 5 volumes of ESP to 1 volume of gel plus in 15 ml polypropylene tubes.

6. Incubate samples in ESP at 50–55°C for 12–24 h. Change the solution by gently pouring off the old ESP and replacing it with freshly prepared ESP. Incubate for a second 12–24 h period. The samples should clear during this step.

7. Store the samples at 4°C in 0.5 M EDTA in the polypropylene tubes.

[a] Protocols for preparing high molecular weight DNA from protoplasts must begin with a method of protoplast isolation optimized for the plant to be studied. For lettuce, we use a modification of procedures developed by Crucefix *et al.* (90).

[b] Modified from reference 79.

[c] Recently, we have tried a method involving embedding protoplasts in agarose microbeads. This has provided high quality large DNA which is readily digested by restriction enzymes. The method requires less proteinase K and restriction enzyme, as well as shorter digestion periods. The protocol will be published shortly (R. Wing and S. Tanksley, in preparation).

Protocol 4. Restriction digestion of high molecular weight DNA in agarose

A. *Single digests*[a]

1. Prepare the following solutions:

TE:
- 10 mM Tris, pH 7.5
- 1 mM EDTA

TE + PMSF: (prepare fresh)
- Add 1/100 vol. of 100 mM PMSF stock prepared in isopropyl alcohol to TE

1 mg/ml BSA:
- (sterile filtered, and stored −20°C)

100 mM Spermine–HCl:
- (sterile filtered, and stored −20°C)

10 × Restriction buffer:
- (specific for each enzyme, supplied by manufacturer).

Protocol 4. *Continued*

Stop dye:
- 40% sucrose
- 100 mM EDTA
- 2.5 mg/ml bromophenol blue
- 2.5 mg/ml xylene cyanol

2. Dialyse the samples in TE + PMSF prior to digestion to inhibit residual proteinase K.

3. Pour off the 0.5 M EDTA storage buffer and add 5 volumes of freshly prepared TE + PMSF per volume of agarose plugs.

4. Incubate the DNA samples for 20 min at 50°C. Repeat 3 times to remove all traces of proteinase K, and store in TE prior to digestion.

5. Carry out the digests of the agarose plugs in a volume of 250–300 μl under buffer conditions recommended by the restriction enzyme supplier with the addition of 1/10 vol. BSA and 1/20 vol. of spermine stock. Manipulate the agarose plugs (100 μl) into the reaction tubes using a sterile spatula. For each digestion, at least 50 units of enzyme is required due to the large volume of the reaction.

6. Leave the samples on ice for 30 min to allow the enzyme to diffuse into the agarose plugs prior to incubation at the desired temperature. Allow the digestions to proceed for 8–12 h.

7. Stop the reaction by adding 25–50 μl of stop dye and then refrigerate the samples until further use.

B. *Double digests*[b]

1. After the first digest dialyse the samples in the reaction tubes for 20 min each with two changes of TE.

2. Add the second reaction mix (as described in step 5 above) to the tube containing the agarose embedded DNA, and let it sit on ice for 30 min prior to digestion at the desired temperature.

C. *Partial digests*

1. Prepare multiple reactions as described above (part A) in 300 μl with 1 unit, 5 units, 10 units, 25 units, and 50 units of enzyme per reaction. Allow the reactions to sit on ice for 30 min to permit the enzyme to diffuse into the sample plugs before incubating at the desired temperature for 8 h.

[a] Exceptions to the conditions described in A may be necessary for some enzymes. For example, we usually omit the spermine from *Not*I digests; *Bss*HII digests are allowed to proceed for only 2–3 h.

[b] Assumes that the two endonucleases require different reaction conditions.

[c] Partial digests in agarose may be made by varying the concentration of enzyme or the time of digestion. We have found that varying the concentration of enzyme gives the most consistent results.

6.3.3 Integration of genetic and physical data

Recombination events between linked markers in the target region are identified by classical genetic analysis of large segregating populations. Individuals with non-parental combinations of linked markers are the only informative members of the population and therefore only these individuals are analysed at the physical level. Loci are scored by conventional Southern analysis of genomic DNA restricted with frequently cutting enzymes for RFLPs close to sequences homologous to the probe, and by PFGE and Southern analysis of DNA restricted with rare cutters for the RFLP distant from the sequences homologous to the probe. Recombination at a site between the two RFLP loci will result in the generation of either novel combinations of alleles or new fragment sizes (*Figure 5*; *Table 1*; reference 89). Detection of a single recombinant allows the orientation of the long-range map relative to the RFLPs and resistance genes in the region. Detection of several events provides an estimation of the frequency of recombination occurring over the physical distance separating the RFLP loci.

The identification of many recombinant individuals over a short genetic distance requires the analysis of large populations, either at the RFLP level or with flanking phenotypic markers. Phenotypic markers that are readily scored allow the analysis of large populations. If resistance genes are clustered, recombination between the resistance genes themselves can provide many

Table 1. Expected allele combinations for recombination events occurring within the long range map surrounding locus CL922 in lettuce

Recombination Event	CL922 Digest	Chromosome 1		Chromosome 2
(1)	HindIII	a		A
	EagI	B		b
	SalI	C		c
(2)	HindIII	a		A
	EagI	b		novel
	SalI	C		c
(3)	HindIII	a		A
	EagI	b		B
	SalI	C		c
(4)	HindIII	a		A
	EagI	b		B
	SalI	c		novel
None	HindIII	a		A
	EagI	b		B
	SalI	c		C

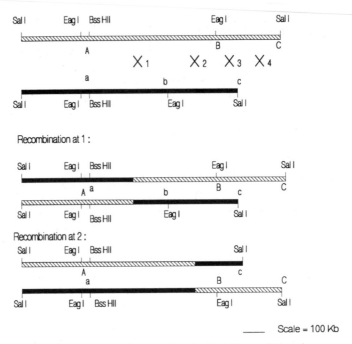

Figure 5. Genetic consequences of recombination in lettuce within a long-range restriction map developed from a single probe. Long-range restriction maps are shown for two inbred parents, one hatched and the other black. The two maps differ in the distance of the rightward *Eag*I and *Sal*I sites from the region homologous to the probe (A). Polymorphism within 20 kb of A was detected by RFLP analysis of conventional Southern blots. Crossovers at any of the four positions shown would generate novel restriction patterns that could be analysed by Southern blots of pulsed-field gels. The novel fragments that would result from recombination at sites 1 and 2 are illustrated. This schematic shows the basis of the analysis described in *Table 1*.

informative recombinants in the region. Recombinants from different segregating populations can be analysed, although the genetic distance and possibly the gene order may vary in different crosses. In addition, near-isogenic lines and recombinant inbreds that have cross-overs within the target region will also be useful.

Long-range RFLP analysis can be performed by analysing recombinant F_2 individuals. However, for several reasons it is more effective to analyse F_3 families. The F_2 plants can be first screened to identify recombinants. F_3 individuals then provide large amounts of material for analysis, grown under appropriate conditions to yield high quality large DNA. Also, the recombinant genotype of the progenitor F_2 plant can be confirmed by analysing the segregation of flanking markers in the F_3 family. Furthermore, analysis of F_3 families allows the identification of individuals that are homozygous for each recombinant chromosome for analysis by PFGE. This is critical as it allows the distinction between heterozygotes and partial digests that may complicate analsyis of pulsed field gels (89).

Deletion mutants can also be very useful in locating the target gene on the physical map. This approach has proved powerful in the cloning of the gene for Duchenne muscular dystrophy (13). Physical perturbations of the map may provide much greater precision than genetic analysis, particularly if only a short sequence is deleted, because the standard errors associated with estimates of genetic distances are avoided.

6.3.4 Sources of error

There are several potential sources of error in estimating genetic and physical distances. For genetic distance it must be remembered that recombination is a biological process whose frequency will vary both with the genotype and with the environment. Within a single segregating population, the absolute size of the standard error of the genetic distance decreases with increasing linkage; however, the ratio of the standard error to the genetic distance increases because the number of informative recombinants becomes limiting at small genetic distances. The standard error can be minimized by selecting as many recombinants as possible from large populations.

There are several aspects of physical mapping with PFGE that can be sources of error. Many of the 'rare cutting' restriction enzymes are sensitive to methylation. Partial digestions are, therefore, possible due to differential methylation. To provide complete and accurate data, it is desirable to build overlapping maps based on a number of physically linked single copy markers. However, target regions may not be saturated with a density of markers sufficient to coalesce the physical maps. Alternatively, the analysis of F_3 families allows the detection of each of the homozygote classes; therefore heterozygous progenitor F_2 genotypes can be distinguished from partial digests. The electrophoresis conditions can also be a source of error. Optimal resolution for any one size range is accomplished with a specific combination of switch interval and voltage. Accurate sizing for fragments which span a range of sizes can only be accomplished using multiple gel runs with different pulse/voltage combinations. Non-linearity and resonance (migration in an order which does not reflect size) in the gels prevents extrapolation of sizes beyond the boundaries of the linear region. Confirmation that only a single size class exists in the region may be accomplished by hybridization of mapped yeast clones to the yeast chromosome size markers.

6.4 Chromosome walking
6.4.1 Introduction

Once a sequence which is physically close to the target locus has been identified, isolation of the genomic fragment containing the resistance gene can be attempted. Unless one is fortunate and the clone selected initially contains the gene of interest, cloning will probably involve multiple rounds of selection and characterization of overlapping clones (chromosome walking). From the precedence of human studies, several different types of libraries will have to be screened. The types of library used will depend on the genetic distance to be walked as determined by PFGE.

Unfortunately, there is currently no substitute for some 'brute-force' effort at this step. At present, it seems necessary to construct complete genomic libraries, then to select sequences from them. Techniques are not sufficiently refined to allow the isolation of specific large fragments of genomic DNA prior to cloning. For example, use of pulsed-field gels preparatively does not result in sufficient enrichment to warrant the extra effort involved. However, techniques are in the developmental stage (see, for example, references 68, 91) that in the future may allow significant enrichment for specific genomic fragments and the generation of minilibraries. These will greatly reduce the effort required to localize the targeted gene.

6.4.2 Genomic libraries

Several types of cloning vectors are available, the main ones being: lambda, cosmid, P1, jumping, and yeast artificial chromosome vectors. Each have their advantages and disadvantages. It is likely that no one library will suffice and several will have to be utilized.

Lambda and cosmid vectors have the advantage that they are the easiest to generate, screen, and manipulate. The technology is the most developed and reliable. However, they suffer the disadvantage that they carry relatively small genomic fragments. Therefore, such libraries may not be suitable for the initial stages of chromosome walking. The small insert size is an advantage in later stages of analysis as it makes identification of the target gene easier.

P1 libraries offer the potential for the rapid generation of genomic libraries with inserts up to 100 kb in *E. coli* (92, 93). Such libraries are intermediate between cosmid and YAC libraries in both transformation efficiency and insert size. Genomic DNA is partially digested, ligated to two vector arms, packaged using P1 head and tail proteins in a two-step process, and transfected into *E. coli*. In a bacterial strain that expresses the *cre* recombinase, the cloned DNA is circularized by recombination between the *lox* sites in each vector arm and maintained as a circular plasmid. However, procedures for making P1 libraries are still being refined. Informal communication with several groups who are experimenting with commercially available vectors suggests that the average insert size may be well below the 100 kb target, and in some cases smaller than in cosmid libraries. In the commercially available vectors, there is no selection for large insert sizes and frequently many clones may contain small or no inserts. Recently developed P1 cloning vectors have positive selection for the presence of inserts, and procedures have been devised to maximize insert size (93, 94). Therefore, generation of genomic libraries with 75–100 kb inserts in P1 vectors may become routine. It remains to be shown how representative and stable such libraries will be. The longer the genomic fragment sequence, the greater the chance of unclonable or unstable sequences.

Jumping libraries provide the possibility of progressing along the chromosome without having to isolate contiguous clones. Therefore, they may be very useful if the starting point of the walk is distant from the target gene. They may also be useful if the walk is blocked either by apparently unclonable DNA (contiguous sequences not represented in libraries) or by a stretch of repeated sequences in which no unique landmarks can be detected by hybridization or PCR. The strategy for jumping libraries has been described in detail by Collins (95). A jumping library contains short fragments that represent sequences separated in the genome by 100 kb or more. The strategy involves:

- generating large fragments of genomic DNA with rare-cutting endonucleases, either by partial or complete digestion
- DNA size-fractionation on pulsed-field gels
- circularization of the size-selected DNA under dilute conditions (0.2 μg/ml) to favour inclusion of a selectable marker (*E. coli* tRNA *supF*) and intramolecular ligation to bring together the two ends of individual large fragments
- a further digestion of these large circles with an enzyme that does not cut within the selectable marker and subsequent ligation into a lambda vector
- cloning of junction fragments is accomplished by infection of *supF⁻* strains using vectors with multiple amber mutations.

Jumping libraries have been used successfully in the cloning of the human cystic fibrosis gene (96) and to obtain markers linked to the gene for Huntington's disease (97). To our knowledge, they have not been used extensively in plants. One of the problems is the large amount of effort required for each step. Recently, a PCR-based jumping approach has been developed (98). This may increase the speed and application of chromosome jumping.

Yeast artificial chromosome (YAC; 99, 100) vectors currently offer the most reliable method of cloning large fragments of genomic DNA. Over the last four years, techniques for generating and manipulating YAC libraries have become increasingly efficient and the average insert sizes are increasing as the methods become refined (100–103). A library with an average insert size of at least 150 kb has been reported for *Arabidopsis* (104). Average insert sizes for some human libraries have reached 620 kb with maximum insert sizes of over 1600 kb (103). Yeast seems tolerant of repeated sequences although the long-term stability of YAC clones is still under investigation (102). There are several disadvantages of YAC libraries. YAC libraries are currently very time-consuming to generate and manipulate. One of the limiting steps is the low transformation efficiency of yeast. YAC libraries representing several genome equivalents cannot be generated in a matter of weeks, maintained as bulks, and screened. Most YAC libraries are generated over many months or years. The colonies grow slowly relative to bacteria, and are usually maintained as ordered arrays of colonies due to the very different growth rates of individual transformants. Also, the large clones are difficult to analyse.

The construction of YAC libraries now usually consists of several distinct steps:

- the isolation of high molecular weight genomic DNA
- partial digestion
- size fractionation to eliminate small fragments
- ligation to YAC vector arms
- a second size fractionation of recombinant molecules
- transformation of yeast
- selection, storage, and characterization of transformants

There are several variants of each of these steps and before embarking on making a YAC library, the latest literature (for example 100, 103, 106, 107, 123) and laboratories that are continuing to refine the technology should be consulted. YAC cloning is still in the developmental phase and further increases in efficiency and insert size are expected. We present the protocols that we have used to generate a partial YAC library of lettuce (*Protocols 5–7*) as a guide to what can be effective in plants and as a discussion of some of the problems that can be encountered. However, we stress the *caveat* that we are still modifying these protocols and that they will likely benefit from the incorporation of refinements recently developed with mammalian DNA as indicated.

The largest libraries with an average insert size of 620 and 700 kb have been obtained from human and mouse DNA, respectively, by the preparation of large molecular weight DNA embedded in agarose, combined with size selection both before and after ligation with the YAC vector arms (103). Larin *et al.* (103) also indicated that addition of polyamines at all stages when the DNA is melted in the agarose, protects the DNA from degradation. We are testing this modification with plant DNA. Our initial attempts to clone large plant DNA from agarose plugs did not seem to be as effective as for mammalian DNA. This was possibly due to the presence of the remains of the plant cell wall inhibiting DNA digestion, or to the degradation of the DNA during isolation from agarose. Therefore, we developed a method (*Protocol 5*) for the preparation of large plant DNA in solution. However, the sizes of our inserts were less than optimal (*circa* 100 kb) and, therefore, the two methods should be reassessed for plant DNA.

Protocol 5. Isolation of high molecular weight plant genomic DNA[a]

1. Prepare the following solutions:

   POM buffer:
   - 13% mannitol
   - 4 μM FeSO$_4$
   - 0.2 mM KH$_2$PO$_4$
   - 0.2 mM KNO$_3$
   - 0.2 mM CaCl$_2$
   - 0.2 mM MgSO$_4$
   - Adjust pH to 5.8 with 200 mM morpholino-ethane sulphonic acid (MOPS)

   Protoplast lysis solution:
   - 3% Na-lauroyl sarcosine
   - 0.5 M Tris–HCl, pH 9.0
   - 0.2 M EDTA

   Gradient buffer:
   - 1.6 M NaCl
   - 40 mM Tris–HCl, pH 8.0
   - 20 mM EDTA

2. Cut young leaves of lettuce plants (3 to 4-weeks-old) into thin (1 mm) strips and wash once with POM buffer.

3. Add cellulase (63 units/mg, Worthington) and macerase pectinase (3.5 units/mg, CalBiochem), at 700 mg/100 ml and 350 mg/100 ml, respectively, in POM buffer. Incubate the leaf strips at room temperature with very gentle agitation (60 r.p.m.) for 6–8 h.

4. Assess the extent of cell wall digestion and the yield of protoplasts by counting with a haemocytometer; 5–6 g fresh tissue yields at least 10^7 protoplasts.

5. Filter the suspension of protoplasts through 70 μm grid nylon mesh (Spectrum) to remove debris and spin at 800 g for 5 min.

6. Resuspend the pellet gently in 2 ml of the POM buffer and add 4 ml of the protoplast lysis solution. Lyse the protoplasts at 65°C for 15 min.

7. Decant the viscous liquid directly on top of a sucrose gradient consisting of 7 ml 50%, 4 ml 20%, and 4 ml 10% (w/v) sucrose in the gradient buffer.

8. Centrifuge the gradient at 11 000 r.p.m. for 18 h at 18°C in a SW28 rotor (Beckman).

Protocol 5. *Continued*

9. Recover the DNA as a loose pellet trapped in the 50% sucrose layer; remove the top layers of sucrose by aspiration.

10. Decant the viscous fraction into dialysis tubing and dialyse against 10 mM Tris–HCl, pH 8.0, and 1 mM EDTA for 6 h.

11. Store the DNA at 4°C in a small Petri dish.

12. Check the size of the recovered DNA by pulsed-field gel electrophoresis (PFGE). Large DNA prepared by this method migrates slower than the largest chromosome of *Saccharomyces pombe* (7 Mb) or does not enter the gel.

[a] Adapted from a protocol developed for yeast by Olson *et al.* (108).

Most of the early YAC libraries were constructed in pYAC4 or derivatives (99, 100). Our vector for YAC construction (*Protocol 6*) utilizes a minor modification of pYAC-RC (109) that includes T7 and T3 promoters either side of the *Eco*R1 cloning site (*Figure 6*). Recently, YAC vectors have been developed that allow a selective increase in copy number (112). The low efficiency of yeast transformation is a major limitation to the generation of YAC libraries. Even though reasonable frequencies can be obtained in small scale experiments, the transformation efficiency tends to drop as the experiment is scaled up. Spheroplast transformation, as described by Burgers and Percival (105) and Lunblatt (110), yields the highest transformation efficiencies and is widely used. With plasmid DNA, this protocol provides efficiencies of up to 10^6 transformants/µg DNA. With large DNA, however, the frequency drops considerably which selects for smaller inserts. Even in good experiments, the rate of transformation is approximately 10^2 transformants/µg DNA. Therefore, the generation of a YAC library of several genome equivalents requires large amounts of size-selected, large DNA.

A critical parameter is the extent of digestion of the yeast cell wall during generation of spheroplasts. The enzyme used considerably affects cell survival. We have routinely used Novozym 234 (CalBiochem); good results have also been reported with a Glusulase preparation (NEN, DuPont), although some batch variation has been observed by others. The concentration of enzyme relative to cells and extent of digestion is critical. The reaction should be monitored for cell wall loss by examining small aliquots added to an equal volume of 10% SDS under the microscope. The spheroplasting reaction should be stopped by centrifuging and washing the cells when the proportion of protoplasts approaches 90%. The reaction can also be monitored spectrophotometrically (105). The reaction is stopped when the OD_{800} has decreased to a few per cent of the original reading. Both methods work well. Over-digestion of the cell wall results in reduced cell viability and consequently low transformation rates.

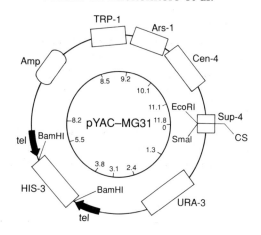

Figure 6. Yeast artificial cloning vector, pYAC-MG31. This was derived from pYAC-RC (109) by the addition of T3 and T7 promoters either side of the cloning site (M. Fortin, unpublished data). CS, cloning site; *HIS-3, URA/3, TRP-1*, yeast selectable markers, histidine, uracil, and tryptophan auxotrophy, respectively; *SUP-4*, suppressor of amber mutations; *Amp*, bacterial ampicillin resistance; *Ars-1*, yeast autonomously replicating sequence; *Cen-4*, centromeric sequence from *Saccharomyces cerevisiae*; tel, telomeric sequences from *Tetrahymena*. T3 and T7, prokaryotic promoters from bacteriophages T3 and T7.

Protocol 6. Cloning plant genomic DNA into YAC vectors

Solutions:

Recovery medium (per 100 ml):

- sorbitol   18.2 g
- yeast extract (Difco)   0.25 g
- peptone (Difco)   0.5 g
- 1 M CaCl$_2$ 0.65 ml

Minimal selective medium:

- sorbitol   182 g/l
- agarose (BRL)[a]   20 g/l
- ammonium sulphate   5 g/l
- dextrose   20 g/l
- yeast nitrogen base   1.8 g/l

275

Protocol 6. *Continued*

- amino acids, to give a final concentration (μg/l) of:

arg	20	asp	100	glu	100	his	20
leu	60	lys	30	met	20	phe	50
ser	375	thr	200	trp	40	tyr	30
val	150						

A. *Partial digestion*

1. Partially digest the high molecular weight DNA from lettuce with *Eco*R1 (1 unit) in a 100 μl reaction containing:
   - 40 mM Tris–HCl, pH 8.0
   - 8 mM MgCl$_2$
   - 40 mM NaCl$_2$

   at 37°C for approximately 15 min to give average fragment sizes greater than 200 kb.[b,c]

2. Inactivate the enzyme by increasing the temperature to 68°C for 15 min.

B. *Ligation with YAC arms*

1. Prepare vector arms of pYAC-MG31 or pYAC-4 by complete digestion with *Bam*H1 and *Eco*R1 followed by treatment with calf alkaline phosphatase.

2. Reduce the NaCl concentration in the solution of genomic DNA to 20 mM by dilution and add ATP to 1 mM.

3. Add the partially digested genomic DNA to the prepared YAC arms in a 1:100 molar ratio (in practice this represents an equal number of μg of each as genomic DNA is approximately 100-fold larger than the vector). Add 10 units of T4 DNA ligase and ligate at 15°C overnight.

4. Adjust the ligated solution to 1 M sorbitol.[d]

[a] To facilitate plating, low melting point agarose (BRL) is added instead of agar to the medium containing the yeast cells after transformation. The lower temperature also facilitates survival of the transformants.

[b] The optimal digestion times for partial digestion varies slightly between different DNA preparations; this is probably because it is difficult to quantify the amount of DNA present. Therefore, pilot digestions with varying amounts of enzyme and length of digestion should be carried out on each preparation, and their products monitored by PFGE.

[c] Construction of recent mammalian YAC libraries utilized a competition reaction between *Eco*R1 and *Eco*R1 methylase to give more controlled partial digestion (75, 103); slower reaction rates allowed better control over the extent of digestion.

[d] When we assayed the effect of size fractionation after ligation, we observed no dramatic increase in insert size but a severe reduction in the number of clones obtained. Therefore, we did not size fractionate after ligation. However, our observations probably resulted from the degradation of DNA during size fractionation that might be prevented by the addition of polyamines (103). Current experiments include size fractionation both before and after ligation with polyamines added to minimize DNA degradation.

Protocol 7 describes yeast transformation and storage of YAC transformants. The only selection used with the pYAC4 vector and derivatives is uracil prototrophy. Even though there is a *TRP1* gene for tryptophan prototrophy on the YAC vector, it is not fully functional and the trp1–1 mutation is a point mutation that reverts at too high frequency to be useful as a selectable marker (106). The addition of adenine sulphate is required to allow transformants to survive in genetic backgrounds such as AB1380. The *ADE* marker allows the selection of clones with inserts; those with inserts develop a red colour due to the interference of expression of the mutant *ADE* gene. New vectors and strains (106) are now available that should allow selection for both arms of the YAC using URA and TRP markers.

An important consideration when embarking on making a YAC library is not to underestimate the time required to manually pick the 10 000s of colonies required for a representative library and the considerable freezer space needed to provide duplicated storage for several genome equivalents. Replica plating using a 40 000 multipin transfer device that avoids the necessity of picking individual colonies, has recently been reported (103). Also, replicas of the YAC library can be stored on filters with considerable saving of space (111).

Protocol 7. Yeast transformation and storage of YAC transformants

A. *Yeast transformation*

1. Prepare mid-log phase yeast cells for spheroplast transformation (105, 110). The culture density is important as it influences the relative concentrations of spheroplasting enzyme and cells.

2. Add DNA to the prepared spheroplasts; we routinely used a ratio of approximately 10 μg of DNA per 100 μl of spheroplasts. It is important not to add a volume of DNA larger than 10 μl since it would then dilute significantly the sorbitol present in the regeneration medium. A larger volume of DNA may be added by first adjusting the ligation mixture to 1 M sorbitol to preserver the osmoticum. The DNA is then precipitated with the spheroplasts in presence of PEG and $CaCl_2$ (105, 110).

3. Wash the cells, resuspend them in recovery medium (sorbitol complete medium) for 1 h at 28°C.

4. Concentrate the cells by centrifuging at 800 *g*, and remove the excess medium.

5. Resuspend the cells in warm (45°C) minimal selective medium (see *Protocol 6*) containing low melting point agarose.

6. Plate the cells immediately as a top agar layered on to agar plates containing the same medium as the top agar.

Protocol 7. *Continued*

7. Incubate at 28°C. The first transformants can be seen at 3 days. However, colonies grow at very different rates and slower growing transformants may not be visible until after 6–7 days. Recombinant YAC clones have a red phenotype due to the interruption of *Sup-4*.

B. *Preservation of the YAC library*

YAC transformants grow embedded inside the regeneration media and, therefore, cannot be replica plated by overlaying with a membrane.

1. Pick up cells of individual recombinant YAC colonies with a toothpick, and place in a microtitre plate well containing 250 μl of minimal selective medium lacking sorbitol and agar but containing 25% glycerol.

2. Incubate the plates for several days until growth is visible in all wells.

3. Store the plates at −70°C.

6.4.3 Selection of overlapping clones

Chromosome walking involves the identification of overlapping genomic clones. The marker identified as being physically the closest to the target gene is used to select the first genomic clone. Terminal sequences are obtained from this clone and used to select contiguous genomic sequences from the library. The precise strategy depends on the type of library being used. As insert size increases, the number of clones required decreases, and the number of walking steps is minimized. The larger the insert the faster the progress, but the more difficult the clone is to analyse.

Screening YAC libraries remains a major task. If yeast transformation efficiencies could be increased sufficiently, it would become possible to generate and screen unordered libraries on a routine basis (103), rather than having to screen ordered arrays as is done at present. The latest generation of YAC vectors allow selective amplification of the YAC vector copy number for easier analysis of the insert DNA; this should facilitate selection of clones by hybridization (112). The identification of contiguous clones requires sequences from the end of the cloned insert. Methods have been developed for obtaining the terminal sequences from YAC inserts by IPCR (113) and by the vectorette method (114). Methods are still required for efficient subcloning from YAC clones (107).

For screens involving hybridization, YAC clones are replica plated on to multiple filters. If ordered colonies are used, colonies from many microtitre plates can be replica plated using a 96-pronged transfer device to a single filter (Sigma). Optimally, at least four genome equivalents should be screened to have nearly a 99% chance of having a clone containing the sequence. Assuming an average size of 250 kb and colonies from 24 plates per filter, this would involve approximately 14 filters for a plant with a genome size of 2×10^9 bp.

YAC libraries can also be screened using PCR-based markers. This allowed the efficient identification of YAC clones in the region containing the cystic fibrosis gene (115). Pooling of YAC clones into groups of nearly 2000 allows efficient screening of YAC libraries (116). If the YAC clones are pooled in several different combinations, the pattern of positive pools can indicate which clones contain the targeted sequence even when the primers detect sequences in multiple clones (128; P. de Jong, personal communication). Such approaches with PCR-based markers should allow chromosome walking through complex regions. This may be a prerequisite for map-based cloning of some resistance genes.

6.4.4 Monitoring progress

The progress and direction of the walk can be monitored both genetically and physically. Terminal sequences of each walking step should be placed on both the genetic and physical map. Genetic analysis involves screening the inheritance of polymorphisms detected by each terminal sequence. This will be most efficiently done by screening the genotypes of recombinants from several crosses. This will show when the walk has progressed over a point of recombination. The use of recombinants from several crosses will maximize the chances of detecting polymorphism. End clones should also be placed on the physical map (see Section 6.3). This may help coalesce previously non-overlapping maps. Analysis of clones for rare-cutting restriction sites will also help localize the cloned sequences on the map. Chromosome walking should proceed until sequences that absolutely genetically co-segregate with the target locus are identified. Preferably walking should continue until sequences have been identified either genetically or physically that are distal to the target. This would indicate that the target locus was definitely within the contiguous clones isolated.

6.4.5 Potential problems

If there are extensive regions of high copy DNA, the progress of chromosome walks based on hybridization to identify contiguous sequences will be blocked. However, PCR-based markers offer the opportunity to select clones even if most or all of the sequence is highly repeated. Providing the spacing of the oligonucleotide primer sequences is unique in the genome, such markers will detect unique genomic fragments. Also, if the walk is initiated from both sides of the target, the chances of being blocked are reduced. Jumping libraries also may be useful in traversing regions that contain high copy sequences.

Not all regions of the genome will be equally represented in a library. Some sequences may be recalcitrant to cloning in *E. coli* (96). It is still not known how generally representative YAC libraries will be (102). The use of multiple types of libraries should minimize this problem. Jumping libraries may be required if screening of multiple libraries fails to identify contiguous clones.

The cloned genomic DNA may be chimaeric due to ligation of more than one fragment into the vector. Apparently this can be a significant proportion in YAC and jumping libraries; one third of YAC clones studied in the vicinity of the cystic fibrosis gene were the result of multiple ligations (115). Chimeric clones should be reduced by ligating genomic and vector sequences under appropriately dilute conditions. The consequence of chimeric clones is that libraries have to be larger than the theoretical minimum for a three- to four-fold redundancy to compensate for the chimeric clones. The clones used in each walking step should be checked on the genetic and physical maps so that diversion of the walk into a different region will be prevented.

There may be insufficient polymorphism or recombination in the region to allow the direction and progress of the walk to be monitored genetically. Overcoming both problems may necessitate making several walking steps without being able to map the end clones used in each step. Continued localization on the physical map should prevent walking into unlinked regions. After multiple steps there should be sufficient recombination to determine the direction of the walk. If there is not, it may be very difficult to localize the target gene genetically and unless there are other indications of genomic position (for example small deletions), the walk will identify a genomic region too long to be tested for function by complementation.

7. Gene identification

7.1 Introduction

Whatever the approach used to isolate a resistance gene, the ultimate confirmation is the introduction of the putative sequences into a susceptible cultivar to test for function by complementation of the susceptible genotype to resistance. The precise transformation strategy to confirm which sequences encode the resistance gene will depend on the species under study and the strategy used to generate the candidate clones. Dicotyledonous species that are amenable to transformation by *A. tumefaciens* will allow more extensive complementation efforts than those species such as many monocots that are difficult to transform. Some strategies, such as transposon tagging, may provide precise information on gene position; others, such as chromosome walking, may only localize the resistance gene to several hundred kilobases.

While shotgun transformation of all potential sequences could be attempted, there are several approaches that have been useful for cloning human genes that may help reduce the number of sequences that require introduction. None of these approaches will likely, unambiguously, identify the location of the resistance gene, and some may be totally uninformative in some systems. However, most are relatively quick and straightforward to perform and could considerably reduce the time and effort spent on time-consuming transformation experiments.

7.2 Pre-selection of sequences for transformation

Mutants of the resistance gene may be highly informative. Mutants should be screened for changes in genomic sequences in the targeted region and for changes in patterns of gene expression relative to their wild-type progenitor. Mutants which have been generated with mutagens that tend to produce deletion mutants should be favoured for this analysis. However, Southern analysis will be obstructed if there are highly repeated sequences in the region. Also, Northern analysis may be uninformative due to the low expression of resistance genes. Clones detecting either deletions in genomic DNA or inactivation of gene expression in a mutant would be given priority for testing by complementation.

It would be useful to focus on those clones containing transcribed sequences. Subclones could be digested, separated by gel electrophoresis and probed with total genomic DNA. Those clones containing only high-copy DNA would be given low priority in transformation experiments. Low-copy fragments could be used as probes to Northerns of mutants and near-isogenic lines to seek for correlations with resistance gene expression. Analysis of candidate regions may also provide evidence for transcribed sequences. In human studies, hypomethylation of CpG rich regions has been indicative of the 5' region of some genes (96). Hypomethylation in plants is less well documented, although several lines of evidence suggest that hypomethylation may also be prevalent in plants (117). Hypomethylation could be detected by differential sensitivity to such isoschizomers as *Hpa*II and *Msp*I.

Conservation of sequences between resistance genes in different species may also provide indications to the physical locations of resistance genes. If some domains of resistance genes are conserved, then hybridizations to genomic DNA of related species may indicate those sequences that contain resistance genes. This approach has proved useful in the identification of genes for human inherited diseases (96, 118). In tomato, cDNA sequences tend to be conserved across related species, while random genomic clones do not cross hybridize within the same species (119). As with most of the approaches at this stage, these experiments are speculative but readily performed and potentially highly informative.

7.3 Complementation

Overlapping fragments that potentially contain the resistance gene, should be tested for function by introduction into a susceptible line. It may be possible to test bulks of candidate clones. However, the amount of effort and time required to regenerate and test whole plants is such that we favour testing candidate sequences individually to ensure each sequence has been subjected to a valid test.

There are several potential problems. The size of the introduced DNA should be as large as possible to maximize the chances that a complete gene will be introduced and to minimize the number of clones requiring introduction. Many plant transformation systems are mediated by *Agrobacterium tumefaciens*. Most chimeric T-DNAs have been approximately 25 kb or less, although over 50 kb have been successfully transferred (120). The sizes of resistance genes are unknown. Most plant genes cloned to date have been relatively small. However, this may, in part, be a function of the types of genes cloned. Some mammalian recognition genes span several 100 kb of genomic DNA. If this is the case with plant resistance genes, then complementation experiments with genomic sequences will not be successful. If there are indications that the resistance gene is larger than the insert sizes available, then a cDNA approach will have to be followed but this could be very complex if there are several related sequences in the genome. Alternatively, methods will have to be developed for the transformation of plants with large pieces of DNA. Recently, mammalian cell lines have been transformed with YAC clones and the introduced DNA was stably integrated in the chromosome (121).

If the transformation is mediated by *A. tumefaciens*, a problem with introducing large inserts in Ti vectors will be that not all the inserted DNA may be transferred. Premature termination due to recognition of false left border sequences could result in partial transfer of T-DNA inserts (122). This will be aggravated in large T-DNA inserts. Multiple transformants should be generated for each candidate clone and checked to ensure they are a valid test of the cloned sequences. Similar problems may be encountered with direct DNA transformation procedures (for example electroporation or microprojectile bombardment) except that the problem may be more severe as the linearity of the DNA may not be retained. The generation of multiple transformants will also minimize the chances of position effects obscuring the expression of the resistance gene.

Once a clone that complements a susceptible genotype to resistance has been identified, standard techniques can be used to identify the precise location of the coding and regulatory sequences. Further confirmation that the correct sequence has been identified should be made by performing the complementation experiments with mutated sequences; such sequences should not confer resistance. Again, analysis of radiation- and chemically-induced mutants will help localize the gene.

8. Postscript

8.1 Multifaceted approaches

It is impossible a priori to predict which strategy will ultimately prove successful for isolating disease resistance genes from plants. Therefore, several different approaches should be pursued when possible, although due to the amount of effort required by each approach, not necessarily by the same research group. Different strategies are likely to generate complementary information. For example, chromosome walking may localize the resistance gene to a large fragment; transposon mutagenesis may generate several putative insertional inactivations. If the sequences flanking the transposon hybridize to the large fragment identified by chromosome walking, one would have high confidence of the position of the targeted gene. Conversely, if it does not cross-hybridize, one would carefully check the clones generated by each approach before investing in extensive further efforts.

8.2 Cloning additional resistance genes

Different plant–pathogen interactions present different challenges and opportunities. Each will have to be judged on a case-by-case basis and the optimal strategy chosen accordingly. It is unlikely that all genes identified phenotypically by their ability to prevent disease, will have related mechanisms of action and sequence similarity. Therefore, it will be necessary to clone resistance genes from several different species to several different types of pathogens. In many cases this will involve utilizing at least one of the time-consuming approaches discussed above.

A major question will be: is there sequence similarity between some resistance genes as the genetic data suggests? Sequence similarity could cause problems during the cloning of the initial resistance genes as it may be difficult to distinguish among related genes, and, therefore, to identify which sequence encodes the targeted specificity. However, sequence similarity between resistance genes should allow the rapid isolation and testing of related sequences. Therefore, once one resistance gene has been cloned, many of the preceding, time-consuming strategies may not be needed to isolate related genes.

Even if there is not sufficient sequence conservation between resistance genes for cloning by cross-hybridization, isolation of a few such genes should provide new options. For example, characterization of the gene products and patterns of expression of several resistance genes should allow cloning strategies for other resistance genes based on this information.

Acknowledgements

We thank Steve Briggs, Ken Feldman, Howard Judelson, Rod Wing, and John Yoder for their helpful comments on various sections of the chapter. Our studies have been supported in part by grants from the DOE Energy BioSciences Program, DE-FG03-88ER13904, and the UDSA CRGO Plant Genetic Mechanisms and Molecular Biology Program, USDA-88-CRCR-37262-3522.

References

1. Lamb, C. J., Lawton, M. A., Dron, M., and Dixon, R. A. (1989). *Cell*, **56**, 215–24.

2. Dixon, R. A. and Lamb, C. J. (1990). *Annu. Rev. Plant Physiol. Plant Mol. Biol.*, **41,** 339–67.
3. Crute, I. R. (1986). In *Mechanisms of plant diseases*, (ed. R. S. S. Frazer), pp. 80–142. Martinus Nijhoft/Dr W. Junk.
4. Shepherd, K. W. and Mayo, G. M. E. (1972). *Science, 175,* 375–80.
5. Michelmore, R. W., Hulbert, S. H., Landry, B. S., and Leung, H. (1987). In *Genetics and plant pathogenesis* (ed. P. R. Day and G. J. Ellis), pp. 221–31. Blackwell Scientific.
6. Pryor, T. (1987). *Trend Genet., 3,* 157–61.
7. Paran, I., Kesseli, R. V., and Michelmore, R. W. (1992). *Genome.* (In press).
8. Hulbert, S. H. and Bennetzen, J. L. (1991). *Mol. Gen. Genet.,* **226,** 377–82.
9. Geliebter, J. and Nathenson, S. G. (1987). *Trends Genet., 3,* 107–12.
10. Nasrallah, J. B., Nishio, T., and Nasrallah, M. E. (1991). *Annu. Rev. Plant Physiol. Plant Mol. Biol.,* **42,** 393–422.
11. Ebert, P. R., Anderson, M. A., Bernatzky, R., Altschuler, M., and Clarke, A. E. (1989). *Cell, 56,* 255–62.
12. Farrara, B., Ilott, T. W., and Michelmore, R. W. (1987). *Plant Pathol.,* **36,** 499–514.
13. Cremers, F. P. M., van der Pol, D. J. R., Wieringa, B., Collins, F. S., Sankila, E.-M., Siu, V. M., Flintoff, W. F., Brunsmann, F., Blonden, L. A. J., and Ropers, H.-H. (1989). *Proc. Natl. Acad. Sci. (USA),* **86,** 7510–14.
14. Kerem, B.-S., Rommens, J. M., Buchanan, J. A., Markiewicz, D., Cox, T. K., Chakravaki, A., Buchwald, M., and Tsui, L.-C. (1989). *Science,* **245,** 1073–80.
15. Keen, N. T., Tamaki, S., Kobayashi, D., Gerhold, D., Stayton, M., Shen, H., Gold, S., Lorang, J., Thordal-Christensen, H., Dahlbeck, D., and Staskawicz, B. (1990). *Mol. Pl.–Microbe Interact, 3,* 112–21.
16. Welsh, J. and McClelland, M. (1990). *Nucl. Acids Res.,* **18,** 7213–18.
17. Culver, J. N. and Dawson, W. O. (1989). *Mol. Pl.–Microbe Interact., 2,* 209–13.
18. Yoder, O. C. (1980). *Annu. Rev. Phytopathol.,* **18,** 103–29.
19. Mayama, S., Tani, T., Ueno, T., Midland, S. L., Sims, J. J., and Keen, N. T. (1986). *Physiol. Mol. Pl. Pathol.,* **29,** 1–18.
20. Wolpert, T. J., Macko, V., Acklin, W., and Arigoni, D. (1988). *Plant Physiol.,* **88,** 37–41.
21. Wolpert, T. J. and Macko, V. (1989). *Proc. Natl. Acad. Sci. (USA),* **86,** 4092–6.
22. Wolpert, T. J. and Macko, V. (1991). *Plant Physiol.,* **95,** 917–20.
23. Dewey, R. E., Siedow, J. N., Timothy, D. H., and Levings III, C. S. (1988). *Science,* **239,** 293–5.
24. Huang, J., Lee, S.-H., Lin, C., Medici, R., Hack, E., and Myers, A. M. (1990). *EMBO J., 9,* 339–47.
25. Lawton, M. A., Yamamoto, R. T., Hanks, S. K., and Lamb, C. J. (1989). *Proc. Natl. Acad. Sci. (USA),* **86,** 3140–4.
26. Walker, J. C. and Zhang, R. (1990). *Nature,* **345,** 743–6.
27. Straus, D. and Ausubel, F. M. (1990). *Proc. Natl. Acad. Sci. (USA),* **87,** 1889–93.
28. Gurr, S. J., McPherson, M. J., Scollan, C., Atkinson, H. J., and Bowles, D. J. (1991). *Mol. Gen. Genet.,* **226,** 361–6.
29. Kohne, D. E., Levison, S. A., and Byers, M. J. (1977). *Biochemistry,* **16,** 5329–41.

30. Kush, G. S. and Rick, C. M. (1968). *Chromosoma*, **23**, 452–84.
31. Phaler, P. L. (1967). *Gentics*, **57**, 523–30.
32. Reardon, J. T., Liljestrand-Golden, C. A., Dusenbury, R. L., and Smith, P. D. (1987). *Genetics*, **115**, 323–31.
33. Staskawicz, B. J., Dahlbeck, D., and Keen, N. T. (1984). *Proc. Natl. Acad. Sci. (USA)*, **81**, 6024–8.
34. Swanson, J., Kearney, B., Dahlbeck, D., and Staskawicz, B. (1988). *Mol. Plant Microbe Interact.*, **1**, 5–9.
35. de Bruijn, F. J. and Lupski, J. R. (1984). *Gene*, **27**, 131–49.
36. Nevers, P., Shepherd, N. S., and Seadler, H. (1986). *Adv. Bot. Res.*, **12**, 103–203.
37. Peterson, P. A. (1987). *CRC Critical Rev. Pl. Sci.*, **6**, 104–208.
38. Shepherd, N. S. (1988). In *Plant molecular biology: a practical approach* (ed. C. Shaw), pp. 187–220. IRL Press, Oxford.
39. Chandlee, J. M. (1990). *Physiologia Plantarum*, **79**, 105–15.
40. Haring, M. A., Rommens, C. M. T., Nijkamp, H. J. J., and Hille, J. (1991). *Plant Mol. Biol.*, **16**, 449–461.
41. Ellis, J. G., Lawrence, G. J., Peacock, W. J., and Pryor, A. J. (1988). *Annu. Rev. Phytopathol.*, **26**, 245–63.
42. Feldmann, K. A., Marks, M. D., Christianson, M. L., and Quatrano, R. S. (1989). *Science*, **243**, 1351.
43. Feldmann, K. A. (1991). *Plant Journal*, **1**, 71–82.
44. Marks, M. D. and Feldmann, K. A. (1989). *Plant Cell*, **1**, 1043–50.
45. Yanofsky, M. F., Ma, H., Bowman, J. L., Drews, G. N., Feldmann, G. N., and Meyerowitz, E. M. (1990). *Nature*, **346**, 35–9.
46. Bingham, P. M., Leavis, R., and Rubin, G. (1981). *Cell*, **25**, 693–704.
47. Cooley, L., Kelley, R., and Spradling, A. (1988). *Science*, **239**, 1121–8.
48. Doring, H.-P. (1989). *Maydica*, **34**, 73–88.
49. Pryor, T. (1987). *Maize Newsletter*, **61**, 37–8.
50. Bennetzen, J. L., Qin, M. M., Ingels, S., and Ellingboe, A. H. (1988). *Nature*, **322**, 369–70.
51. Baker, B., Schell, J., Lorz, H., and Federoff, N. (1986). *Proc. Natl. Acad. Sci. (USA)*, **83**, 4844–8.
52. Pereira, A. and Sadler, H. (1989). *EMBO J.*, **8**, 1315–21.
53. Martin, C., Prescott, A., Lister, C., and Mackay, S. (1989). *EMBO J.*, **8**, 997–1004.
54. Martin, C. and Lister, C. (1989). *Devel. Genet.*, **10**, 438–51.
55. Greenblat, I. M. (1984). *Genetics*, **108**, 471–85.
56. Dooner, H. K. and Belachew, A. (1989). *Genetics*, **122**, 447–57.
57. Dooner, H. K., Keller, J., Harper, E., and Ralston, E. (1991). *Plant Cell*, **3**, 473–82.
58. Masterson, R. V., Furtek, D. D., Grevelding, C., and Schell, J. (1989). *Mol. Gen. Genet.*, **219**, 461–6.
59. Jones, J., Carland, F., Maliga, P., and Dooner, H. K. (1989). *Science*, **244**, 204–7.
60. Jorgensen, J. H. (1977). *Euphytica*, **26**, 55–62.
61. Weil, C. F. and Wessler, S. R. (1990). *Annu. Rev. Physiol. Plant Mol. Biol.*, **41**, 527–52.

62. Coen, E. S. and Carpenter, R. (1986). *Trends Genet.,* **2**, 292–6.
63. Belzile, F., Lassner, M. W., Tong, Y., Khush, R., and Yoder, J. I. (1989). *Genetics,* **123**, 181–9.
64. Nelsen-Salz, B. and Doring, H.-P. (1990). *Mol. Gen. Genet.,* **223**, 87–96.
65. Robertson, D. S. (1985). *Mol. Gen. Genet.,* **200**, 9–13.
66. Ludecke, H. J., Seuger, G., Claussen, U., and Horsthemke, B. (1989). *Nature,* **338**, 348–50.
67. Van Dilla, M. A., Deaven, L. L., Albright, K. L., Allen, N. A., Aubuchon, M. R., Barthold, M. F., *et al.* (1986). *Biotechnology,* **4**, 537–52.
68. Strobel, S. A., Doncette-Stamm, L. A., Riba, L., Houseman, D. E., and Dervan, P. B. (1991). *Science,* **254**, 1639–42.
69. Williams, J. G. K., Kubelik, A. R., Livak, K. J., and Rafalski, J. A. (1990). *Nucl. Acids Res.,* **18**, 6531–5.
70. Bernatzky, R. and Tanksley, S. D. (1986). *Theor. Appl. Genet.,* **72**, 314–21.
71. Olsen, M., Hood, L., Cantor, C., and Botstein, D. (1989). *Science,* **248**, 1434–5.
72. Young, N. D., Zamir, D., Ganal, M., and Tanksley, S. D. (1988). *Genetics,* **120**, 579–85.
73. Martin, G. B., Williams, J. G. K., and Tanksley, S. D. (1991). *Proc. Natl. Acad. Sci. (USA),* **88**, 2336–40.
74. Hinze, K., Thompson, R. D., Ritter, E., Salamini, F., and Schulze-Lefert, P. (1991). *Proc. Natl. Acad. Sci. (USA),* **88**, 3691–5.
75. Stam, P. and Zeven, C. (1981). *Euphytica,* **30**, 227–38.
76. Muehlbauer, G. J., Sprecht, J. E., Thomas-Compton, M. A., Staswick, P. E., and Bernard, R. L. (1988). *Crop Sci.,* **28**, 729–35.
77. Michelmore, R. W., Paran, I., and Kesseli, R. V. (1991). *Proc. Natl. Acad. Sci. (USA),* **88**, 9828–32.
78. Dooner, H. K. (1986). *Genetics,* **113**, 1021–36.
79. Ganal, M. W., Young, N. D., and Tanksley, S. D. (1989). *Mol. Gen. Genet.,* **215**, 395–400.
80. Steinmetz, M., Stephen, D., and Lindahl, K. F. (1986). *Cell,* **44**, 895–904.
81. Clarke, L. and Carbon, J. (1980). *Nature,* **287**, 504–9.
82. Bender, W., Sperier, P., and Hogness, D. S. (1983). *J. Mol. Biol.,* **168**, 17–33.
83. Gemmill, R. M., Coyle-Morris, J. F., McPeek, F. D., Ware-Uribe, L. F., and Hetch, F. (1987). *Gene Anal. Tech.,* **4**, 119–31.
84. Hermann, B. C., Barlow, D. P., and Lehrach, H. (1987). *Cell,* **48**, 813–25.
85. Lawrance, S. K., Smith, C. L., Sirvastava, R., Cantor, C. R., and Weissman, S. M. (1987). *Science,* **235**, 1387–90.
86. van Ommen, G. J. B., Verkerk, J. M. H., Hofker, M. H., Monaco, A. P., Kunkel, L. M., Ray, P., Worton, R., Wieringa, B., Bakker, E., and Pearson, P. L. (1986). *Cell,* **47**, 499–504.
87. Meagher, R. B., McLean, M. D., and Arnold, J. (1988). *Genetics,* **120**, 809–18.
88. Guzman, P. and Ekker, J. R. (1988). *Nucl. Acids Res.,* **16**, 11091–105.
89. Francis, D. M. (1991). PhD thesis. University of California, Davis.
90. Crucefix, D. N., Rowell, P. M., Street, P. F. S., and Mansfield, J. W. (1987). *Physiol. Molec. Plant Pathol.,* **30**, 39–54.
91. Koob, M. and Szybalski, W. (1990). *Science,* **250**, 271–3.
92. Sternberg, N. (1990). *Proc. Natl. Acad. Sci. (USA),* **87**, 103–7.
93. Sternberg, N. (1990). *Gene Anal. Tech.,* **7**, 126–32.

Richard W. Michelmore et al.

94. Pierce, J. C. and Sternberg, N. (1991). *Abstr. Genome Mapping and Sequencing*, Cold Spring Harbour Meeting, p. 13.
95. Collins, S. F. (1988). In *Genome analysis: a practical approach* (ed. E. K. Davies), pp. 73–94. IRL Press, Oxford.
96. Rommens, J. M., Iannuzzi, M. C., Kerem, B. S., Drumm, M. L., Melmer, G., Dean, M., Rozmahel, R., Cole, J. I., Kennedy, D., Hidaka, N., Zsiga, M., Buchwald, M., Riordan, J. R., Tsui, L.-C., and Collins, F. S. (1989). *Science*, **245**, 1059–65.
97. Richards, J. E., Gilliam, T. C., Cole, J. L., Drumm, M. L., Wasmuth, J. J., Gusella, J. F., and Collins, F. C. (1988). *Proc. Natl. Acad. Sci. (USA)*, **85**, 6437–41.
98. Kandpal, R. P., Shukla, H., Ward, D. C., and Weissman, S. M. (1990). *Nucl. Acids Res.*, **18**, 3081.
99. Burke, D. T., Carle, G. F., and Olson, M. V. (1987). *Science*, **236**, 806–12.
100. Burke, D. T. and Olson, M. V. (1991). *Methods in Enzymology*, **194**, 251–70.
101. Imai, T. and Olson, M. V. (1990). *Genomics*, **8**, 297–303.
102. Albertsen, H. M., Abderrahim, H., Cann, H. M., Dausset, J., Le Paslier, D., and Cohen, D. (1990). *Proc. Natl. Acad. Sci. (USA)*, **87**, 4256–60.
103. Larin, Z., Monaco, A. P., and Lehrach, H. (1991). *Proc. Natl. Acad. Sci. (USA)*, **88**, 4123–7.
104. Ward, E. R. and Jen, G. C. (1990). *Pl. Mol. Biol.*, **14**, 561–658.
105. Burgers, P. M. J. and Percival, K. J. (1987). *Anal. Biochem.*, **163**, 391–7.
106. McCormick, M. K., Antonarakis, S. F., and Hieter, P. (1990). *Gene Analysis Techniques*, **7**, 114–18.
107. Nelson, D. L. (1990). *Gene Anal. Tech.*, **7**, 100–6.
108. Olson, M. V., Loughney, K., and Hall, B. D. (1979). *J. Mol. Biol.*, **132**, 387–410.
109. Marchuk, D. and Collins, F. S. (1988). *Nucl. Acids Res.*, **16**, 7743.
110. Lundblatt, S., Ausubel, F., Brent, R., Kingston, R. E., Moore, D. D., Seidman, J. G., Smith, J. A., and Struhl, K. (1991). In *Current protocols in molecular biology*, Section 13. Wiley, London.
111. Arnand, R., Riley, J. H., Butler, R., Smith, J. C., and Markham, A. F. (1990). *Nucl. Acids Res.*, **18**, 1951–6.
112. Smith, D. R., Smyth, A. P., and Moir, D. T. (1990). *Proc. Natl. Acad. Sci. (USA)*, **87**, 8242–6.
113. Silerman, G. A., Ye, R. D., Pollock, K. M., Sadler, J. E., and Korsmeyer, S. J. (1989). *Proc. Natl. Acad. Sci. (USA)*, **86**, 7485–9.
114. Riley, J., Butler, R., Ogilvie, D., Finniear, R., Jenner, D., Powell, S., Anand, R., Smith, J. C., and Markham, A. F. (1990). *Nucl. Acids Res.*, **18**, 2887–90.
115. Green, E. D. and Olson, M. V. (1990). *Science*, **250**, 94–8.
116. Green, E. D. and Olson, M. V. (1990). *Proc. Natl. Acad. Sci. (USA)*, **87**, 1213–17.
117. Antequera, F. and Bird, A. P. (1988). *EMBO J.*, **7**, 2295–9.
118. Monaco, A. P., Neve, R. L., Colletti, C., Bertleson, C. J., Kurnit, D. M., and Kunkel, L. M. (1986). *Nature*, **323**, 646–50.
119. Zamir, D. and Tanksley, S. D. (1988). *Mol. Gen. Genet.*, **213**, 254–61.
120. David, C., Petit, A., and Tempe, J. (1988). *Plant Cell Rep.*, **7**, 92–5.

121. Eliceiri, B., Labella, T., Hagino, Y., Srivastava, A., Schlessinger, D., Pilia, G., Palmieri, G., and D'Urso, M. (1991). *Proc. Natl. Acad. Sci. (USA)*, **88**, 2179–83.
122. Zambryski, P. (1988). *Annu. Rev. Genet.*, **22**, 1–30.
123. Burke, D. T. (1990). *Gene Analysis Techniques*, **7**, 94–9.
124. Smith, C. L., Klco, S. R., and Cantor, C. R. (1988). In *Genome analysis: a practical approach* (ed. K. E. Davies), pp. 41–72. IRL Press, Oxford.
125. van Kan, J. A. L., van den Ackerveken, G. F. J. M., and de Wit, P. J. G. M. (1991). *Mol. Pl. Microbe Interact.*, **4**, 52–9.
126. Sun, T., Goodman, H. M., and Ausubel, F. M. (1992). *Plant Cell*, **4**, 119–28.
127. Paran, I., Kesseli, R. V., and Michelmore, R. W. (1991). *Genome*, **34**, 1021–7.
128. Amemiga, C. T., Alegria-Hartmann, M. J., Aslanidis, C., Chen, C., Nikohic, J., Gingrich, J. C., and de Jong, P. I. (1992). *Nucl. Acids. Res.* (In press).
129. Meeley, R. B., Johal, G. S., Briggs, S. P., and Walton, J. D. (1992). *Plant Cell*, **4**, 71–7.

A1

List of suppliers

Apin Chemicals Ltd, Unit 29D, Milton Park, Abingdon, Oxford, UK.

Agar Aids Ltd, PO Box 101, Hemel Hempstead, Herts, UK.

Aldrich Chemical Co., 940 West Saint Paul Avenue, Milwaukee, WI 53223, USA; The Old Brickyard, New Road, Gillingham, Dorset, SP8 4JL, UK.

Alltech, Deerfield, ILL, USA.

Amersham International, Amersham Place, Little Chalfont, Buckinghamshire HP7 9NA, UK; 2636 S. Clearbrook Drive, Arlington Heights, IL 60005, USA.

Amicon Corporation, 17 Cherry Hill Drive, Danvers, MA 01923, USA; Upper Mill, Stonehouse, Gloucester GL10 2J, UK.

Anachem, Anachem House, 20 Charles Street, Luton, Bedfordshire LU2 OE6, UK.

Applied Biosystems Inc., 850 Lincoln Center Drive, Foster City, CA 94404, USA; Kelvin Close, Birchwood Science Park North, Warrington, Cheshire WA3 7PB, UK.

Bayer, UK Supplier, Dalgetty Agriculture Ltd, Green Lane West, Rackheath, Norwich, NR13 6N4, UK.

BDH, Broom Road, Poole, Dorset, BDH12 4NN, UK.

Beckman Instruments Ltd, Progress Road, Sands Industrial Estate, High Wycombe, Buckinghamshire, HP12 4YL, UK.

Bio 101 Inc., Box 2284, La Jolla, CA 92038-2284, USA; Stratech Scientific Ltd., 61/63 Dudley Street, Luton, Bedfordshire, LU2 0NP, UK.

Bio-Rad, 1414 Harbour Way South, Richmond, CA 94804, USA; Claxton Way, Watford Business Park, Watford, Hertfordshire, WD1 8RP, UK.

Biosupplies, PO Box 835, Parkville, Australia.

Boëhringer-Mannheim GmbH, Postfach 310120, D-6800 Mannheim 31, Germany; PO Box 50816, Indianapolis, IN 46250, USA; Bell Lane, Lewes, Sussex, BN7 1LG, UK.

Branson Ultrasonic Corporation, Danbury CT, USA.

BRL. See **Gibco-BRL.**

British Biotechnology Ltd, 4–10 The Quadrant, Barton Lane, Abingdon, Oxon, OX14 3YS, UK.

Brownlee Laboratories, Santa Clara, CA 95050, USA.

BTX, 3742 Jewell Street, San Diego, CA 92109, USA.

Calbiochem, 10933 North Torrey Pines Road, La Jolla, CA 92037, USA; Novobiochem (UK) Ltd, Freepost, Nottingham NG7 1BR, UK.

Cambio, 34 Millington Road, Newnham, Cambridge, CB3 9HP, UK.

Cambridge Bioscience Ltd, 25 Signet Court, Stourbridge Business Centre, Cambridge, CB5 8LA, UK.

Canberra Packard Ltd, Brook House, 14 Station Road, Pangbourne, Berkshire, RG8 7DT, UK.

Cen Saclay, Molecules Marquees, Gif-Sur-Yvette, France.

Cetus. See **Perkin Elmer Cetus.**

Chempack Products Ltd, Geddings Road, Hoddesdon, Hertfordshire, UK.

Ciba Corning Diagnostic Corporation, Oberlin, Ohio, USA.

Costar. See **Northumbrian Biologicals Ltd.**

Difco Laboratories Ltd, PO Box 14B, Central Avenue, East Moseley, Surrey, England.

Dionex Ltd, Camberley, Surrey, GU15 2PL, UK.

DuPont Company, Biotechnology Systems Division, BRML, G-50986, Wilmington, DE 19898, USA; Wedgewood Way, Stevenage, Hertfordshire, SG1 4QN, UK.

Fisher Scientific, Pittsburgh, PA, USA; Zurich, Switzerland.

FMC Bioproducts, 5 Maple Street, Rockland, ME 04841-2994, USA; Flowgen Instruments Ltd, Broad Oak Enterprise Village, Sittingbourne, Kent, ME9 8AQ, UK.

IBF Biotechnics, Colombia, MD, USA.

ICN/FLOW, High Wycombe, Bucks, HR13 7DL, UK.

Genetic Research Instruments Ltd, Gene House, Dunmow Road, Felsted, Dunmow, CM6 3LD, UK.

Gibco-BRL, Grand Island, NY, USA; PO Box 35, Trident House, Renfrew Road, Paisley, PA3 4EF, UK.

Invitrogen, San Diego, CA 92121, USA (see also British Biotechnology Ltd).

Janssen Life Sciences and J. W. Scientific Products Merck Darnstadt, Germany.

J. and W. Scientific. See **Sera Lab UK Ltd.**

Karlan Chemical Corporation, 23875 Madison Street, Torrance, CA 90505, USA.

Kinematica UK, Philip Harris Scientific, 618 Western Avenue, Park Royal, London W3 0TE, UK.

Kodak (Eastman Kodak), Acorn Field Road, Knowsley Industrial Park North, Liverpool, LS3 72X, UK; Rochester, New York 14650, USA.

LEP Scientific, Sunrise Parkway, Linford Wood East, Milton Keynes, Bucks, MK14 6QF, UK.

Marine Colloids, FMC Corporation, Bioproducts Division, 5 Maple Street, Rockland, ME 04841, USA; 1 Risingevej, DK-2665 Vallensbaek Strand, Denmark.

Merck, Shaw Road, Speke, Liverpool, L24 9LA, UK.

Millipore Corp., 80 Ashby Road, Beford, MA 01730, USA; The Boulevard, Blackmoor Lane, Watford, Hertfordshire WD1 8YW, UK.

MJ Research, Kendall Square, Box 363, Cambridge, MA 02142, USA.

NBL. See **Northumbrian Biologicals Ltd.**

New England Biolabs, 32 Tozer Road, Beverley, MA 01915, USA; Postfach 2750, 6231 Schwalbach/Tanus, FRG; CP Laboratories, PO Box 22, Bishop's Stortford, Herts, CM23 3DM, UK.

New England Nuclear. See **DuPont.**

Northrop-King Co, PO Box 959, MIN, USA.

Northumbrian Biologicals Ltd, (NBL), South Nelson Industrial Estate, Cramlington, Northumberland, NE23 9HL, UK.

Novo Biolabs, Novo Industri A/S, Novo Allé, Ak-2880 Bagsvaerd, Denmark.

Onozuta, Karlan, Torrance, CA, USA (see also Yakult Honsha Ltd).

Operon Technologies, Almeda, CA, USA.

Organomotion Associates Inc., Berlin, MA, USA.

Oxoid Ltd, Wade Road, Basingstoke, RG24 0PW, Hants, England.

Canberra Packard Ltd, Brook House, 14 Station Road, Pangbourne, Berkshire, RG8 7DT, UK.

Parr Instrument Co., Moline, Illinois, USA.

Perkin Elmer Cetus, Main Avenue, Norwalk, CT 06859-0012, USA; Postfach 101164, 7770 Ueberlingen, FRG; Maxwell Road, Beaconsfield, Buckinghamshire HP9 1QA, UK.

Polysciences, Paul Valley Industrial Park, Warrington, PA 18976, USA.

Pharmacia LKB Biotechnology, AB, S-75182 Uppsala, Sweden; 800 Centennial Avenue, Piscataway, NJ 08854, USA; Southampton SO1 7NS, UK.

Pierce, Rockford, ILL, USA.

Pierce and Warriner (UK) Ltd, 44 Upper Workgate Street, Chester, CH1 4EF.

Promega Corporation, 2800 South Fish Hatchery Road, Madison, WI 537-5305, USA; Episcon House, Enterprise Road, Chilworth Research Centre, Southampton, SO1 7NS, UK.

Rainin Instrument Company, Woburn, MA, USA.

Raymond A. Lamb, 6 Sunbeam Road, London, NW10 6YL, UK.

Sartorius Ltd, Belmont, Surrey, SM2 6JD, UK.

Seishin Pharmaceuticals Co. Ltd, 4.13, Koamicho, Nihoubashi, Tokyo, Japan.

Sera Lab UK Ltd, Crawley Down, Sussex, RH10 4FF, UK.

Serva Feinbiochemica GmbH & Co., D-6900, Heidelberg 1, PO Box 105260.

Stewart Plastics, Purley Way, Croydon, UK.

Sigma Chemical Co., PO Box 14508, St. Louis, MS 63178, USA; Fancy Road, Poole, Dorset, BH17 7NH, UK.

Spectrum Medical 44 Upper Northgate Street, Chester CH1 4EF, UK.

Stratagene Ltd, Cambridge Innovation Centre, Cambridge Science Park, Milton Road, Cambridge, CB4 4GF, UK; 11099 North Torrey Pines Road, La John, CA 92037, USA.

Sunkist Growers, Corona, California, USA.

Supelco, Bellefonte, USA.

Tetko Inc., New York, USA.

US Biochemicals. See **British Biotechnology Ltd.**

Water Associates Inc. See **Millipore.**

Whatman Biosystems Ltd, Springfield Mill, Maidstone, Kent ME14 2LE, UK; 9 Bridewell Place, Clifton, NJ 07014, USA.

Weck & Co. Inc., Research Triangle Park, NC, USA.

Yakult Honsha Co. Ltd, Medicine Department, Enzyme Division, 1-1-19, Higashi-Shinbashi, Minatokv, Tokyo 105, Japan.

Zeiss, Welwyn Garden City, Hertfordshire, AL7 1LU, UK.

Index

Contents of Volume I

SECTION 2.   DEFENCE RESPONSES (GENES)